Despite more than one century of observational stellar spectroscopy, the resulting data are not available in an easily accessible format. The necessity of such basic information is greater than ever, because new wavelength regions (ultraviolet, infrared) are now accessible and because modern receivers can only analyze short stretches of spectra, so that a careful preselection of strategic elements is mandatory.

This book presents a summary of our knowledge of the behavior of all chemical elements identified in stars, based on observations rather than on their interpretations. Whenever possible, the behavior is described quantitatively, with the help of equivalent widths in different types of stars, or different ionization stages, for both absorption and emission features. It will provide an authoritative reference book for the astrophysical community.

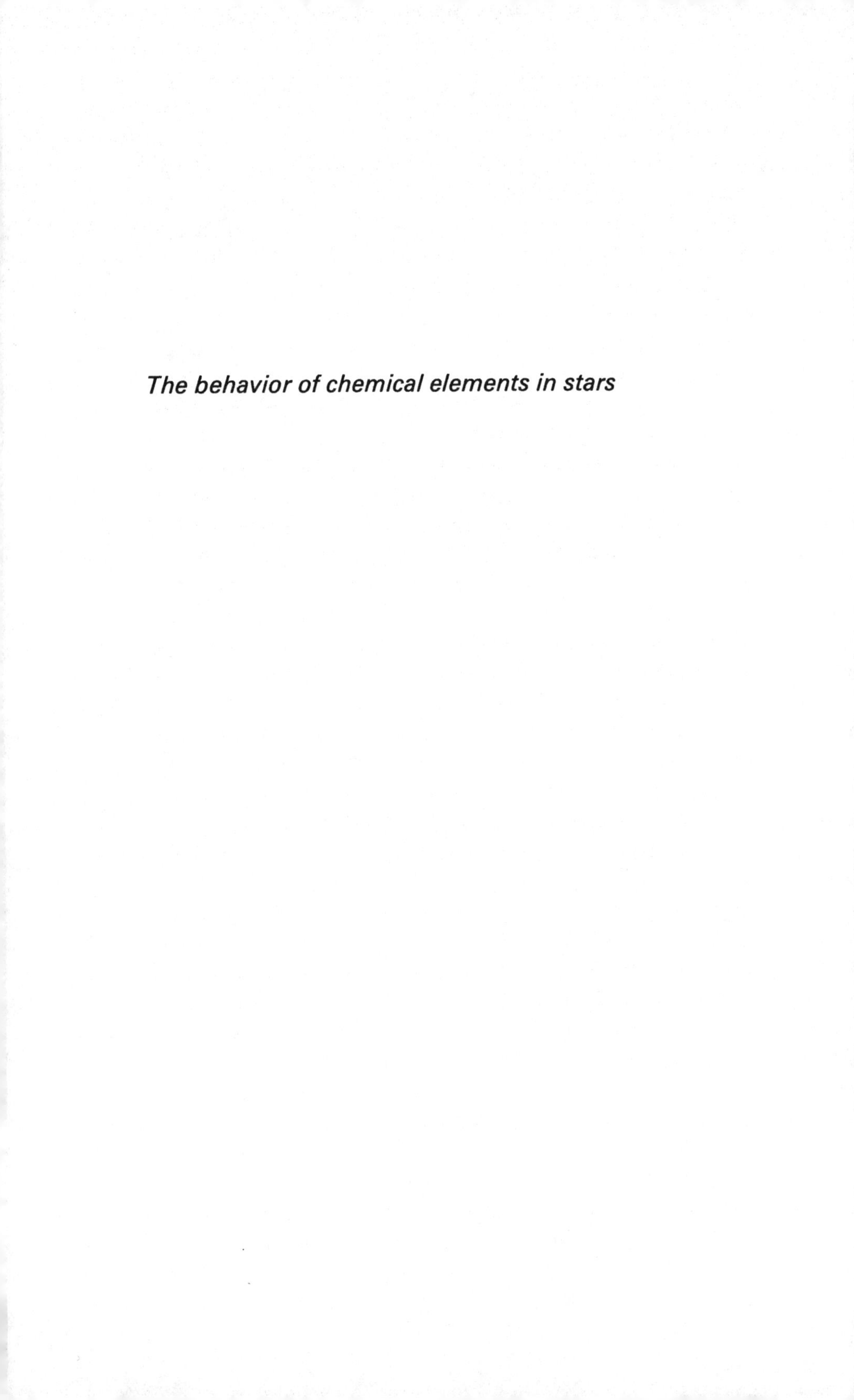

The behavior of chemical elements in stars

The behavior of chemical elements in stars

CARLOS JASCHEK and MERCEDES JASCHEK
formerly of Observatoire de Strasbourg, France

CAMBRIDGE UNIVERSITY PRESS
Cambridge, New York, Melbourne, Madrid, Cape Town, Singapore, São Paulo, Delhi

Cambridge University Press
The Edinburgh Building, Cambridge CB2 8RU, UK

Published in the United States of America by Cambridge University Press, New York

www.cambridge.org
Information on this title: www.cambridge.org/9780521102407

First published 1995
This digitally printed version 2009

A catalogue record for this publication is available from the British Library

Library of Congress Cataloguing in Publication data

Jaschek, Carlos.
The behavior of chemical elements in stars/ Carlos Jaschek and
Mercedes Jaschek.
p. cm.
ISBN 0 521 41136 X
1. Stars – Spectra. 2. Cosmochemistry. 3. Chemical elements.
I. Jaschek, Mercedes, 1926– . II. Title.
QB871.J36 1995
523.8 – dc20 94-34688 CIP

ISBN 978-0-521-41136-3 hardback
ISBN 978-0-521-10240-7 paperback

Contents

Part Four

Preface

The purpose of this book is to provide an outline of our knowledge about the behavior of the chemical elements in stars. As every observational spectroscopist knows, one is often confronted with essentially simple questions of the following kinds. What is the behavior of a given element in a given group of stars, for example, europium in metallic line stars or in S-type stars? Are the neutral lines of this element visible, are they strengthened or weakened with regard to those of normal dwarfs? Questions like these are often difficult to answer even for specialists and we have thus thought that it would be useful to collect the available information and to present it in such a way as to be useful for others.

We have reviewed the literature for both normal and non-normal stars, in the classical wavelength region (3800–4800 Å) as well as in the ultraviolet and the infrared (when available) for both absorption and emission lines. We have tried to stick as closely as possible to observations and to refrain from interpretation; this means for instance that we quote equivalent widths rather than abundances, whenever possible. The separation of observations from interpretation is especially useful in fields that are in a constant state of flux. This alludes for instance to interpretations of observed abundances in terms of the thermonuclear processes going on in the stars, or to interpretations involving physical processes like diffusion in stellar atmospheres or mechanisms for heating of the corona.

For completeness and for reasons detailed in the introduction to that chapter, we have also summarized our knowledge about the behavior of molecules in stars.

In order to keep this book down to a reasonable size we have avoided discussing individual objects like zeta Aurigae or eta Carinae. We have preferred instead to deal with the main characteristics of stellar groups and of groups of variable stars.

There exists, however, one important exception to our rule that no individual stars are discussed, namely the sun. In fact the sun is the cornerstone of stellar spectroscopy. On the other hand, many things that can be studied in great detail in the sun (like the chromosphere and the corona) can only be seen in a global and cursory way in other stars. We have thus retained only that part of our knowledge of the sun that can be applied to stellar studies.

We excluded from the book non-stellar objects such as nebular objects (planetary nebulae, nebulae, HII regions) and interstellar features (such as star clouds).

Stellar spectroscopy has existed for more than a century and produces an ever increasing flow of papers, which at present is of the order of several hundred per year. A complete coverage of the literature is thus impossible, the more so since we do not want to write the history of the subject. The bibliographic coverage is thus limited to essentials – whatever that means – providing usually (but not always) only the first and the latest paper on the subject. The selection of what we felt important leads unavoidably to a strong personal bias and we ask the pardon of authors who have not been quoted.

The book is divided into four parts. Part one constitutes the main body of the book. It starts with a content description followed by a separate section for each chemical element and finishes with a figure providing the general behavior of the elements in stars of different types. Part two contains in the first chapter a summary of our knowledge of molecules in stellar atmospheres and circumstellar envelopes. It finishes with a table providing the general behavior of molecules in stars. The second chapter deals with groups of elements, namely metals and rare earths. The third chapter deals with stellar chromospheres and coronas. Part three contains five short chapters on major topics such as the terminology of spectral lines, the selection of stars included in the book, line identification, equivalent widths and abundances and a last section called 'afterthoughts'. These sections do not provide a complete treatment of the subject, but provide only information necessary for a better understanding of the book. Part four contains auxiliary tables to aid use of the book. The bibliographic references are given at the end of the volume ordered by (first) author and year of publication.

This book can be regarded as a late successor to Merrill's book on the *Lines of the Chemical Elements in Astronomical Spectra* published in 1956. The book is a complement to our book on *The Classification of Stars*. In the latter the reader can find a general discussion of spectral types, stellar groups and other matters, which is not repeated in this book. For instance the uses of certain lines of certain elements for stellar classification purposes are discussed in our earlier book.

The book is the result of a long-standing involvement of the authors with the subject, which started at the La Plata Observatory with Professor L. Gratton, forty years ago, and continued at Córdoba, Lick, Mt Wilson, Leuschner, Yerkes, Ohio State, Geneva, Lausanne and Strasbourg observatories. Over the years we have had the privilege to know personally many of the astronomers who produced the results contained in this book and to work with some of them.

We would like to thank all the colleagues who have read and commented critically on parts of this book and/or have provided bibliography. These were Y. Andrillat, W. Balfour (Chemistry Department of the University of Victoria), H. Behrens, W. P. Bidelman, G. Cayrel de Strobel, R. Cayrel, Ch. Cowley, R. Freire Ferrero, N. Grevesse, M. Holtzer, L. Houziaux, P. C. Keenan, J. Koeppen, A. Maeder, A. Slettebak, V. Trimble, J. B. Tatum, A. J. Sauval and Mme Thierry (Chemistry Department of the University of Strasbourg). We also thank Dr M. Creze, director of the Strasbourg Observatory, and the staff of the Strasbourg Observatory for their help. Our special gratitude goes to Mr J. Marcout, who made all the drawings, Mrs M. J. Wagner, who helped with the bibliography, and Mrs M. Hamm, who provided much of the non-astronomical bibliography. Finally, we thank Miss B. del Huerto, who contributed to the indices and checked the bibliography.

Strasbourg

Part One

Content description

This part is divided into 80 sections, each devoted to one element. The elements are ordered by their English names. Cross references between names, formulae and atomic numbers are given in the appendices (part four, tables 2–4).

For the description of the behavior of each element we have kept a constant order, which is as follows.

1 The formula, the English name and the atomic number.
2 A short history of the discovery of the element and an explanation of its name and/or abbreviation. If not indicated otherwise, these data are taken from Elmsley (1989).
3 The ionization energies, in electron-volts (eV), rounded off to the nearest tenth of an electron-volt. The data were taken from Samsonov (1968) and Elmsley (1989). We have only quoted those ionization stages that are observed in stars, plus the next one.
4 We then start a description of the behavior of the lines of each ionization stage (i.e. each species), beginning with the neutral lines. In order to be as precise as possible, we have given the behavior of the equivalent widths of one or more absorption lines with spectral type. Wavelengths and equivalent widths are given in ångström units (1 Å$=10^{-8}$ cm$=0.1$ nm), except when stated otherwise. These values are listed in tables and, if sufficient values are available, plotted in figures. The figures provide only smoothed curves, based upon the values listed in the tables. All figures give the spectral types on the abscissae and the equivalent width (in ångström units) on the ordinates.

   If data exist in the literature, separate curves have been plotted for dwarfs and supergiants and sometimes also for giants. Uncertain trends (few data points or large dispersions) are indicated by broken lines. Readers interested in more details should consult chapters 2–4 in part three. The multiplet to which the line belongs is indicated in parentheses after the wavelength.

   After the tables and figures follows a summary of the behavior of the species in the HR diagram, with mention of luminosity effects. The term 'positive luminosity effect' expresses the fact that the line is stronger in stars of higher luminosity, i.e. stronger in supergiants than in dwarfs. 'Negative luminosity effect' implies on the contrary that the line is stronger in dwarfs than in supergiants.

If possible we comment upon the behavior of the species in different wavelength ranges – ultraviolet, photographic, visual and infrared.

5 For each species then follows information on emission lines, both permitted and forbidden. (For more details on the terminology see chapter 1 of part three.)

After the information for neutral lines (both in absorption and in emission) we proceed to do the same for the singly and doubly ionized lines.

6 After describing the behavior of the lines in normal stars, the next section deals with the behavior of the element in non-normal stars. As explained in the preface, we have refrained from going into detail concerning the behavior of lines in individual peculiar stars, so we only discuss the behavior of the species in groups of stars. The groups quoted (which are italicized in the text) are those discussed in our book *The Classification of Stars* (Jaschek and Jaschek 1987a) plus additional groups. Short definitions of all groups mentioned in the book are given in chapter 2 of part three. Wherever possible the behavior of the lines is quantified in terms of equivalent widths (*W*), taken from the sources indicated in the references. In those cases in which the behavior of an element is described differently by different authors we have inserted the note 'under discussion'.

Readers should be attentive to the fact that the behavior of some of the non-normal groups is also described under the heading 'emission lines' of the different species.

7 A short summary is provided of the key isotopes. The data are taken from the compilation of R. L. Heath in the CRC Handbook, 63rd edition. Two modifications in the terminology have been introduced in this book. The first is that, if an isotope has a half life shorter than 10^4 years, it was simply called 'short-lived' and further comment omitted. The second modification is that we have also called 'stable' those isotopes whose half life is longer than 10^9 years, because their life time is then comparable to the age of the sun.

The percentages of occurrence of the different isotopes are quoted for the solar system. The data have been taken from Anders and Grevesse (1989). The values quoted have been rounded off. Their sum should always be 100%, except if one isotope is present in a minute fraction (like 0.1%). In such cases the sum is slightly greater than 100. Besides the information for isotopes in the solar system, we have also summarized observations of isotopes of the element in stars.

Whenever an isotope exists with a half life longer than 10^7 years, we have added the comment 'can be used for radioactive dating'.

8 Origin. We provide a short note on the processes through which the element can be produced from a nucleosynthetic point of view. This information was also taken from Anders and Grevesse (1989). The following processes are mentioned, either with the complete name or with an abbreviation:

cosmological nucleosynthesis
hydrogen burning
hot hydrogen burning
helium burning
carbon burning
oxygen burning
neon burning
explosive nucleosynthesis
nuclear statistical equilibrium (the so-called e process)
slow neutron process (s process)
rapid neutron process (r process)
proton process (p process)
cosmic spallation
actinide-producing r process (ra)

For a general overview of which elements are visible in given types of stars, see figure 1 at the end of this part (page 209).

Readers should be aware that as an exception we have dealt with the 'collective' behavior of some elements in chapters 2 and 3 of part two.

In some places in the text we have used the notation 'dex'. This notation provides the decimal logarithm of the quantity. For instance -3 dex means 10^{-3}.

ALUMINUM Al $Z=13$

The element was discovered in 1827 by F. Woehler. The name comes from the Latin *alumen* (alum). The element is also called aluminium.

Ionization energies

Al I 6.0 eV, Al II 18.8 eV, Al III 28.4 eV, Al IV 120 eV.

Absorption lines of Al I

Table 1. *Equivalent widths of Al I*

	W(3944) (resonance line UV M.1)		W(6699)(5)		
Group	V	Ib	V	III	Ib
B 7	0.006				
B 9	0.051				
A 0	0.085	0.05			
A 1	0.098				
A 2	0.14	0.085(Ia)			
A 3		0.050(0)			
A 7	0.18				
F 0		0.30(Ia)			
F 4	0.29				
F 5					0.015
F 6	0.47				
F 8	0.35	0.28			
G 1	0.4				0.056
G 2	0.50		0.018		0.073, 0.092
S	0.488		0.021		
G 5					0.093
G 6					0.142
G 8			0.022(IV)		0.119
K 0			0.006	0.052	
K 2				0.062	0.188
K 3				0.11	
K 5			0.063		0.157
M 2					0.117

Al I (for instance the line at 3944) is present in dwarfs from late B-type stars onwards, increasing steadily in strength toward G-type stars. This line shows no luminosity effect. The line at 6699 shows a positive luminosity effect from late F-type onwards.

Emission lines of Al I

That at 3944 (1) appears in emission from some *T Tau stars* (Joy 1945)

Absorption lines of Al II

Al II has its resonance line at 1671 (UV M.2). Sadakane *et al.* (1983) provide the following values.

Table 2. *Equivalent widths of Al II 1671(2)*

Group	V
B 5	0.59(IV)
B 7	1.02
B 9	1.60

The ultraviolet blend at 1723(6) shows a positive luminosity effect in B-type stars (Heck *et al.* 1984).

In the sun a line at 3900.66 is probably Al II

Al III is present in B-type and more weakly in A-type stars.

Absorption lines of Al III

Al III has its resonance lines in the ultraviolet (1854 and 1863 – UV M.1). Sadakane *et al.* (1983) provide the following values.

Table 3. *Equivalent widths of Al III*

	W(1863)(1)	W(3612)(1)	W(4512)(3)		
Group	V	V	V	III	Ia
0 7			0.004		
0 9			0.015		
B 0		0.012	0.022		
B 0.5			0.032	0.030	
B 2		0.027(IV)	0.036	0.046	
B 3		0.012	0.017		0.031
B 5	0.48(IV)		0.004		0.039
B 7	0.40				
B 9	0.20				

Al III (for instance the line at 4512) is present in B-type stars with a maximum around B5 and a positive luminosity effect. Al III 4149(5) is present in the solar chromosphere, besides Al I and Al II (Pierce 1968).

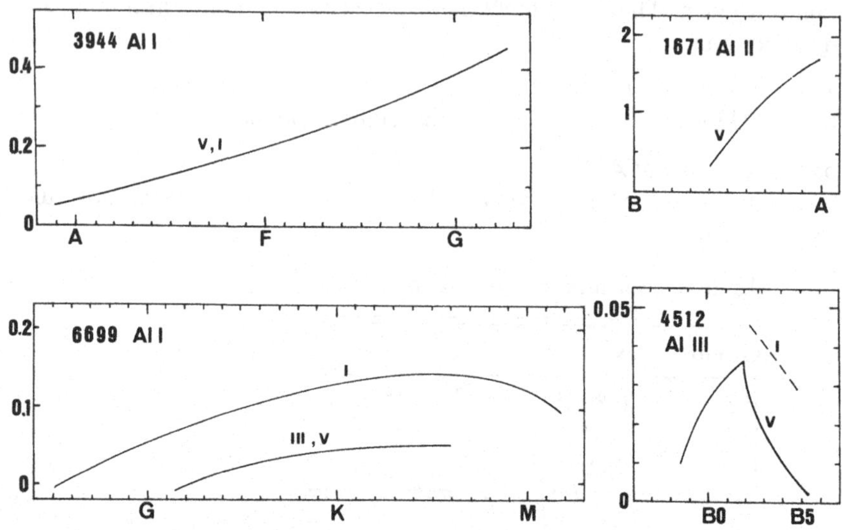

Behavior in non-normal stars

Sadakane *et al.* (1983) found that both Al II and Al III are weak in *Bp* stars of the Hg–Mn, Si and He-weak subgroups. The extent of weakening varies from star to star, with equivalent widths lesser by factors two to six than in normal stars.

Al I is normal or slightly weak in *Ap* stars of the Cr–EU–Sr type (Adelman 1973b).

Al is either slightly strengthened in *Am* stars (Burkhart and Coupry 1991) or normal (Smith 1973, 1974).

Al is probably underabundant with respect to Fe in the most *metal-weak stars* ($Fe/H \times 10^{-4}$) by factors of the order of 4–5 (Molaro and Bonifacio 1990). This is also true for *horizontal branch stars* (Adelman and Philip 1990, 1992a). Magain (1989) suggests that Al has a normal (i.e. solar) abundance in normal stars, whereas it is increasingly deficient in increasingly metal-weak dwarfs, a fact that was also pointed out by Gratton and Sneden (1988). Since not all authors use the same lines, the abundances differ between authors more than would seem reasonable (Spite 1992).

Al apparently has an erratic behavior from star to star in *globular cluster stars*, but seems to be overabundant with respect to iron (Francois 1991). However, Gonzalez and Wallerstein (1992) find it underabundant in one F-type *globular cluster supergiant*.

In *novae* Al II and Al III lines usually appear in emission in the ultraviolet region during the 'principal spectrum' phase. Lines of [Al IV] and [Al VIII] sometimes appear during the nebular stage (Warner 1989).

Al is very overabundant in the spectra of novae of the O–Ne–Mg subgroup (Andreae 1993).

Isotopes

Al has eight isotopes and isomers. The only stable one is Al^{27}. One of the isotopes, Al^{26}, has a half life of 7.4×10^5 years. Branch and Peery (1970) derive $Al^{26}/Al^{27} < 0.15$ from the AlH bands in one S-type star.

Origin

The element can be produced by neon burning or by explosive nucleosynthesis.

AMERICIUM Am Z=95

This transuranium element was discovered by T. Seaborg, R. James, L. Morgan and A. Ghiorso in Chicago in 1944. Its name alludes to America.

Ionization energies

Am I 6.0 eV

Several lines of Am II were found by Jaschek and Brandi (1972) in the spectrum of one *Ap star* of the Cr–Eu–Sr subgroup. The authors regarded this identification as dubious because of the lack of agreement with laboratory line intensities.

Later work did not confirm the presence of this transuranium element.

Isotopes

The longest lived isotope of this unstable element (Am^{243}) has a half life of 7.4×10^3 years. A total of 13 isotopes and isomers exist.

Origin

Am is produced by the r process.

ANTIMONY Sb *Z*=51

This element was already known in antiquity. The name alludes to the Greek words *anti* and *monos* (not alone). In Latin this element is called *stibium*.

Ionization energies
Sb I 8.6 eV, Sb II 18.6 eV.

Absorption lines of Sb I
The equivalent width of Sb I 3267(2) in the sun is 0.011.

Isotopes
There exist 29 isotopes and isomers. The two stable isotopes are Sb^{121} and Sb^{123}. In the solar system their abundances are respectively 57% and 43%.

Origin
Sb^{121} can be produced by the r and s processes whereas Sb^{123} is a pure r process product.

ARGON Ar Z=18

This element was discovered in 1894 by D. Rayleigh and W. Ramsay in Great Britain. The name comes from the Greek *argos* (inactive). The name of the element has also been abbreviated as A instead of Ar (for instance by Moore (1945)).

Ionization energies

Ar I 15.8 eV, Ar II 27.6 eV, Ar III 40.7 eV, Ar IV 59.8 eV.

Absorption lines of Ar II

Table 1. *Equivalent widths of Ar II*

	4371(1)	4431(1)		4590(21)	
Group	V	V	Ia	IV	III
B 0.5		0.004			
B 1				0.0025	
B 2				0.0065	
B 3	0.006	0.008	0.052		
B 6					0.004

Source: Data are from Keenan *et al.* (1990).

Ar II (for instance the line at 4590) is represented by weak lines in early type B-stars. It has a maximum around B2. The line at 4431 (see table) shows a positive luminosity effect.

Emission lines of Ar II

The lines at 920 and 932 (UV M.1) are seen in emission in the ultraviolet solar spectrum (Feldman and Doschek 1991). *Supernova* 1987A also showed Ar II lines in emission (Arnett *et al.* 1989).

Emission lines of Ar III

The 7135 line of [AR III] (M.1F) has been observed in at least one *nova* (Thackeray 1953).

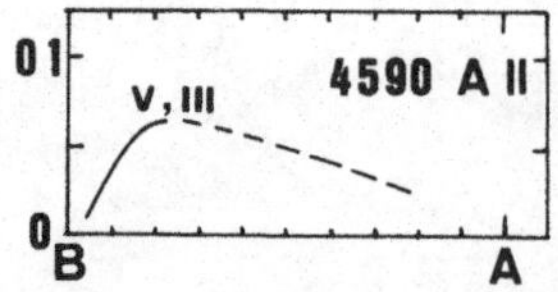

Behavior in non-normal stars

Ar II lines are strong in the spectrum of the *extreme He star* upsilon Sgr (Morgan 1935).

Forbidden lines of Ar III, Ar IV and Ar V are often visible in *symbiotic objects* (see for instance Freitas Pacheco and Costa (1992)). In at least one *recurrent nova* Joy and Swings (1945) found forbidden lines of Ar V, Ar X and Ar XI.

Isotopes

There exist eight isotopes. The three stable isotopes are Ar 36, 38 and 40. In the solar system 99.6% of Ar is in the form of Ar^{40}.

Origin

Ar^{36} produced by explosive nucleosynthesis, Ar^{38} by the same process or by oxygen burning and Ar^{40} by the s process or neon burning.

ARSENIC As Z=33

This element was probably first isolated by Albertus Magnus in the thirteenth century. Its name comes from the Greek *arsenilon* (yellow orpiment). Orpiment comes from *auri pigmentum* (gold pigment).

Ionization energies

AsI 9.8 eV, AsII 18.6 eV.
This element is not observed in the sun.

Leckrone *et al.* (1991a) detected an absorption line at 1937 in the spectrum of one *Ap star* of the Hg–Mn subgroup, which they identify with an AsI line from multiplet 1.

Isotopes

As has one stable isotope, As^{75} and 13 short-lived isotopes and isomers. In the solar system only As^{75} is observed.

Origin

As can be produced by the r process and the s process.

BARIUM Ba Z=56

This element was first isolated in 1808 by H. Davy in London. The name comes from the Greek *barys* (heavy).

Ionization energies

Ba I 5.2 eV, Ba II 10.0 eV, Ba III 37.3 eV.

Absorption lines of Ba I

The ultimate line of Ba I at 5535(2) has been detected in the sun as a weak blended line. Smith and Lambert (1985) observed Ba I at 7392(10) as a weak line (*W* about 0.025) in K5III to M5III stars. Other infrared Ba I lines have also been observed in M-type stars (Fay *et al.* 1968).

Absorption lines of Ba II

Table 1. *Equivalent widths of Ba II*

	4554(1)			5853(2)		
Group	V	III	Ib	V	III	Ib
B 9	0.018					
A 0	0.076		0.02(Ia)			
A 2	0.14					
A 7	0.17					
F 0	0.28					
F 4	0.20					
F 5	0.19, 0.21		0.38	0.103		0.305
F 6	0.21					
F 8	0.19		0.55	0.050		0.380
G 0	0.18					0.340
G 1	0.27			0.074		
G 2	0.19			0.047		0.288
S	0.159			0.055		
G 5				0.085		0.263
G 8	0.21(IV)				0.107	0.316
K 0		0.27		0.068	0.155	
K 2		0.30		0.062	0.130	0.295
K 3					0.138	0.340
K 5				0.106		0.295
M 0					0.174	
M 2						0.348
M 2.5					0.177	

Ba II (for instance the 4554 and 5853 lines) appears in early A-type dwarfs and increases in intensity toward F-type. From there on the lines increase asymptotically in strength. The lines have a strong positive luminosity effect.

Behavior in non-normal stars

Ba II lines are weak ($W(4554)=0.07$) in *Ap stars* of the Cr–Eu–Sr subgroup (Adelman 1973b) and are also weak in the Hg–Mn subgroup (Adelman 1992). Cowley (1976) finds Ba to behave erratically in *Ap* stars.

Ba II lines are strengthened in general in *Am* stars by factors of less than two in *W* when compared with stars of the same temperature (Smith 1973, 1974). There exist, however, stars in which Ba lines are about normal.

Ba II lines are also slightly strengthened in *delta Del stars* (Kurtz 1976). Ba is overabundant in *blue stragglers* of an old open cluster (Mathys 1991) and underabundant in *HB stars* (Adelman and Philip 1992a). Ba II lines are very strong in the spectra of the so-called *Ba stars*. These are giants (G 2–K 4) showing a very strong Ba II 4554 line (Bidelman and Keenan 1951). Fernandez-

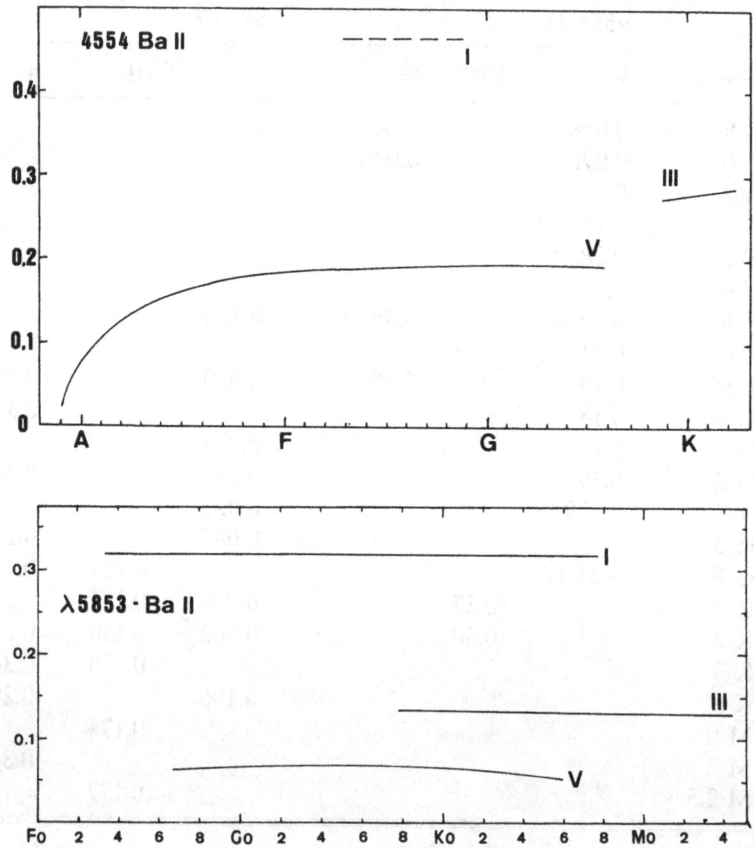

Villacañas *et al.* (1990) provide measures of W(5853) for a sample of Ba stars, with W varying between 0.103 and 0.259. Danziger (1965) found that W(4554) may be up to 1.0. This implies overabundances of Ba of more than one order of magnitude.

The coolest Ba stars have weak carbon compounds and weak Sr lines (Sowell 1989).

Ba II lines are also strengthened in the spectra of *subgiant CH stars* (also called hot Ba stars) (Luck and Bond 1982, Krishnaswamy and Sneden 1985).

Ba lines are probably strengthened in *R CrB stars* (Cottrell and Lambert 1982).

Ba II lines are very strong in *late C-stars* (>C 5), with W(5853) typically 1.5 Å (Fujita *et al.* 1966). The overabundance of Ba in some, but not all, C-stars was confirmed by Kilston (1975), who suggests that abundance of this element is correlated with that of Zr. Dominy (1985) suggests that the C-stars in which Ba is not enhanced are predominantly of type J. Utsumi (1895) finds Ba overabundant with respect to Ti in most late C-type stars.

Ba I is strengthened in *S-type stars*. According to Smith and Lambert (1986), typically W values are larger by a factor of two than in normal stars of the same temperature. Ba is also enhanced in *SC stars* (Kipper and Wallerstein 1990).

The behavior of Ba in *metal-weak stars* has been studied by several authors, with conclusions that do not always agree completely. Ba seems to be underabundant with respect to iron in metal-weak stars according to Luck and Bond (1985). Molaro and Bonifacio (1990) found the same result in the most extreme metal-weak stars (Fe/H$=10^{-4}$), but in less pronounced metal-weak stars its behavior probably parallels that of Fe (Magain 1989). This behavior is also followed in *globular cluster stars* (Wheeler *et al.* 1989, Francois 1991). However, Gonzalez and Wallerstein (1992) found Ba to be overabundant with respect to iron in one *globular cluster supergiant*. In general Ba seems to behave in metal-weak stars in a manner parallel to that of Eu (Spite 1992). In some *halo stars* Ba lines are strengthened.

Ba is enhanced in the *Magellanic Cloud stars* (Luck and Lambert 1992).

Isotopes

There exist 25 isotopes and isomers, of which seven are stable, namely Ba 130, 132, 134, 135, 136, 137 and 138. In the solar system 72% and 11% of Ba is in the form of Ba^{137} and Ba^{138} respectively.

Origin

Ba is produced by a variety of processes: Ba^{130} and Ba^{132} by the p process, Ba 134, 136 and 138 by the s process and Ba^{135} and Ba^{137} by either the s or the r process.

BERYLLIUM Be *Z*=4

This element was isolated independently by F. Woehler in Berlin and by A. Bussy in Paris in 1828. The name comes from the gem beryl, in Greek *beryllos*.

Ionization energies
Be I 9.3 eV, Be II 18.2 eV, Be III 153.9 eV.

Absorption lines of Be I
The equivalent width of Be I 3321(1) in the sun is 0.008 Å.

Absorption lines of Be II

Table 1. *Equivalent widths of Be II 3130(1) (resonance line)*

Group	V	III
F 0	0.043	
F 5	0.083	
G 0	0.100	
S	0.085	
K 0		0.035

Source: The stellar data are from Boesgaard (1976b).

Be II (the 3130 line) appears in A-type stars, increases up to G-type stars and decreases thereafter.

Behavior in non-normal stars
Sadakane *et al.* (1985) found a large range of Be strengths in *Bp stars* of the Hg–Mn subgroup (up to *W*(3130)=0.38) and in *Ap* stars of the Cr–Eu–Sr subgroup. In late Ap stars Gerbaldi *et al.* (1986) found no strengthening of Be II.

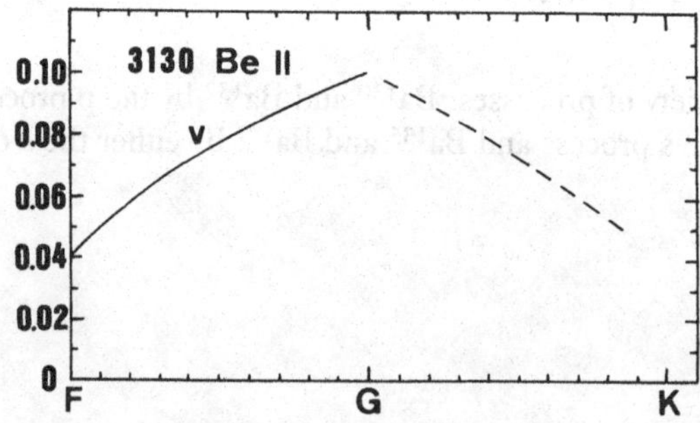

Boesgaard (1976b) remarks that, in late type F dwarfs, there exists also a 'gap' like the Li gap. Such a gap is, however, absent in the Hyades (Boesgaard and Budge 1989). Stars that are deficient in Be are also deficient in Li (Boesgaard and Lavery 1986).

Rebolo (1991) and Ryan *et al.* (1992) found that, in *metal-weak* stars, Be is weakened with respect to Fe, but that it is not absent. Spite (1992), from an analysis of *halo dwarfs*, finds a positive correlation between Be and Fe/H.

Isotopes

Be has six isotopes. The more important are Be^{10} with a half life of 2.7 million years and the stable Be^9. Apparently no short-lived isotope exists in stars (Boesgaard 1976a).

Ryan *et al.* (1992) analyzed the presence of Be^9 in four weak-lined dwarfs. They found that the deficiency in Be^9 accompanies that of other metals. Furthermore the deficiency of Be is not as uniform as that of the neighboring light elements like Li^7.

Be is more stable than Li; it burns at 3.5×10^5 instead of 2×10^6 K and both are less stable than B.

Origin

Be is produced mostly by cosmic ray spallation but a small amount is (or may be) primeval.

BISMUTH Bi $Z=83$

This element has been known since the Middle Ages. Its name comes from the German *weiße Masse* (*bisemutum*).

Ionization energies

Bi I 7.3 eV, Bi II 16.7 eV, Bi III 25.6 eV.

Absorption lines of Bi I and Bi II

Bi has not been detected in the sun. Guthrie (1972) detected Bi I in one *Ap star* of the Cr–Eu–Sr subgroup. Guthrie (1984) found the Bi II 4259 line (W=0.046) in one *Bp star*, of the Hg–Mn subgroup, and Cowley (1987) found this line in one Ap star of the Cr–Eu–Sr subgroup.

Isotopes

Bi has one stable isotope (Bi^{209}), one (Bi^{208}) with a half life of 3×10^5 years, an isomer with half life 3×10^6 years (Bi^{210}) as well as 17 short-lived isotopes and isomers.

Origin

Bi can be produced by either the r or the s process.

BORON B Z=5

This element was discovered in 1808 by L. J. Lussac and L. J. Thenard in Paris and independently by H. Davy in London. The name comes from Arabic *buraq*.

Ionization energies
BI 8.3 eV, BII 25.2 eV, BIII 37.9 eV.

Absorption lines of BI
The BI 2496(1) resonance doublet has been observed in the sun as a faint feature (Kohl *et al.* 1977). The line was also observed by Lemke *et al.* (1993) in three F-type stars. In two of them B is normal (and Li and Be are normal too), whereas in one B is weakened by at least −0.5 dex, whereas Li and Be are highly depleted (−1.7 dex).

Absorption lines of BII
Beosgaard and Heacox (1978) observed the resonance line at 1362(1) in 16 B- and A-type stars. The feature has W between 0.100 for B1 III and 0.060 for B 3 V.

Behavior in non-normal stars
Sadakane *et al.* (1985) analyzed the ultraviolet resonance line 1362 of BII in *Bp stars* of the Hg–Mn subgroup and found a wide variety of B strengths, from $W=0$ to strong lines with $W=0.150$.

Molaro (1987) set an upper limit for B in one *subdwarf* of one tenth of the solar value.

Duncan *et al.* (1993) observed three *halo dwarfs* and found B/Be ≃ 10, which corresponds approximately to the solar value.

Isotopes
B has two stable isotopes, B^{10} and B^{11}. The solar ratio B^{10}/B^{11} is about 0.25. There exist four shorter lived isotopes.

The isotopic ratio could be derived from the BI resonance line, where the isotopic shift is 0.025 Å.

Origin
B is produced by cosmic ray spallation.

CADMIUM Cd Z=48

This element was discovered by F. Stromeyer in 1817 in Göttingen, Germany. The name comes from the Latin *cadmia* (calomine, a zinc carbonate). The element was isolated from impurities in calomine.

Ionization energies

Cd I 9.0 eV, Cd II 16.9 eV, Cd III 37.4 eV.

Absorption lines of CdI

The equivalent width of the Cd I line 5085(2) in the sun is 0.0007.

Behavior in non-normal stars

Jaschek and Brandi (1972) detected Cd I and Cd II in one *Ap star* of the Cr–Eu–Sr subgroup.

Cd II was detected in two *Am stars* by Boyartchuk and Snow (1978) with *W*(2265)=0.018. Sadakane (1991) found two Cd II lines (2144 and 2265, both of M.1) in the spectrum of another Am star.

Isotopes

Cd has eight stable isotopes – Cd 106, 108, 110, 111, 112, 113, 114 and 116, as well as 14 short-lived isotopes and isomers. In the solar system one finds respectively Cd 106 (1%), 108 (1%), 110 (12%), 111 (13%), 112 (24%), 113 (12%), 114 (29%) and 116 (7%).

Origin

Cd is produced by several processes: Cd^{106} and Cd^{108} by the p, Cd^{110} by the s, Cd^{116} by the r and Cd 111, 112, 113 and 114 by either the r or the s process.

CALCIUM Ca $Z=20$

This element was first isolated by H. Davy in London in 1808. The name comes from the Latin *calx* (lime).

Ionization energies

CaI 6.1 eV, CaII 11.9 eV, CaIII 50.9 eV.

Ca is an element of great astrophysical importance. Lines of this element are seen in stars of practically all spectral types, except the hottest ones. If CaII lines are visible in O- and B-type stars then they are due to the interstellar medium, or, more infrequently, to circumstellar envelopes.

Absorption lines of CaI

Table 1. *Equivalent widths of CaI*

	4226(2)		6439(18)		
Group	V	Ib	V	III	Ib
B 9	0.034				
A 0	0.06				
A 1	0.07				
A 2	0.10, 0.09				
A 7	0.27				
F 0	0.43	0.575(Ia)			
F 2		0.562			
F 4	0.51				
F 5	0.43, 0.50	0.42			
F 6	0.69		0.15		
F 8	0.63	0.88	0.19		0.24
G 0	1.09		0.185		0.31
G 1	1.15		0.23		
G 2	1.48		0.22, 0.16		0.35
S	1.476		0.156		
G 5	1.80		0.31		0.28
G 8				0.23(IV)	0.364
K 0	3.36		0.40		0.22, 0.20(III)
K 2			0.43	0.20	0.38
K 3					0.38
K 5	9.42		0.65		0.39

Ca I lines (see for instance 4226, the resonance line), appear at about A 0 and grow steadily toward later types. At K 5 the line has an equivalent width of about 10 Å and is still growing in intensity toward later types. It thus constitutes an important temperature indicator for spectral classification.

Before type K a slight positive luminosity effect exists. For stars cooler than K 2 the luminosity effect is negative. It becomes more pronounced in type M (the 6439 line). For detailed comments see Burwell (1930) and Keenan and McNeil (1976). This behavior closely parallels that of the NaI line.

The Ca I lines 4435 and 4455(4) also show a negative luminosity effect for late type stars (Adams and Pease 1914).

Emission lines of Ca I

4226 (M.2) appears in emission in *T Tau stars* (Joy 1945).

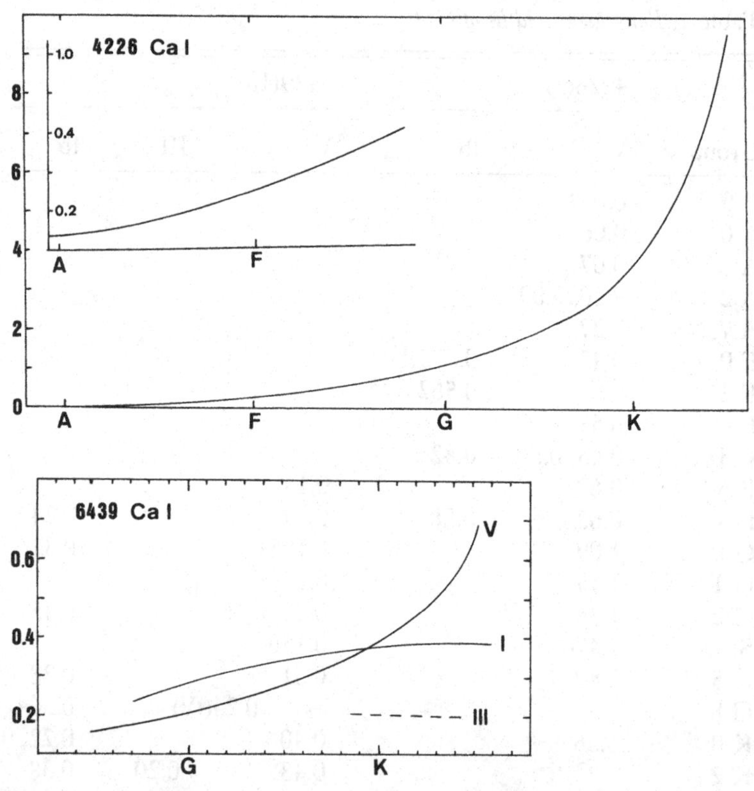

Absorption lines of Ca II

Table 2. *Equivalent widths of Ca II*

	3933(1)			8542(2)	
Group	V	III	I	V	III
B 3			0.40(Ia)		
B 5	0.11		0.483(Ia)		
B 6	0.22	0.28			
B 8			0.67(Ia)		
B 9	0.54				
A 0			0.60(Ib)		
A 1	1.2				
A 2	1.0, 1.5		1.17(Ia)		
			1.54(0)		
A 3			1.400(0)		
A 7	3.6			1.19	0.82
F 0		4.9(II)		1.32	2.02
F 2				2.11	2.17
F 5	6.3		7.6(Ib)	2.65	2.40
F 6				2.69	
F 7				2.96	
F 8			22.1(Ib)		
G 2	14.20			3.25	
S	20.26			3.67	
G 5				3.45	4.20
G 7					4.41
G 8				3.46	4.41
K 0				3.60	4.62
K 3				3.75	4.90
K 5				3.63	5.07
M 0				3.04	5.12
M 2				2.87	5.68

The values for 8542 from A 7 to G 2 are from Jaschek and Jaschek (unpublished) and those for the later type stars from Zhou Xu (1991).

The most important Ca II lines in the classic region are the so-called Fraunhofer H and K lines, which correspond to M.1 (resonance lines), 3968 and 3933. In late type stars (types K, M, C and S) in the infrared region the dominating feature is the triplet 8498, 8542 and 8662(2). The triplet lines have been used to determine physical parameters like gravity and metal abundance of late type stars (see for instance Diaz *et al.* (1989)).

Ca II lines are seen faintly at mid-B-type. They increase steadily toward later types, with a maximum at about K 0 (for example 8542). A positive luminosity effect exists.

In late type giants and supergiants there very often exist circumstellar components in these lines, which can only be seen at high dispersion. The velocity and strength of these lines can be used to establish the mass loss of the stars (see for instance Sanner (1976)). The strength of the CA II line can also be measured photoelectrically (Lockwood 1968) which can provide a rapid assessment of the Ca abundance.

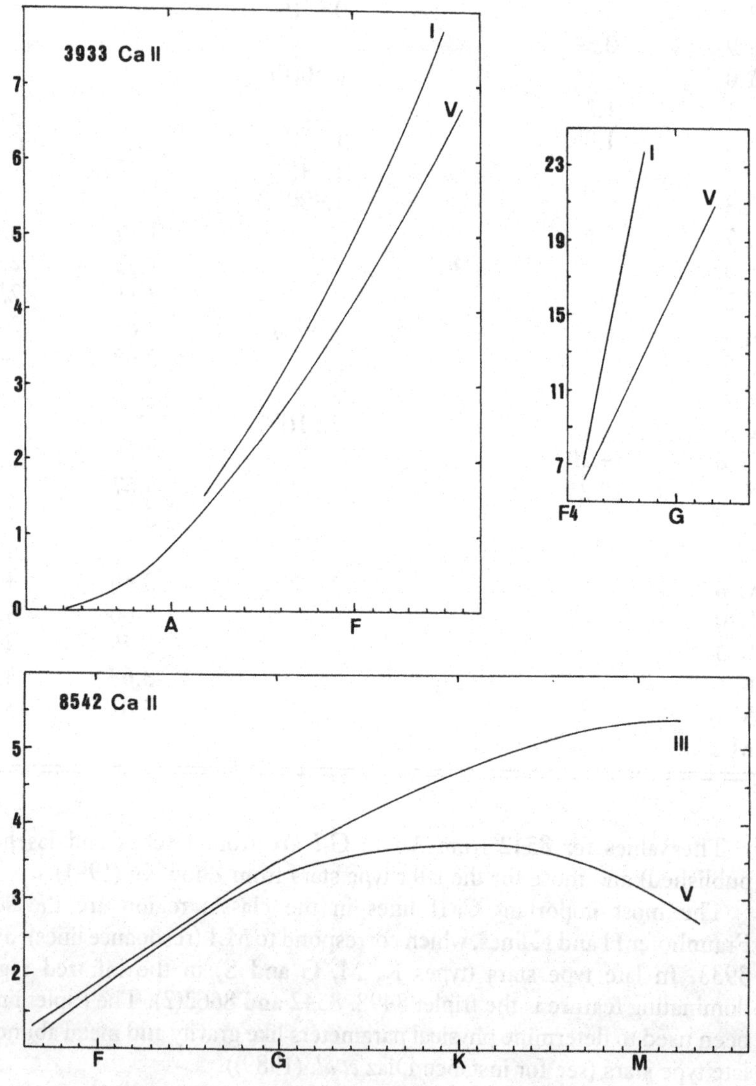

Emission lines of Ca II

Ca II emissions originate in the chromospheres of late type stars (see also the discussion on chromospheres) and are frequently used to identify mass motion and the presence of circumstellar material. The emissions appear in the center of the deep absorption lines 3933 and 3968 (see figure). The emission itself may have a central reversal and be shifted with respect to the center of the absorption line. The emission peaks may also have different intensities. Ca II emissions are a

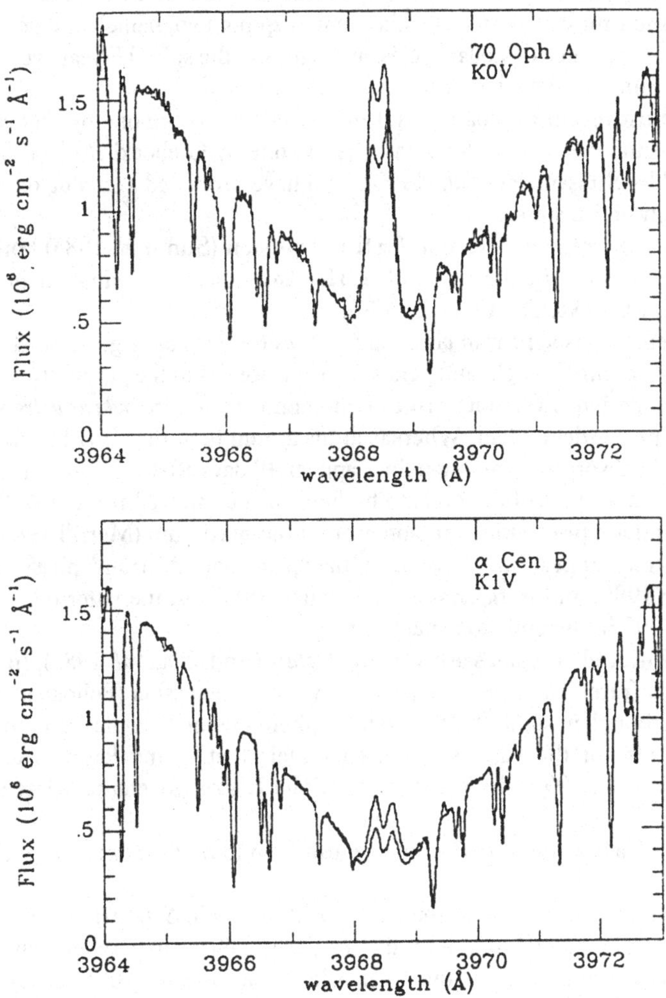

Profiles of Ca II emissions.
Abscissa: wavelength in ångström units; ordinate: absolute flux. The two profiles correspond to the extreme profiles observed within the period covered by the observations. From Rebolo *et al.* (1989). (Courtesy of *AA Suppl.*)

fairly common phenomenon. Broadly speaking they are rare in F-type, become more frequent in G-type and are very common in K- and M-type stars. Bidelman (1954) published a very useful catalog of Ca II emission objects.

The separation of the emission peaks (located within the broad absorption line) is related to the absolute magnitude of stars, in the form

$$M_v = -14.94 \log w + 27.6$$

This is the so-called Wilson–Bappu effect.

The Ca II emission line intensity has been used extensively by Wilson and co-workers studying chromospheres (for details see Jaschek and Jaschek (1987a)). Among their most important results we may quote two, namely the demonstration of the existence of stellar cycles, analogous to the solar 11-year cycle and the demonstration of the (slow) rotation of late type stars.

The most recent catalog of Ca II measurements is the one by Duncan *et al.* (1991). Another useful older catalog is the one by Glebocki *et al.* (1980), now replaced by Strassmeier *et al.* (1990), who have produced a catalog of chromospherically active stars.

T Tau stars frequently exhibit Ca II in emission (Sun *et al.* 1985) both in the photographic and in the infrared. Ca II also appears in emission in *compact infrared sources* (McGregor *et al.* 1984).

Ca II emissions appear in classical *Cepheids* (of periods larger than about four days) after minimum light and persist typically for 0.4 of the period (Kraft 1960).

Ca II emission is a characteristic phenomenon of *long-period variables*, which is strongly phase-dependent. Whereas at maximum light the Ca II H and K lines are seen in absorption, emission lines appear 40 days after maximum and persist during a major part of the cycle. The lines of the infrared triplet (M.2) on the other hand are present in emission around the maximum (Merrill 1960).

Ca II lines appear in emission in the 'principal spectrum' phase of *novae* (Warner 1989) and in *supernovae* (Fillipenko 1988). In the latter one sees also (Ca II) 7290 in the emission spectrum.

Ca II infrared emissions are seen in *Oe stars* (Andrillat *et al.* 1982). In early *Be stars* the infrared Ca II triplet is frequently seen in emission (Polidan and Peters 1976), although in stars later than B 6 the phenomenon becomes rare. In general the behavior of the Ca II emissions parallels that of the Paschen emissions (Andrillat *et al.* 1990). In *B[e] stars* the triplet is strongly emitted (Jaschek *et al.* 1992).

Strong Ca II infrared emissions are also seen in *cataclysmic variables* (Persson 1988).

It should be added that in some *Be stars* (but also in *S-type stars*), the infrared Ca triplet and the H and K lines do not always behave in parallel – one may be found in emission, whereas the other is in the absorption (Bretz 1966).

Forbidden lines of higher ionization stages

Lines of Ca XII (3328), Ca XIII and Ca XV, with ionization potentials between 589 and 814 eV are seen in the spectrum of the solar corona (Zirin 1988) (see also the discussion on coronae). Ca V, VI, VII and XIII forbidden lines have been observed in the *recurrent nova* RS Oph (Joy and Swings 1945). Forbidden Ca V lines are present in the *C star* RX Pup (Swings and Struve 1941b).

Behavior in non-normal stars

Ca is weak in *Ap stars* of the Cr–Eu–Sr subgroup (Adelman 1973b).

Ca II is very weak in *Am stars*, which is easily visible and has led many authors to classify Am stars according to the Ca II line strength only. Since Ca is weak, the spectral type corresponding to these lines is always earlier than the other spectral types adjudicated on the basis of the hydrogen or metallic lines. The *W* values of Ca II are smaller by factors about 3–4 than in normal stars of the same temperature.

Ca lines are also weak in *delta Del stars* (Kurtz 1976).

Ca II is weak also in *lambda Boo stars*. Typically *W* values are smaller by factors 1.5–3 than in normal stars of the same temperature (Venn and Lambert 1990).

Ca lines are weakened in *F-type HB stars* by factors up to six with respect to normal stars of the same colours (Adelman and Hill 1987). Adelman and Philip (1992a) later extended this result to all *HB stars*.

Ca II lines are present in the spectra of *supernovae*. Ca II and (Ca II) emission lines are prominent in the nebular phase of supernovae of class II (Branch 1990).

The infrared Ca lines can be used to separate *RV Tau stars* from normal supergiants (Mantegazza 1991).

Ca II lines exist in the spectra of many, if not all, *DZ degenerates* (Sion *et al.* 1990), with *W*(3933) up to 73 Å. Ca II lines were also found in *DB* and *DA stars* (Sion *et al.* 1988), where their presence was not expected, since DB stars had been defined as possessing only He lines and DA stars as showing only hydrogen lines. (For more information see Jaschek and Jaschek (1987a).) On the other hand, Ca I is usually weak in DZ degenerates $W<2.5$ Å (Sion *et al.* 1990).

Ca is overabundant with respect to iron in *metal-weak stars* (Luck and Bond 1985, Magain 1989) by a factor of the order of two and this seems also to be true for *globular cluster stars* (Wheeler *et al.* 1989, Francois 1991).

Ca I lines are weakened in the spectra of *CH stars* (Yamashita 1967).

Forbidden lines of Ca V, Ca VI and Ca VII are sometimes observed in the spectra of *symbiotic objects* (Freitas Pacheco and Costa 1992).

Isotopes

Ca has 14 isotopes. Among these six are stable, namely Ca 40, 42, 43, 44, 46 and 48. In the solar system most (97%) is in the form of Ca^{40}. Of the unstable isotopes, Ca^{41} has a half life of 8×10^4 years.

Origin

Ca^{40} is produced by explosive synthesis and other stable isotopes by this process or others, namely Ca^{42} by oxygen burning, Ca^{43} by carbon burning or the s process, Ca^{44} by the s process and Ca^{46} by carbon or neon burning.

CALIFORNIUM Cf Z=98

This element was produced in 1950 by S. G. Thompson, K. Street, A. Ghiorso and G. T. Seaborg in Berkeley, California, USA. Its name comes from the state of California.

Ionization energies

CfI 6.3 eV.

Cf has not been observed in stars. Marginal evidence for its production in supernova explosions comes from the supernova light curves (Burbidge *et al.* 1957).

Isotopes

Cf is a short-lived element. It has 12 isotopes, of which the longest lived have a half life of 900 years (Cf^{251}) and 60.5 days (Cf^{254}).

Origin

Cf is produced by the r process.

CARBON C Z=6

This element has been known since remote times. The name comes from the Latin word *carbo* (charcoal).

Ionization energies

CI 11.3 eV, CII 24.4 eV, CIII 47.9 eV, CIV 64.5 eV, CV 392 eV.

Absorption lines of CI

Table 1. *Equivalent widths of CI*

	4771(6)		10 685 + 10 691(1)	
Group	V	Ib	V	III
B 6			0.05	
A 0			0.76	
A 2	0.003		0.96,1.16	
A 3			1.12,1.25	
A 4			1.39	
A 5			1.41	1.64
A 7	0.06		1.38	1.75
F 0		0.158(Ia)	1.22	
F 1				1.86
F 5	0.05,0.021	0.06,0.031		
F 6				1.84
F 8		0.045		
G 0				
S	0.015		0.18	
K 0		0.015		
K 2		0.013(III)		

Note that the resonance line is 8335(10).

Many CI lines are present in the ultraviolet spectrum – among others 1657 (UV 2), 1931(33) – a few weak ones in the photographic region and some stronger lines in the infrared.

The infrared lines (10685+91(1)) are seen from mid-B-type stars, with a maximum around A 6. In the sun these lines are very weak. A positive luminosity effect is present.

In C-type stars the [CI] line 8727 is prominent (Fujita 1992).

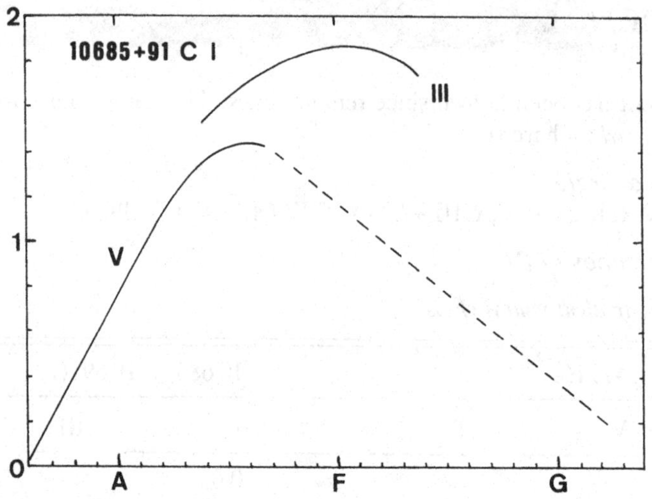

Emission lines of CI

When the C I line 1993(UV 32) is seen in emission from late type stars it indicates the presence of a chromosphere; the line is, however, never very intense. It is observed mostly in G- and K-type dwarfs. See also Part Two, section 3.1.

Absorption lines of CII

Table 2. *Equivalent widths of CII 4267(6)*

Group	V	III	Ia
O 9	0.04		
O 9.5	0.067		
B 0	0.082		0.095
B 1	0.134		0.135
B 2		0.219	0.155
B 3	0.161		0.22(Iab) 0.427(Ia)
B 5	0.118		0.29(Ib)
B 6	0.145	0.132	
B 7	0.071	0.11	
B 8	0.062		0.190,0.170
B 9	0.02		
A 0			0.05,0.080
A 1	0.003		

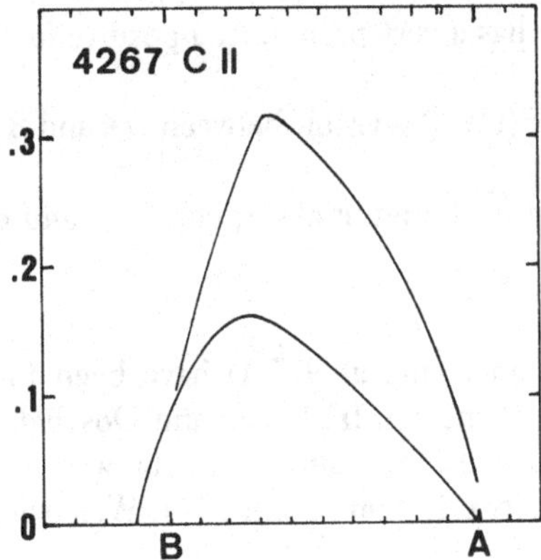

C II lines are present in the ultraviolet region (for instance 1335(UV 1)) and in the photographic region, where there exist some medium strength lines. The 4267(6) line appears in late O-type stars and disappears at late B-type, with a maximum around B 5 and a positive luminosity effect. C II 6578–6583(2) lines are very strong from B 2 to B 5 supergiants (Walborn 1980). Barnett and McKeith (1988) analyzed a series of measurements of these lines in B-type stars and concluded that they should be preferred for abundance analysis over the lines in the photographic region.

The ultraviolet C II line 1335 decreases from early B-type to B 8, where it disappears (Heck *et al.* 1984).

Emission lines of C II

For *WR stars* see later.

When the C II 1335(1) line is seen in late type spectra it indicates the existence of a chromosphere. See the discussion on chromospheres. The line is prominent in F, G and K dwarfs.

Absorption lines of C III

Table 3. *Equivalent widths of C III 4647(1)*

Group	V	III	I
O 7	0.04		0.15(f)
O 8	0.15		
O 9	0.225		
B 0	0.155		0.530
B 1		0.215	
B 2	0.020		

C III lines (for instance 4647) appear at about O 7 and disappear at B 0.5, with a maximum around O 9.5. This line has a very pronounced positive luminosity effect.

The ultraviolet C III feature at 1175(UV 4) is visible between O 4 and B 6, with a maximum at B 1.

The infrared C III line 8500 is always found in absorption in O and *Of stars* (Mihalas *et al.* 1975).

Emission lines of C III

The 1175(4) feature and the resonance line at 977(1) have been found in emission in the ultraviolet spectrum of the sun (Feldman and Doschek 1991).

The C III 5696(2) line is seen in emission in supergiants between O 6 and O 9.5 and constitutes a good luminosity criterion (Conti 1974, Walborn 1980).

For *WR stars* see later.

Absorption lines of C IV

Table 4. *Equivalent widths of C IV*

	5801(I)		1550(UV 1)		
Group	V	I	V	III	I
O 6	0.2	0.25(f)	10	12	12
O 7	0.400		9.5	11.5	12
O 9	0.090		6.5	10	11
O 9.5	0.065	0.33(f)			
B 0			4	6.5	9
B 2			1	2.5	4
B 4				0.5	2.5

Note: Since the scatter in measurement is very large, mean values have been give.

Source: The values for 1550 were taken from Sekiguchi and Anderson (1987).

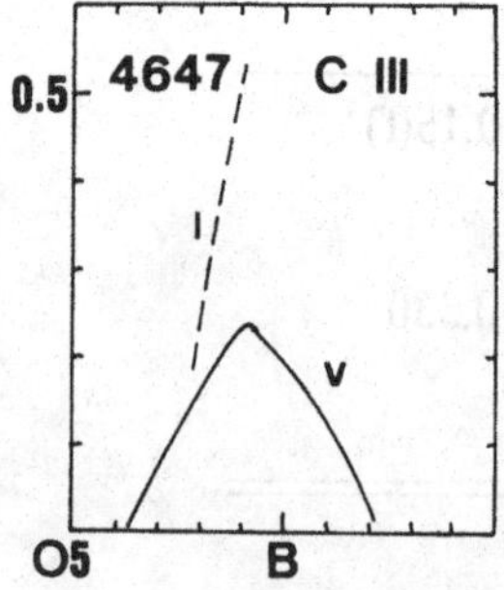

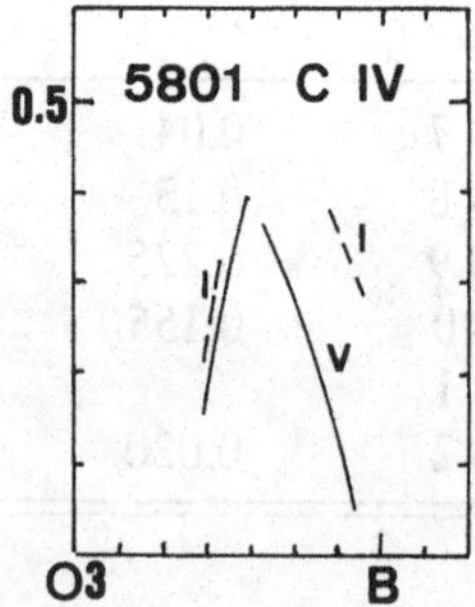

C IV (see 5801) characterizes O-type stars, with a maximum around O 7. It has a positive luminosity effect. The same behavior is observed for C IV 5812 (Walborn 1980).

The ultraviolet C IV lines (resonance doublet 1548 and 1550(1)) decrease from O-type toward B 2, where they disappear (in dwarfs). The feature has a strong positive luminosity effect (Heck *et al.* 1984). The feature is frequently seen in emission before B 0, independently of the luminosity.

C IV lines are characteristic of the hottest stars so far known, the *pre-degenerates* (Werner *et al.* 1991).

In many O-type stars, 1550 has a P Cygni profile (or extended blue wings). This fact is interpreted as indicating mass loss from the stellar atmosphere (Howarth and Prinja 1989). In *Be Stars* the line is visible up to B 5V (Marlborough 1982).

Emission lines of C IV

A strong 5801–5812 (M.1) doublet in emission and a pronounced 4646 absorption of C IV are characteristic of the *pre-degenerates* (PG 1159 stars) (Motch *et al.* 1993).

For *WR stars*, see below.

When the 1550 line appears in emission in late type stars, it indicates the presence of a chromosphere, or, more properly speaking, of a transition region. The emission is visible in a domain extending from F to K 4 dwarfs and giants (see the discussion on coronal stars in Part Two, section 3.2). The limit at K 4 coincides approximately with what is found from other chromospheric indicators (Haisch 1987).

The carbon sequence in Wolf Rayet stars

A *WR star* exhibiting strong emission lines of carbon is called a WC star. Carbon is present in the form of C III and C IV. The most important emission features are C IV 5801–12 (whose average wavelength is often quoted as 5808), which has

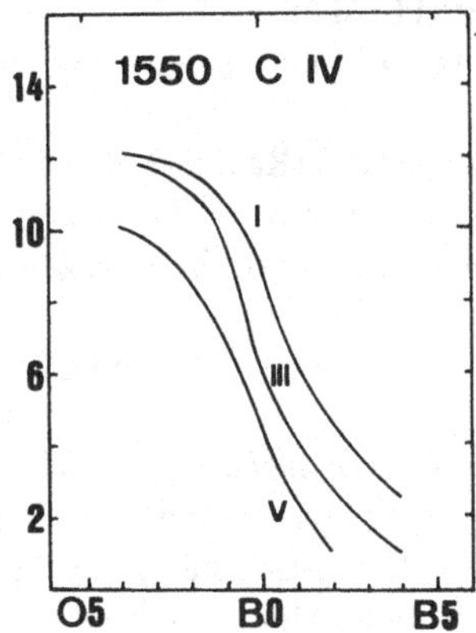

equivalent widths between 2300 and 10 Å, with a clear tendency to exhibit the largest *W* in early type (WC 5 and 6) stars.

Another feature, which appears in strong emission, is the blend of C III 4650, C IV 4658 and He II 4686, with average wavelength 4650. This feature has *W* between 1300 and 10 Å. Another feature present in strong emission is C III 5696 (*W* between 1000 and 30 Å). Usually, but not always, He I at 5876 (*W* between 170 and 7 Å) and O III + O V at 5592 (*W* between 400 and 5 Å) are also present.

It should be added that no nitrogen lines are seen in WC stars (Conti and Massey 1989).

In the infrared spectral region one also finds the same atomic species in emission. One thus finds He II 10124 (*W* up to 260 Å) in practically all stars, except a few late WC stars. C II is present at 7236, 9234 and 9891; the first line being the most intense one (*W* up to 500 Å). This species is absent in early WC stars. C III is represented by 6742, 8500, 8665 and 9711, the last one being the most intense (*W* up to 1000 Å). This species is present in practically all WC stars. C IV is represented by 7061, 7726 and 8859, the most intense being 7726 (*W* up to 280 Å). These lines are present in practically all WC stars (Conti *et al.* 1990). The presence of C III and C IV was confirmed in the 1–3 μm region by Eenens *et al.* (1991), who detected several features corresponding to both species. Smith and Hummer (1988) add also emission lines of Si IV, C II and O V and O VI in the 1–3 μm region.

In the ultraviolet, finally, one finds again the same species in emission: C IV (very strong emissions), C III and C II (the last two preferentially in the later subclasses) and He II (1640 in very strong emission). In addition one also sees emissions of C V in the earlier subclasses, O V and O IV, Si IV (1402) and Fe V, Fe IV and Fe III. For details see Willis *et al.* (1986). Nitrogen is again absent in the ultraviolet spectral region.

Behavior in other non-normal stars

C is usually weakened in *sdO* and *sdB* stars in which He is weak. For instance C II (4267) has *W* smaller by a factor of two or more (Baschek and Norris 1970). When He is strong, C is normal or strong (Husfeld *et al.* 1989). A strong ultraviolet C II 1335 line is characteristic of *He-strong objects* (Jaschek and Jaschek, unpublished).

sdO stars with strong He may also have C enhanced (Rauch *et al.* 1991). C IV lines (1548 and 1550) are strongly enhanced in *CNO stars* of the carbon variety (by factors of 2–3), whereas they are weakened in those of the nitrogen variety (Walborn and Panek 1985).

C III lines are weakened by factors of 2–3 in nitrogen-enhanced CNO stars (Schoenberner *et al.* 1988).

C lines show a variety of behaviors in the *central stars of planetary nebulae*. Many but not all of the H-deficient subgroup show large C overabundances and there exist also indications of a variety of C strengths in the H-rich subgroup. The lines can be observed both in emission and in absorption (Mendez 1991).

C is very overabundant in *pre-degenerate stars* (Werner *et al.* 1991, Werner 1991).

C I and C II have been analyzed in early type stars by Roby and Lambert (1990). On average C is about normal in *Bp stars* of the Hg–Mn and Si subgroups, but seems slightly weak in *Ap stars* of the Cr–Eu–Sr subgroup (*W* values less by a factor of four or thereabouts).

C is about normal in *Am stars* (Boyartchuk and Savonov 1986).

The ultraviolet lines of C I (1657 and 1931) are very strong in *lambda Boo stars* (Baschek *et al.* 1984).

C is weak in most of the *HB stars* (Lamontagne *et al.* 1985).

The abundance of C has been investigated by Boesgaard and Friel (1990) in F-type stars covering a wide range in age. No trend of C with Fe is apparent.

C I absorption lines are strong in the principal spectrum phase of *novae*, whereas C II lines are prominent in the Orion spectrum phase (Payne-Gaposchkin 1957). Emission lines of C II, C III and C IV are present in the ultraviolet region (Warner 1989). In general C is overabundant by one order of magnitude in novae of the C–O subgroup (Andreae 1993).

In late type stars C is not easily observable, except as the molecules C_2, CH, CO and CN. These molecules have numerous strong bands in the classical region and dominate the infrared region.

Carbonated molecules (in warmer stars C_2 and CN, in cooler stars HCN and C_2H_2) are especially strong in the so-called *C stars*, formerly also called *R* (warmer) and *N* (cooler) stars, see Jaschek and Jaschek (1987a). In fact the explanation of the existence of C stars invokes the fact that more C atoms than O atoms exist in the atmospheres of these giant stars. The O atoms are then used to form CO molecules, with no O atoms left for oxygenated molecules like TiO, VO and ZrO, which are characteristic of M-type stars. The explanation that, in oxygen-rich giants, all C is locked up in CO is true in general, except in the coolest and densest atmospheres, where the C is locked up mostly in CH_4 (Tsuji 1964).

M stars and C stars are found over the same temperature interval, but M-type stars are about 100 times more frequent than C stars. C and M stars are the result of the (different) evolution that the stars have undergone. The C/O ratio is, however, not the same in all C stars and it may vary in the range 1–5 (Kilston 1975, Dominy 1984). In the sun this ratio is 0.6. Later work by Lambert *et al.* (1986) has shown that the high ratios in C stars are probably spurious and that the C/O ratio is less than 1.6. Green *et al.* (1991) and Warren *et al.* (1993) have discovered a small number of *dwarf carbon stars*.

Usually the C giants are surrounded by circumstellar shells which are rich in carbon, characterized by an emission feature at 11.0–11.5 μm attributed to silicon carbide. Only a few exceptions to this rule are *J-type stars* (Lloyd Evans 1991, Barnbaum *et al.* 1991, Lorenz Martins and Codina 1992). For a definition of J-type stars, see the discussion under carbon isotopes.

A cautionary remark must, however, be made. If the relative abundances of C and O were the factor determining the existence of carbonated or oxygenated molecules, then it would be impossible to explain how some variables can change from S to C stars (Lambert 1989). An *S star* is a star of the oxygen sequence, where the TiO bands have been replaced by those of ZrO. It is thus not certain that what we see on the surface of the star is in all cases what nucleosynthesis has fabricated in the interior of the star. In *Ba stars*, the CH and C_2 bands are moderately enhanced (Bidelman and Keenan 1951), which leads to a moderate overabundance of C. C I lines are strengthened in the spectrum of *R CrB stars* (Cottrell and Lambert 1982) and also in the so-called *hot R CrB stars* (Jeffery and Heber 1993).

C has been analyzed in normal (i.e. not metal-weak) K 5–M 6 giants by Lazaro *et al.* (1991) using the 2.3 μm band of CO. They find a systematic depletion of C with respect to the sun, the average deficiency being −0.6 dex.

C is weak in the *weak G-band stars* (Cottrell and Norris 1978).

C seems to behave in a manner parallel to that of Fe in *metal-weak stars* (Carbon *et al.* 1987, Wheeler *et al.* 1989) although this is not established. Some authors think that C becomes overabundant with respect to Fe in extremely metal-weak stars.

In *globular cluster giants*, measurement of CN strength has shown that the behavior of C varies from one cluster to another, even at the same Fe/H ratio (Langer *et al.* 1992). Several authors have suggested correlations of C with either the Fe/H or the C^{12}/C^{13} ratio, or an anticorrelation with N. Since most of these studies are based on a small number of objects it seems best to consider this problem as being unsolved.

C seems to vary in a manner parallel to that of Fe in the *stars of the Magellanic Clouds* (Barbuy *et al.* 1981, Spite and Spite 1990).

An interesting group is constituted by the *hydrogen deficient carbon stars*. This is a small group of stars, first defined by Bidelman (1953), which exhibit strong C I lines and strong C_2 and CN bands, as well as very weak or absent H lines and weak or absent CH bands.

Carbon isotopes

Carbon has seven isotopes, among them two stable ones, namely C^{12} and C^{13}, and the short-lived C^{14} (half life 5.7×10^3 years). In the solar system the ratio C^{13}/C^{12} is of the order of 1.1×10^{-2}. The easiest way to study this ratio in stars is to look at the behavior of molecules like CN, CO, C_2 and CH, which may be formed with C^{12} and C^{13} isotopes. The following band heads have frequently been used:

4744 $C^{12}C^{13}$
4753 $C^{12}C^{12}$
6206 $C^{12}N^{14}$
6260 $C^{13}N^{14}$
8004 $C^{13}N^{14}$
8010 $C^{12}N^{14}$

Sanford (1929) was the discoverer of the $C^{12}C^{13}$ bands and he concluded correctly that this implied anomalies in the isotope ratio C^{12}/C^{13}. This originated a large number of studies in which a wide variety of isotope ratios has been found. For instance in C stars the ratio C^{12}/C^{13} varies in the range 3–20 (Kilston 1975, Dominy 1984). Lambert *et al.* (1986) found that the ratio lies, in general, between 30 and 70. Stars very rich in C^{13} are rather exceptional.

It seems also that there exists a progression of C^{12}/C^{13} ratios from M, MS, S, SC and C stars. Whereas for M stars this ratio is of the order of 13, it is 60 for early *C stars*. However, not all authors agree with this conclusion – see for instance Lazaro *et al.* (1991).

It has also been suggested that the isotope ratio is a function of the strength of the iron peak elements (Sneden 1991).

Jorissen (1988) finds ratios between 8 and 24 for *Ba stars*. Tsuji *et al.* (1991) find very high ratios in some *CH stars* and suggest that these ratios are related to the strength of the C_2 bands, in the sense that ratios are high if C_2 bands are strong.

Cottrell and Lambert (1982) find very high C^{12}/C^{13} ratios in *R CrB stars.*

Schild *et al.* (1992) have investigated the ratio in *symbiotic stars* and find that it is undistinguishable from that in normal red giants.

Fix and Cobb (1987) studied the C^{12}/C^{13} ratio in one *OH master* and found a ratio of about 10.

Other molecules have also been used to derive the C^{12}/C^{13} ratio, like for instance CS and SiC_2 (Cernicharo *et al.* 1991b), and one can add the use of CO and HCN.

All these studies present many difficulties from a theoretical viewpoint – for a discussion of some of the difficulties, see Gustafsson (1989).

Carbon isotope ratios in *globular cluster stars* have also been studied several times – see for instance Suntzeff and Smith (1991).

There exists a group of stars – the so-called J stars – whose isotopic ratios are very anomalous. The J stars were defined by Bouigue (1954) as carbon stars in which the C^{13}/N^{14} band is very conspicuous, which can happen only if the C^{12}/C^{13} ratio is rather low.

Early J stars show weaker Sr, Y, Zr and Ba lines than C stars of the same temperature, whereas this does not happen in late J-type stars (Dominy 1985).

Up to now, no definite evidence for the existence of C^{14} in stars has been found (Gustafsson 1989), despite a careful search made in C, M and SM stars on the basis of the CO band at 4130 cm^{-1} (Harris and Lambert 1987).

Origin

C^{12} is produced by He burning and C^{13} by hydrogen or by hot hydrogen burning.

CERIUM Ce Z=58

This element was discovered by J. Berzelius and W. Hisinger in Sweden, in 1803. It was, however, isolated only much later (1875). The name comes from the goddess Ceres, whose name was given to an asteroid discovered in 1801.

Ionization energies

Ce I 5.5 eV, Ce II 10.8 eV, Ce III 20.2 eV, Ce IV 36.8 eV.

Absorption lines of Ce I

The strongest line of Ce I (5699) is not present in the solar spectrum.

Absorption lines of Ce II

Table 1. *Equivalent widths of Ce II*

	4628(1)		5274(15)		5610(26)	
Group	V		V	III	III	Ib
F 0		0.024(Ib)				
F 5	0.014					
G 0	0.02					0.046
G 2			0.008			0.032
S	0.014		0.006		blend	
G 5						0.038
G 8						0.074
K 0				0.062	0.018	
K 2				0.028		0.095
K 3				0.063		0.098
K 5						0.123
M 2						0.198

Table 2. *Equivalent widths of Ce II 6043(30)*

Group	III	Ib
G 1		0.031
G 2		0.027,0.035
S		0.002
G 5		0.033
G 8		0.042
K 0	0.004	
K 2		0.047
K 5	0.010	0.034,0.068
M 0	0.013	

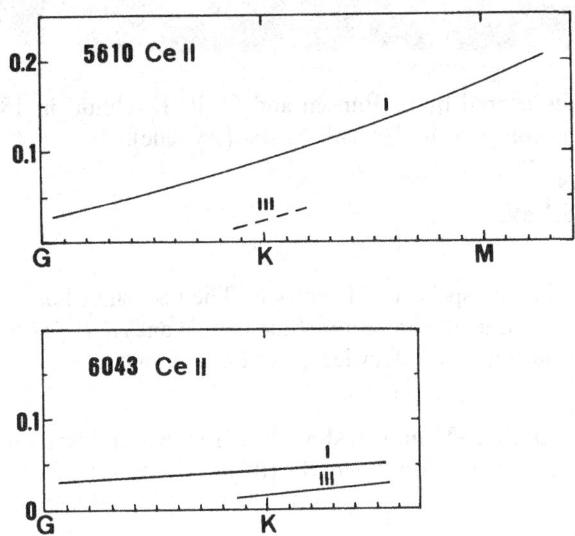

Ce II lines (see for instance the line at 5610) are present in stars later than G-type. A positive luminosity effect is observed.

Behavior in non-normal stars

Ce II is strong in *Ap stars* of the Cr–Eu–Sr subgroup (Adelman 1973b) (*W*(4628) ≃ 0.030–0.100). Of all rare earths, this is the most abundant one in all Ap stars (Cowley 1976). Ce III also has been observed in some of these stars (Aikman *et al.* 1979, Cowley 1984).

Ce II is also strong in *Am stars* (Smith 1973, 1974), where the *W* values are much larger than in normal stars of the same temperature. As an example *W*(4137) is about 0.100 in late Am stars, whereas for normal F 0 dwarfs, *W*(4137) is only 0.080.

As with other rare earths, Ce II lines are intensified in *Ba stars*, which leads to large overabundances (Lambert 1985). A typical value is *W*(4628)=0.16 for a K 0 III Ba star (Danziger 1965).

Ce II is also enhanced in at least one early *C star* (Dominy 1984).

Ce II has been observed in G- and K-type *metal-weak stars* by Gilroy *et al.* (1988). It is overabundant with respect to iron (see also Part Two, section 2.2).

Isotopes

Ce has 19 isotopes and isomers, four of them stable, namely Ce 136, 138, 140 and 142. In the solar system most of the Ce (88%) is in the form of Ce^{140}.

Origin

Ce^{136} and Ce^{138} are produced by the p process, Ce^{142} by the r process and Ce^{140} by either the r or the s process.

CESIUM Cs Z=55

The element was discovered by R. Bunsen and G. R. Kirchhoff in 1806 in Heidelberg. Its name comes from the Latin *caesius* (sky blue).

Ionization energies
CsI 3.9 eV, CsII 25.1 eV.

Behavior in stars
Cs has been observed in the spectrum of sunspots. The resonance line is 8521. CsII was identified as marginally present by Bidelman in one *Ap star* of the Sr–Cr–Eu subgroup according to Ch. Cowley (private communication).

Isotopes
Cs has one stable isotope, Cs^{132}, and 20 short-lived isotopes and isomers. The longest lived one is Cs^{135} with a half life of 3×10^{6} years.

Origin
Cs^{132} is produced by both the r and the s process.

CHLORINE Cl Z=17

This element was discovered by Scheele in Uppsala, Sweden, in 1774. The name comes from the Greek *chloros* (pale green).

Ionization energies
Cl I 13.0 eV, Cl II 23.8 eV, Cl III 39.7 eV.

Absorption lines of Cl I
Artru *et al.* (1989) report some Cl I lines in the ultraviolet spectra of B 7 and B 9 V stars.

A weak line at 8376(2) in the solar spectrum has been attributed to Cl I and some other weak lines have been observed in sunspot spectra. The identification should be confirmed. Nevertheless, the presence of Cl is certain because of the identification of vibration–rotation bands of HCl in sunspots (Hall and Noyes 1972).

Phillips and Keenan (1990) have reported the detection in an X-ray flare of Cl XVI.

Absorption lines of Cl II

Table 1. *Equivalent widths of Cl II*

	3860(25)		4794(1)
Group	V	Ia	V
B 0	0.014		
B 2			0.003
B 5		0.016	

Weak lines of Cl II are seen in early B-type stars.

Weak lines of Cl II are seen in early B-type stars.

Behavior in non-normal stars
Cl II is strong in some early *Bp stars* (Bidelman 1966, Jaschek and Lopez García 1967, Cohen *et al.* 1969), with *W*(4794)=0.031. Sadakane (1992) calls attention to one star with exceptionally strong Cl II lines, with *W*(4253)=0.082, which leads to an overabundance of about 4.0 dex. This star also has highly enhanced Co lines as well as other abundance anomalies.

Cl is also strong in some *HB stars* (Heber 1991), where *W*(4794)=0.035.

Isotopes

There exist 11 isotopes and isomers, including two stable ones, Cl^{35} and Cl^{37}. In the solar system they have respectively 76% and 24% abundances. One short-lived isotope, Cl^{36}, has a half life of 3×10^5 years.

The isotopes Cl^{35} and Cl^{37} have been observed in the circumstellar envelope of the *C-type star* IRC + 10216 through the presence of the Cl isotopes in NaCl and AlCl molecules. The ratio Cl^{35}/Cl^{37} is compatible with the terrestrial value (Cernicharo and Guelin 1987). A similar procedure could be applied to the isotopes of Cl in HCl.

Origin

Cl^{35} is produced by explosive nucleosynthesis and Cl^{37} by this process or by either carbon burning or the s process.

CHROMIUM Cr $Z=24$

The element was discovered by Vauquelin in Paris in 1797. The name comes from the Greek *chromos* (color), because of the colored salts it forms.

Ionization energies

Cr I 6.8 eV, Cr II 16.5 eV, Cr III 31.0 eV, Cr IV 49.1 eV, Cr V 69.3 eV, Cr VI 90.6 eV.

Cr is well represented in stellar spectra, especially those of later types. In the solar spectrum, Cr I lines are more numerous than those of any other neutral element, except Fe.

Absorption lines of CrI

Table 1. *Equivalent widths of Cr I*

	4254(1)		5345(18)	
Group	V	Ib	III	Ib
B 9	0.03			
A 0	0.02			
A 2	0.05			
A 7	0.18			
F 0	0.19	0.22(Ia)		
F 2		0.26		
F 4	0.19			
F 5	0.19	0.35		
F 6	0.24			
F 8	0.24	0.43		
G 0	0.27			0.26
G 1	0.37			0.29
G 2	0.30			0.29
S	0.393		0.107	
G 5	0.45			0.30
G 8				0.29
K 0	1.27		0.14	
K 2			0.40	0.46
K 3				0.33
K 5	2.60			0.62
M 0			0.65	
M 2				0.59
M 2.5			0.38	
M 3			0.54	
M 4			0.47,0.84	
M 5				0.72
M 7			(1.16)	

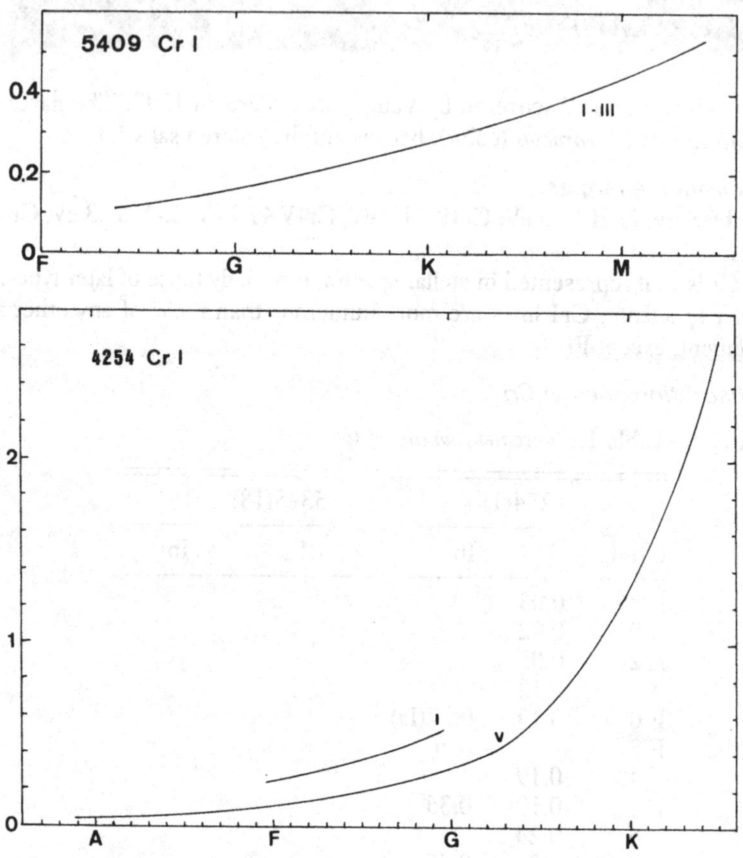

Table 2. *Equivalent widths of CrI 5409(18)*

Group	V	III	Ib
F 4	0.12		
F 5	0.10		0.09
F 6	0.12		
F 8	0.15		0.23
G 0	0.14		
G 1	0.21		
G 2	0.15		
S	0.154		
G 5	0.21		
G 8	0.20(IV)	0.26	
K 0	0.28	0.21	

Table 2. (*cont.*)

Group	V	III	Ib
K 2	0.38		
M 0		0.49	
M 2.5		0.36	

Cr I (for instance the lines at 4254 and 5409) appears in A-type spectra and grows monotonically toward later types. A small positive luminosity effect seems to exist for the line at 4254. According to Keenan and McNeil (1976) the luminosity effect becomes negative after type K.

The most intense infrared line of Cr I is that at 9290(29). In the sun, W=0.065.

Absorption lines of Cr II

Table 3. *Equivalent widths of Cr II*

	4558(44)		5502(50)		
Group	V	Ib	V	III	Ib
B 5	0.005				
B 7	0.023				
B 8		0.036			
B 9	0.080				
B 9.5	0.047,064				
A 0	0.076	0.19,0.15(Ia)			
A 1	0.097				
A 2	0.15,0.10	0.658(0)			
A 3		0.340(0)			
A 7	0.17				
F 0	0.15	0.62(Ia)			
F 2		0.26			
F 4	0.13				
F 5	0.14	0.36			
F 6	0.11				
F 8	0.11	0.46			
G 0	0.09		0.00	0.00	0.072
G 1	0.14				
G 2	0.07		0.017		0.083
S	0.066		0.023		

Table 3. (*cont.*)

	4558(44)		5502(50)		
Group	V	Ib	V	III	Ib
G 5					0.062
G 8	0.095				0.058
K 0	0.055		0.017	0.035	
K 2				0.027	0.078
K 3				0.049	0.095
K 5	0.06				0.129

Table 4. *Equivalent widths of CrII 5237(43)*

Group	V	Ib
S	0.049	
G 5		0.144
G 8		0.161
K 2		0.181
K 5		0.153,0.197

Cr II appears in late B-type dwarfs, increases to a flat maximum in early F-type and declines toward late dwarfs (see for instance the line at 4558). In supergiants the maximum is displaced toward later types. A positive luminosity effect exists.

Emission lines of CrII

Both permitted and forbidden Cr emission lines are observed in *B[e] stars* (see Swings (1973) for the analysis of a typical object). They are also observed in the ultraviolet region of *T Tau stars* (Appenzeller *et al.* 1980) and in *symbiotic stars* (Swings and Struve 1941a).

Absorption lines of CrIV and CrV

CrIV and CrV have been identified in the ultraviolet spectra of O-type stars (Dean and Bruhweiler 1985).

Behavior in non-normal stars

Cr II is strong in *Ap stars* of the Cr–Eu–Sr subgroup. Typically *W*(4558)=0.30. It should, however, be observed that the strongest Cr II laboratory lines are not the Cr II lines that are most enhanced in these stars (Jaschek and Jaschek 1974).

Cr behaves in a manner parallel to that of Fe in *metal-weak dwarfs* (Magain 1989) and in *globular cluster stars* (Wheeler *et al.* 1989).

Cr can be considered to be a typical 'metal'.

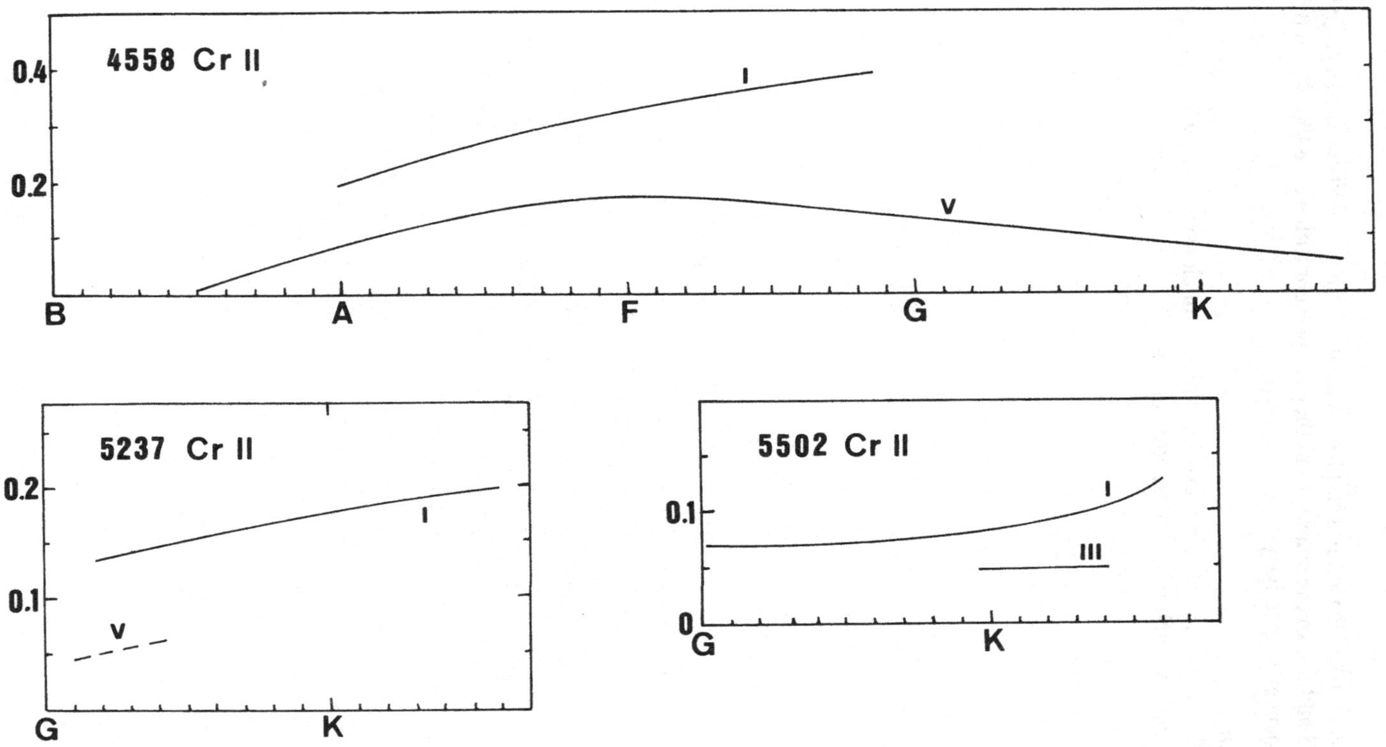
4558 Cr II
0.4
0.2
I
V
B
A
F
G
K
5237 Cr II
0.2
0.1
I
V
G
K
5502 Cr II
0.1
0
I
III
G
K

Isotopes

Cr has four stable isotopes and five short-lived ones. The stable ones are Cr 50, 52, 53 and 54. In the solar system they are present with 4%, 84%, 9% and 3% abundances respectively.

Origin

Cr 50, 52 and 53 are produced by explosive nucleosynthesis, whereas Cr^{54} is produced by nuclear statistical equilibrium.

COBALT Co Z=27

This element was discovered in 1735 by G. Brandt in Stockholm. The name comes from the German *Kobold* (goblin).

Ionization energies
Co I 7.9 eV, Co II 17.1 eV, Co III 33.5 eV.

Absorption lines of Co I

Table 1. *Equivalent widths of Co I*

	4121(28)		5352(172)		
Group	V	Ib	V	III	Ib
A 2	0.003				
F 0		0.050(II)			
F 4			0.012		
F 5	0.135	0.120			0.007
F 6			0.021		
F 8		0.160	0.012		
G 0	0.120				
G 1			0.024,0.022		0.057
S	0.125		0.021		
G 2	0.130				0.070
G 5	0.160		0.016		0.074
G 6					0.123
G 8			0.037(IV)	0.044	0.139
K 0			0.02	0.065	
K 2			0.02	0.066	0.152
K 3				0.114	0.135
K 5	0.240				0.162
M 0				0.083	
M 2					0.153

Co is represented by numerous faint lines in late type spectra. As far as the number of lines are concerned, Co I ranks after Fe I, Cr I and Ti I.

Co I (see the lines at 4121 and 5352) appears in A-type stars and increases toward later types. A positive luminosity effect is present.

In the infrared one important Co I line is that at 8093(189). In the sun, $W(8093)=0.026$.

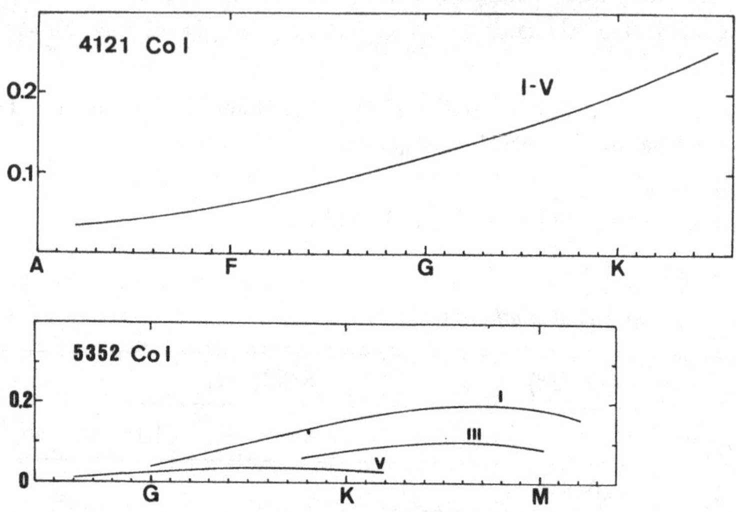

Emission lines of Co I

In *long-period variables* of type S, some Co I lines (those at 3894, 3997, 4118 and 4121) appear in emission around maximum light (Merrill 1952).

Absorption lines of Co II

The equivalent width of Co II 3388(2) in the sun is 0.031.

Co II has been identified in the ultraviolet spectrum of an A 0V star (Rogerson 1989).

Emission lines of Co II

Co II 2202(M.l) is often seen in emission in *T Tau stars* (Appenzeller *et al.* 1980). The *supernova* 1987A presents some strong Co II emissions and it is possible that these are due to radioactive Co^{56} decaying from Ni^{56} (Arnett *et al.* 1989). Varani *et al.* (1990) identified an emission at 1.547 μm with [Co II] and have proved that it is due to Co^{56}.

Behavior in non-normal stars

Co I is enhanced in some *Ap stars* of the Cr–Eu–Sr subgroup (Adelman 1973b). Typically $W(3873) \simeq 0.070$. Co I is also enhanced in many stars of the Hg–Mn subgroup (Takada-Hidai 1991).

Cowley (1979) called attention to a few Ap stars in which the Co II lines are very strong. Sadakane (1992) studied one of these stars in detail and found very strong Co II lines, as well as other anomalies. The Co overabundance is of the order of 3.6 dex.

Co is about normal in *Am stars* (Smith 1973, 1974).

Co behaves in a manner parallel to that of Fe in *globular cluster stars* (Wheeler *et al.* 1989). For field stars the evidence is somewhat contradictory, but on the whole it seems that Co behaves like Fe (Luck and Bond 1985, Gilroy *et al.* 1988).

Strong absorption lines of Co II appear in the spectra of *supernovae* of types Ia and II. [Co II] and [Co III] appear in the nebular stages of supernovae of types II and Ia respectively (Branch 1990).

Isotopes

Co has 14 isotopes and isomers – 13 short-lived ones and one stable one, Co^{59}. In the solar system all the Co is in this form.

Origin

Co^{59} is produced either by the nuclear statistical equilibrium process or by carbon burning.

COPPER Cu Z=29

This element has been known since prehistory. The name comes from the Latin word *Cuprum* (Cyprus), because on this island there existed heavily exploited copper mines.

Ionization energies
CuI 7.7 eV, CuII 20.3 eV, CuIII 36.8 eV.

Absorption lines of CuI

Table 1. *Equivalent widths of CuI 5105(2)*

Group	V	III	Ib
F 0	0.035		
F 4	0.025		
F 5	0.049		0.030
F 6	0.037		
F 8	0.037		0.083
G 0	0.060		0.158
G 1	0.083		0.184
S	0.082		0.203
G 2	0.092		0.178
G 5			0.245
G 6			0.301
G 8	0.112(IV)	0.15	0.286
K 0	0.110	0.200	
K 2		0.182	0.269,0.354
K 3		0.230	
K 5			0.309,0.368,0.394
M 0		0.58	
M 2			0.388,0.58(Ia)
M 3		0.34	

CuI (see the line at 5105) appears in early F-type stars and intensifies toward later types. A positive luminosity effect exists.

Absorption lines of CuII
CuII lines are visible in the ultraviolet spectral region of B- and A-type stars.

Emission lines of CuII
The 3806 line from [CuII] is present in the *luminous blue variable* eta Car (Thackeray 1953) and in one typical *B[e] star* (Swings 1973).

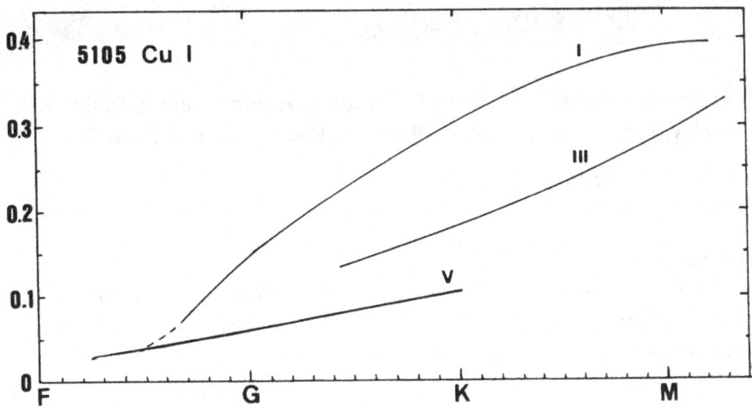

Behavior in non-normal stars

The first detection of the resonance line of Cu II 1358(M.3) line in *Ap stars* was made by Bidelman (1979). Jacobs and Dworetsky (1981) measured the line in several *Bp stars* of the Hg–Mn subgroup and found it to be very strong (*W* up to 0.270). Takada-Hidai (1991) noticed, however, that in some stars of this subgroup Cu can be very weak or absent.

Cu I lines are also strong in *Am stars*. Typical *W* values are twice as large as in normal stars of the same temperature (Smith 1973, 1974).

Sneden *et al.* (1991) find that, in *metal-weak* (disk and halo) *stars*, Cu is weakened with respect to the metals Fe and Ni, to the point that, in extreme metal-weak objects, Cu lines are undetectable. A similar behavior is found in *globular cluster stars* (Wheeler *et al.* 1989). One emission line of [Cu II] is seen in one *VV Cep star* (Rossi *et al.* 1992).

Isotopes

Cu has nine short-lived isotopes and two stable ones, Cu^{63} and Cu^{65}. In the solar system 69% of all Cu is in the form of Cu^{63} and the remainder is in the form of Cu^{65}.

Origin

Cu^{63} and Cu^{65} are both produced by explosive nucleosynthesis, but Cu^{63} can also be produced by carbon burning.

CURIUM Cm $Z=96$

This element was discovered by G. Seaborg, R. James and A. Ghiorso in 1944 in Berkeley, California. Its name alludes to Pierre and Marie Curie.

Ionization energy
Cm I 6.0 eV.

Absorption lines of Cm I
Cm I was identified on the basis of eight lines present in one *Ap star* of the Cr–Eu–Sr subgroup (Jaschek and Brandi 1972). Later work, however, did not confirm this identification.

Isotopes
Cm is unstable. The two longest lived isotopes (there exist in total 13) are Cm^{247} and Cm^{248}, with half lives of 1.6×10^7 and 4.7×10^5 years respectively. Cm^{247} could be used for radioactive dating.

Origin
Cm is produced by the r process.

DYSPROSIUM Dy *Z*=66

This element was discovered by P. E. Lecoq de Boisbaudran in Paris in 1886. The name comes from the Greek *dysprositos* (hard to obtain).

Ionization energies

DyI 5.9 eV, DyII 11.7 eV, DyIII 22.8 eV, DyIV 41.5 eV.

Absorption lines of DyI

DyI is absent in the sun, but present in one M2III star (Davis 1947).

Absorption lines of DyII

Table 1. *Equivalent widths of DyII 4103(–)*

Group	V
S	0.012
K 5	0.050

Several weak lines of DyII were reported in one F 0Ib star, in one F 5V star (Kato and Sadakane 1986) and in late type giants. *W*(4000)=0.047 according to Reynolds *et al.* (1988) for F 5V.

Abundance anomalies

DyII is present in *Ap stars* in general (Cowley 1976, Adelman *et al.* 1979). In several Ap stars with strong DyII lines, Aikman *et al.* (1979) also detected DyIII, which fact was confirmed by Cowley and Greenberg (1987). DyII is also present in at least one late *Am star* (van t'Veer-Menneret *et al.* 1988) with *W*(3944)=0.152.

DyII was observed by Gilroy *et al.* (1988) in *metal-weak* G- and K-type dwarfs. These authors find the element to be overabundant with respect to iron. See also the discussion on rare earths.

DyI lines are enhanced in some *Ba stars* (Lambert 1985).

DyII lines were found in at least one *S-type star* (Bidelman 1953).

Isotopes

Dy occurs in the form of seven stable isotopes – Dy 156, 158, 160, 161, 162, 163 and 164. In the solar system Dy^{161} represents 19% and Dy 162, 163 and 164 represent 25%, 25% and 28% respectively. The rest is made up by the other stable isotopes. There exist 14 unstable isotopes and isomers, among them Dy^{154} with a half life of 10^6 years.

Origin

Dy is produced by a variety of processes; Dy^{156} and Dy^{158} by the p process, Dy^{161} and Dy^{163} by the r process, Dy^{160} by the s process and Dy^{162} and Dy^{164} by either the r or the s process.

ERBIUM Er $Z=68$

This element was discovered by C. Mossander in Stockholm in 1842. The name comes from the city of Ytterby in Sweden. This city is also referred to in the names of the elements yttrium and terbium.

Ionization energies

Er I 6.1 eV, Er II 11.9 eV, Er III 22.7 eV, Er IV 42.6 eV.

Absorption lines of Er II and Er III

Er II 3616 is seen in one F 0Ib star (W=0.015) according to Reynolds *et al.* (1988).

Behavior in non-normal stars

Er II lines are strengthened in the spectra of some *Ap stars* of the Cr–Eu–Sr subgroup. Aikman *et al.* (1979) also observed Er III in the spectra of stars with strong Er II lines, and this was confirmed by Cowley and Greenberg (1987). The Er III line at 4000 has W=0.040. Er II and Er III lines were also detected in the spectrum of one *Bp star* of the Si subgroup (Cowley and Crosswhite 1978).

Er II lines are also seen in at least one *Am star* (van t'Veer-Menneret *et al.* 1988) with W(4009)=0052.

Er II lines are enhanced in at least one *Ba star* (Lambert 1985) and in at least one *S-type star* (Bidelman 1953).

Isotopes

Er has six stable isotopes and ten unstable isotopes and isomers. The stable ones are Er 162, 164, 166, 167, 168 and 170. In the solar system Er^{166} represents 33% and Er 167, 168 and 170 respectively 23%, 27% and 14%.

Origin

Er is made by several processes – Er^{162} by the p process, Er^{167} and Er^{170} by the r process and the others can be made by two processes, namely Er^{164} by the p or the s and Er^{166} and Er^{168} by either the r or the s process.

EUROPIUM Eu Z=63

This element was discovered by E. A. Demarcay in Paris in 1901. The name comes from the continent, which in turn alludes to a Greek mythological figure.

Ionization energies

Eu I 5.7 eV, Eu II 11.2 eV, Eu III 24.9 eV, Eu IV 42.6 eV.

Eu is the rare earth that has been most extensively studied in stars. Whereas Eu I lines are very weak, Eu II lines are more easily detectable.

Absorption lines of Eu I

In the sun Eu I 4627(1) has W=0.004. According to Gopka *et al.* (1990), W(5765)=0.013 in both K 5III and M 0III.

Absorption lines of Eu II

Table 1. *Equivalent widths of Eu II*

	4205(1)		6645(8)		
Group	V	Ib	V	III	Ib
F 0	0.010				
F 5	0.010	0.22			
G 0	0.120				0.083
S blend with VI			0.004		
G 2			0.004		0.063
G 5	0.176				
K 0				0.031	
K 2				0.033	0.098
K 3				0.045	0.089
K 5					0.098

Eu II lines (for instance those at 4205 and 6645) appear at mid-F-type and grow in intensity toward cooler stars. A positive luminosity effect exists.

Behavior in non-normal stars

Eu II lines are very strong in *Ap stars* of the Cr–Eu–Sr subgroup. Typically W(4205)=0.20 (Sadakane 1976). Attention should be paid to the fact that many lines are widened by hyperfine structure (Cowley 1980) and/or by Zeeman broadening. Abundances should thus be corrected for these effects (Hartoog *et al.* 1974). If these corrections are applied, then the abundance of Eu becomes comparable to that of other rare earths. However, Cowley (1984) calls attention to the fact that there exists at least one star in which Eu is the dominant rare earth.

Eu II is also strong in *Am stars*. A typical value for early type Am stars is W(4205)=0.010, and, for late Am stars, W(4205)=0.300.

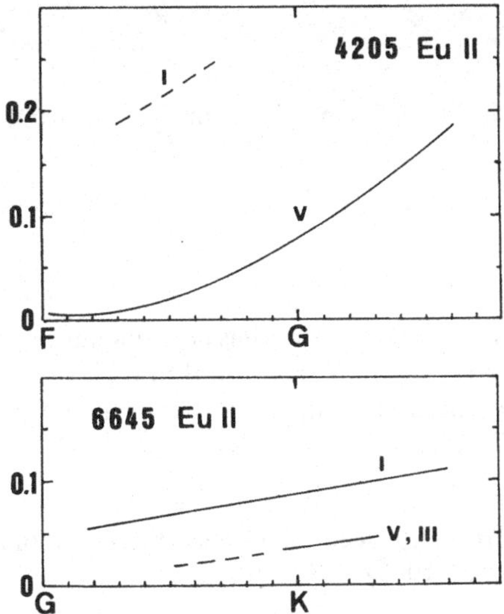

Eu II is also strong in *delta Del stars*. For instance *W*(4205)>0.150 (Kuntz 1976).

Eu I and Eu II lines are enhanced in some *Ba stars* to varying degrees. This leads to overabundances which are, however, smaller than those of other rare earths (Lambert 1985). As a general comment Cowley and Dempsey (1984) observe that, whereas Eu is often very strong in Ap stars, the enhancement in Ba stars is less pronounced.

Eu II lines are present in *S-type stars* (Merrill 1947).

Eu II lines have been observed in G- and K-type *metal-weak dwarfs* by Gilroy *et al.* (1988). These authors find Eu overabundant with respect to Fe. However, Sneden *et al.* (1988) call attention to the large scatter in the Eu/Fe ratio in different stars. Magain (1989) confirmed the overabundance with respect to iron and indicated an overabundance of 1.0 dex. (See also Part Two, section 2.2.) Gilroy *et al.* (1988) pointed out a close correlation between the abundances of Ba and Eu, which seems also to be valid for *globular cluster stars* of our galaxy (Francois 1991). Spite *et al.* (1991) find the same correlation in one cluster of the *Small Magellanic Cloud.*

Isotopes
Eu has two stable isotopes, Eu^{151} and Eu^{153}. In the solar system they represent 48% and 52% respectively of all Eu. There exist 19 unstable isotopes and isomers.

Kato (1987) found from the 4130 line in one F 5 star that the ratio Eu^{153}/Eu^{151} agrees with the solar ratio.

Origin
Eu^{151} and Eu^{153} are produced by either the r or the s process.

FLUORINE F Z=9

This element was first isolated by H. Moissan in Paris in 1886. Its name comes from the Latin *fluere* (to flow).

Ionization energies

F I 17.4 eV, F II 35.0 eV, F III 62.7 eV.

F in stars

The presence of some faint F II lines in B-type stars has been announced, but not confirmed. Beals (1951) observed absorption lines of F II in P Cyg.

The best evidence for the presence of F in stars is the molecule HF, which is present in the sun and in K 5 III to M 4III stars (Jorissen *et al.* 1992).

Behavior in non-normal stars

In *S, SC and C stars* the HF feature is enhanced, which leads to an overabundance of F by up to 0.6 dex (Jorissen *et al.* 1992).

Isotopes

F has only one stable isotope, namely F^{19}, and five short-lived ones.

Origin

F^{19} is produced by hot hydrogen burning.

GADOLINIUM Gd Z=64

This element was discovered by J. Galisard in Geneva in 1880. The name alludes to the Finnish chemist J. Gadolin.

Ionization energies
Gd I 6.1 eV, Gd II 12.1 eV, Gd III 20.6 eV, Gd IV 44.0 eV.

Absorption lines of Gd I

Table 1. *Equivalent widths of Gd I 5803*

Group	III
K 5	0.010
M 0	0.008

Source: Data from Gopka *et al.* (1990).

No Gd I lines are seen in the solar spectrum.

Absorption lines of Gd II

Table 2. *Equivalent widths of Gd II*

	3101(12)	4342(15)	5952(95)
Group	V	III	III
S	0.098		
K 0		0.128	
K 5			0.014
M 0			0.021

Source: Data for 5952 are from Gopka *et al.* (1990).

Gd II appears in late F-type stars and intensifies toward cooler types.

Gd II is also seen in one F 0 Ib star (Reynolds *et al.* 1988).

Emission lines of Gd II
Some Gd II lines, excited by fluorescence from H epsilon, have been observed in emission in one *Mira variable* (Grudzinska 1984).

Behavior in non-normal stars
Gd II is weakly present in some *Bp stars* of the Hg–Mn subgroup, with $W(3850)=0.005$ (Adelman 1989).

Gd II is strong in *Ap stars* of the Cr–Eu–Sr subgroup (Adelman 1973b).

Typical values are $W(4251)=0.04$ (Sadakane 1976). There even exists a star in which, exceptionally of all rare earths, only Gd II is strengthened (Cowley and Henry 1979). Gd III is seen in Ap stars of the Cr–Eu–Sr subgroup (Cohen *et al.* 1969).

Gd II is also strong in *Am stars*. Typical values for $W(4251)$ of early types are 0.020 and of late types, $W=0.100$ (Smith 1973, 1974). Similar values are found in *delta Del stars* (Kurtz 1976).

Gd I and Gd II lines are enhanced in some *Ba stars* (Lambert 1985). In these stars the W values may be twice as large as in normal giants of the same temperature.

Gd II lines are present in *S-type stars* (Merrill 1947).

Gilroy *et al.* (1988) observed Gd II 4130(M.19,49) in a number of *metal weak G- and K-type dwarfs*. These authors find the element to be overabundant with respect to iron. See also Part Two, section 2.2.

Isotopes

Gd has seven isotopes – Gd 152, 154, 155, 156, 157, 158 and 160. In the solar system Gd 155, 156, 157, 158 and 160 are present as 15%, 21%, 16%, 25% and 22% of all Gd. Furthermore, there exist ten other short-lived isotopes, among them Gd^{150}, with a half life of 2×10^6 years.

Origin

Gd is made by several processes: Gd 155, 156, 157 and 158 by either the r or the s process, Gd^{152} by the p or the s process, Gd^{154} only by the s process and Gd^{160} only by the r process.

GALLIUM Ga Z=31

This element was discovered in 1875 by P. Lecoq de Boisbaudran in Paris. The name comes from the Latin designation of France (*Gallia*) or from the translation of Lecoq (*gallus*).

Ionization energies

Ga I 6.0 eV, Ga II 20.5 eV, Ga III 30.7 eV, Ga IV 64.2 eV.

Absorption lines of Ga I

The equivalent width of Ga I 4172(1) in the sun is 0.042. GaI is also seen in one M 2 III star (Davis 1947) and in *long-period variables* of types M and S (Merrill and Lowen 1953).

Emission lines of Ga I

The 4172 line of Ga I is present in emission toward minimum light in *long-period variables* (Merrill and Lowen 1953).

Absorption lines of Ga II

The equivalent width of Ga II 4256 in B 8 V stars is 0.005.

Behavior in non-normal stars

Ga II was first detected by Bidelman and Corliss (1966) in *Bp stars* of the Hg–Mn subgroup. It is enhanced in almost all objects of this type. Typically $W(4256)=0.090$ (Kodaira and Takada 1978).

The ultraviolet Ga II resonance line 1414 is strong in some Hg–Mn stars (Jacobs and Dworetsky 1981), with W up to 1.30. Jaschek and Jaschek (1987a) have introduced a subgroup called '*ultraviolet gallium stars*' characterized by a strong 1414 line. Objects of this group may belong in the photographic region to a variety of subgroups (Si, He-weak, Hg–Mn, etc.).

Takada-Hidai *et al.* (1986) investigated in detail the ultraviolet lines of GaII (1414) and Ga III (1495) in hot Bp stars. They found Ga III lines to be strong in Bp stars of the same subgroups as those in which Ga II lines are enhanced, with $W(1414)$ up to 1.86 in Hg–Mn stars and $W(1495)$ up to 0.067. For these spectral types (B 5–A 0) the Ga lines are below the detection limit (which can be assumed to be 0.002 Å) in normal stars. Lanz *et al.* (1993) have observed the red Ga II lines (6334, 6419 and 6456) in *Ap stars*, finding that Ga II is enhanced both in magnetic and non-magnetic stars. Typically $W(6334)<0.11$.

Isotopes

Ga has two stable isotopes – Ga^{69} and Ga^{71} – and 12 short-lived ones. In the solar system, Ga^{69} is 60% of all Ga and Ga^{71} is 40%.

Origin

The two stable Ga isotopes are produced by s, r or nuclear statistical equilibrium processes.

GERMANIUM Ge Z=32

This element was discovered by C. Winkler in Freiberg, Germany in 1886. The name comes from *Germania* (Germany).

Ionization energies

GeI 7.9 eV, GeII 15.9 eV, GeIII 34.2 eV.

Absorption lines of GeI

Table 1. *Equivalent widths of GeI 4685(3)*

Group	V	III
S	0.002	
K 0		0.034

Source: The value for the sun is from Lambert *et al.* (1969).

Behavior in non-normal stars

GeI lines were announced as probably present in one star of the Cr–Eu–Sr variety of *Ap stars* (Adelman *et al.* 1979). GeII has been observed in one *Bp star* of the Hg–Mn subgroup by Leckrone *et al.* (1991a).

Isotopes

Ge has five stable isotopes, namely Ge 70, 72, 73, 74 and 76 as well as 12 short-lived isotopes and isomers. The five stable isotopes represent in the solar system 20%, 27%, 8%, 37% and 8% respectively of the Ge.

Origin

Ge isotopes are produced by different processes. Ge^{76} is produced only by the nuclear statistical equilibrium process but all the others are produced by this process or the s process. Ge 72, 73 and 74 can also be produced by the r process.

GOLD Au Z=79

Gold has been known since prehistory. The symbol is derived from Latin *aurum* (gold).

Ionization energies
AuI 9.2 eV, AuII 20.5 eV, AuIII 30.0 eV.

Absorption lines of AuI
In the sun, the equivalent width of AuI 3122(1) is 0.005.

Behavior in non-normal stars
The probable detection of AuI was announced by Jaschek and Malaroda (1970) in one *Ap star* of the Cr–Eu–Sr subgroup. Fuhrmann (1989) detected Au through the ultimate line of AuII at 1740(2) in several *Bp stars* of the Si and *Ap stars* of the Cr–Eu–Sr subgroups. The presence of Au seems to be associated with that of platinum and mercury.

Isotopes
Au has one stable isotope, Au^{197} and 20 short-lived isotopes and isomers.

Origin
Au can only be produced by the r process.

HAFNIUM Hf Z=72

This element was discovered in 1923 by D. Coster and G. von Hevesey in Copenhagen, Denmark. The name comes from the Latin name of Copenhagen (*Hafnia*).

Ionization energies
HfI 6.6 eV, HfII 14.9 eV, HfIII 23.3 eV.

Absorption lines of HfI
The equivalent width of HfI 3616 in the sun is 0.001.

Absorption lines of HfII
The equivalent width of HfII 3505(7) in the sun is 0.024. In one F 0 Ib star, *W*(3399)=0.008 (Reynolds *et al.* 1988).

Behavior in non-normal stars
HfI lines are enhanced in some *Ba stars*, which leads to large overabundances of the order of 0.5–1.0 dex (Lambert 1985, Smith 1984).

Isotopes
Hf has six stable isotopes, 12 short-lived isotopes and isomers and one isotope with a long half life. The stable ones are Hf 174, 176, 177, 178, 179 and 180. In the solar system they represent respectively 0.2%, 5%, 19%, 27%, 14% and 35% of all the Hf. The long-lived isotope Hf^{182} has a half life of 9×10^6 years, which could be useful for radioactive dating.

Origin
Hf isotopes can be produced by several nuclear processes. Hf^{174} is a pure p process and Hf^{176} a pure s process product. The other four (Hf 177–180) can be produced by both the r and the s process.

HELIUM He Z=2

This element is the only one found in the sun (1868, by Janssen) before being detected on earth. It was first isolated by W. Ramsey in London in 1895, and independently by P. T. Cleve and N. A. Langlet in Uppsala. The name comes from the Greek *helios* (sun). The history of its discovery is given by Frost (1895).

Ionization energies
He I 24.6 eV, He II 54.4 eV.

Absorption lines of He I

Table 1. *Equivalent widths of He I*

	4471(14) series ^3D			4387(51) series ^1D		
Type	V	III	Ib	V	III	Ia
O 3	0.20		0.13(f)			
O 4	0.15		0.12			
O 6	0.4		0.40			
O 8	0.9		0.6	0.40		0.22(f)
O 9	1.0	0.87	0.738			
B 0	1.12	1.04	0.74(Ia)	0.545		0.415
B 0.5		0.73				
B 1			0.708(Ia)			
B 2	1.47	1.13	1.20	0.95	0.74	
B 3	1.39		0.83(Ia)	0.68		0.419
B 5	0.84	0.68	0.86	0.48		0.418
B 6	0.75	0.61		0.59	0.23	
B 7	0.48	0.35		0.185		
B 8		0.45	0.22			
B 9	0.64		0.1			
B 9.5	0.145					
A 0	0.04					

Equivalent widths of He lines have also been published by Norris (1971).

He lines are used for discriminating the spectral type of the hot stars. If only He I is present, then the star is of type B; if He II is present, then the star is of type O. In the transition region at late O-type, He I and He II lines coexist. When He I lines disappear one has stars of spectral type A (or later).

Usually, for the determination of spectral type, one considers also the hydrogen lines, because hydrogen lines increase in strength monotonically from early O-type to A 2, where they have a maximum. Since both H and He lines are

Table 2. *Equivalent widths of He I 7065 series ^3S*

Group	V	III	Ia
O 4	0.22		
O 6	0.50		
O 8	0.65	0.75	
O 9	0.60	0.74	0.92
B 0	0.58	0.74	0.86
B 2	0.42	0.60	0.74
B 5	0.27		0.60
B 7	0.18		0.50

Source: The data are averaged from Jaschek and Jaschek, unpublished.

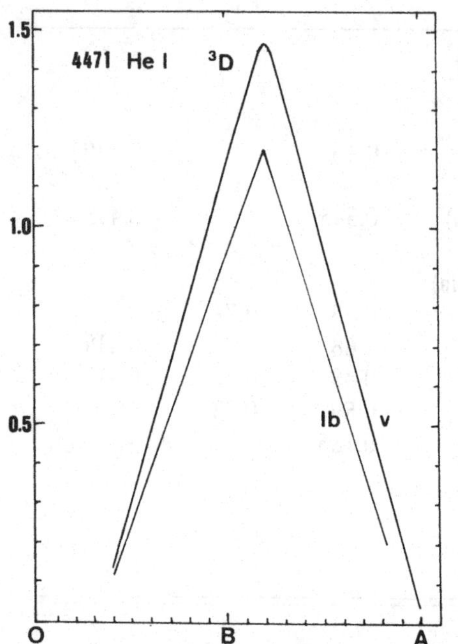

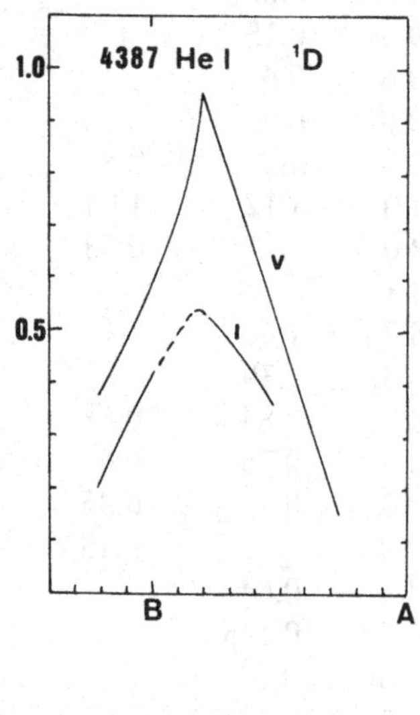

sufficiently strong to be easily measurable, the classification procedure can be carried out photometrically, using interference filters centered on selected He and H lines. For details of the procedures used in practice, see Golay (1974) and Jaschek and Jaschek (1987a). Another use of the photoelectric technique is to single out stars in which He lines are too strong or too weak for the temperature inferred from the H lines. In this sense Nissen (1974) used the 4026 line (18) of He I for singling out *He-rich* and *He-poor* stars.

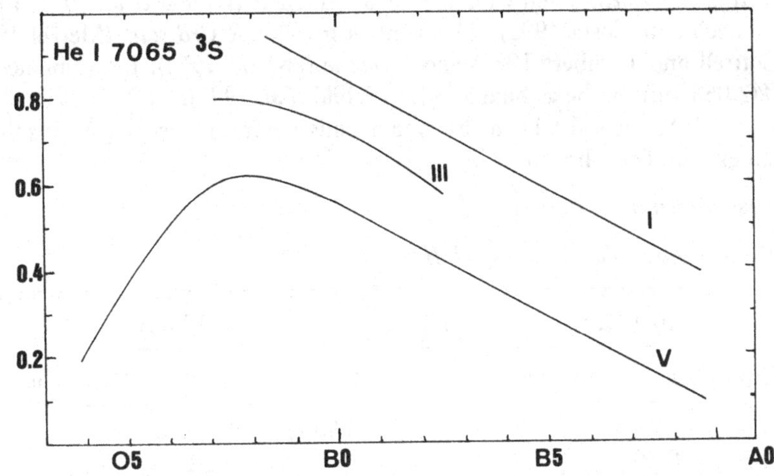

For late type stars two He I lines, 10830(1) and 5876(11), have a considerable importance, since they are indicative of a stellar chromosphere (see Part Two, section 3.1).

Forbidden lines of He I

Several forbidden lines of He I, for instance 4469, 4025 and 4920 (Underhill 1966), are visible (only) in B-type dwarfs. If the rotation is sufficiently low (and the plate factor of the spectrograms is adequate) the forbidden lines can be separated from the permitted nearby He I lines.

Emission lines of He I

Some He I lines are present in emission in *Oe stars* (Conti and Leep 1974, Frost and Conti 1976, Andrillat *et al.* 1982). The 10830(1) line is in emission in most O-type stars outside the main sequence. The emissions are strong (up to 46 Å) and variable in time (Andrillat and Vreux 1979). In the later subtypes of *WN stars* He I 6678(46) and 7065(10) are present in emission, the former being the stronger, with *W* up to 60 Å (Conti *et al.* 1990). No He I in emission has been observed in the ultraviolet (Nussbaumer *et al.* 1982, Willis *et al.* 1986).

Eenens *et al.* (1991) recently found a number of strong He I emission lines in the infrared region (1–3 μm) of *WC stars*.

One finds He I in emission in some *B*[e] *stars* (Ciatti *et al.* 1974), in *high-luminosity stars* (McGregor *et al.* 1988) and in *T Tau stars* (Schneeberger *et al.* 1978).

He I lines are seen in emission in *RV Tau* and *W Vir stars*; they are absent in *RR Lyr, Cepheids* and *Miras* (Gillet 1988). In *long-period variable stars*, 10830 is usually in emission around the light maximum (Querci 1986); 5876 and 6678 appear in emission in the so-called *dMe stars* (see the discussion on hydrogen emission lines, p. 79).

Symbiotic objects often exhibit He I in emission (Baratta *et al.* 1991, Freitas Pacheco and Costa 1992). The same is true for *R CrB stars* (Merrill 1951b, Cottrell and Lambert 1982) and in *novae* (Hyland 1974). In symbiotic stars *W*(2.058 μm) can be as high as 30 Å (Schild *et al.* 1992).

He I 10830 has also been observed in emission in all those *S-type stars* that do not exhibit Tc (Johnson 1992).

Absorption lines of He II

Table 3. *Equivalent widths of He II*

	1640(12)		4200(3)		10123(2)		
Group	V	I	V	I	V	III	Ia
O 4				0.50(f)			
O 5	0.50				1.72		
O 6				0.50(f)			
O 7	0.68	0.42			1.24		
O 8	0.78	0.55	0.50	0.40(f)	0.80		
O 9	0.80	0.85					
O 9.5			0.25				1.10
B 0	0.75	1.15			0.31	0.42	0.41
B 0.5							0.27
B 1	0.68	0.95					
B 2	0.50	0.70					
B 3	0.25	0.40					

Source: Values for 1640 averaged for MK standards are from Prinja (1990). Values for 10 123 are from Jaschek *et al.*, unpublished.

Table 4. *Equivalent widths of He II 4541(2)*

Type	V	Ia
O 3	0.79	0.74(f)
O 4	0.8	
O 5	0.9	
O 6	0.9	
O 7	0.8	
O 8	0.6	
O 9	0.4	
B 0	0.2	0.145

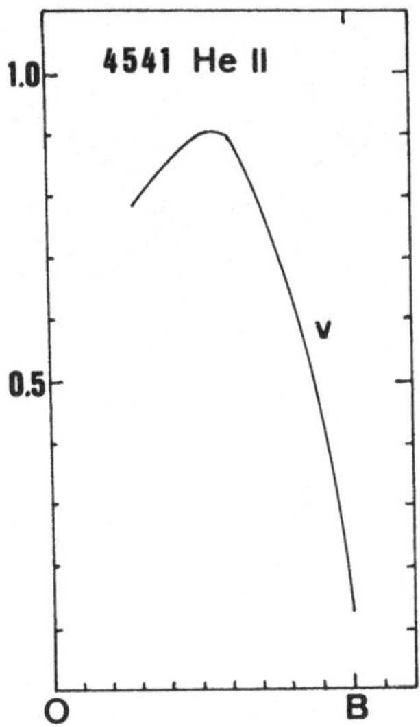

He II lines are only visible in spectral type O, except 1640, which is also seen in early B-type stars. He II lines have a positive luminosity effect, except 1640, which shows a negative luminosity effect before O 8.

Emission lines of He II

An O-type star that shows N III (4630–34) in emission and also exhibits He II 4686 in emission is called an *Of star*. When 4686 appears as a fill-in, the star is classified *O(f)* and when it is an absorption line, *O((f))* (Walborn 1973).

In *WN stars* 5411(2) is in strong emission, with $4<W<70$ Å (Conti and Massey 1989), as also is 4686, with $10<W<150$ Å. In the infrared region WN stars exhibit 6683, 6891, 8237 and 10123 in emission, the latter being the most intense (W up to 380 Å). In *WC stars* 4686 is probably present in emission but strongly blended with C III 4650 and C IV 4658 (Conti and Massey 1989). In the infrared 10123 is seen in strong emission in all subtypes except the latest (W up to 270 Å) (Conti *et al.* 1990). For detailed measurements of the infrared helium emission features, see Vreux *et al.* (1990).

In WN stars one also sees in emission the ultraviolet lines 1640 (W up to 160 Å), 2252, 2306, 2511 and 2733, whereas in WC stars there appear only 1640 and 2733 (W up to 60 Å) (Nussbaumer *et al.* 1982, Willis *et al.* 1986). In the infrared Eenens *et al.* (1991) recently detected the line at 1.344 μm.

The He II line 1640 appears sometimes in *late type stars* in emission. Its appearance indicates the existence of a stellar chromosphere. The line is visible in G and K dwarfs, and disappears in M-type dwarfs. See the discussion on chromospheres.

He II 4686 is always seen in emission from *symbiotic stars* (Allen 1988) and is

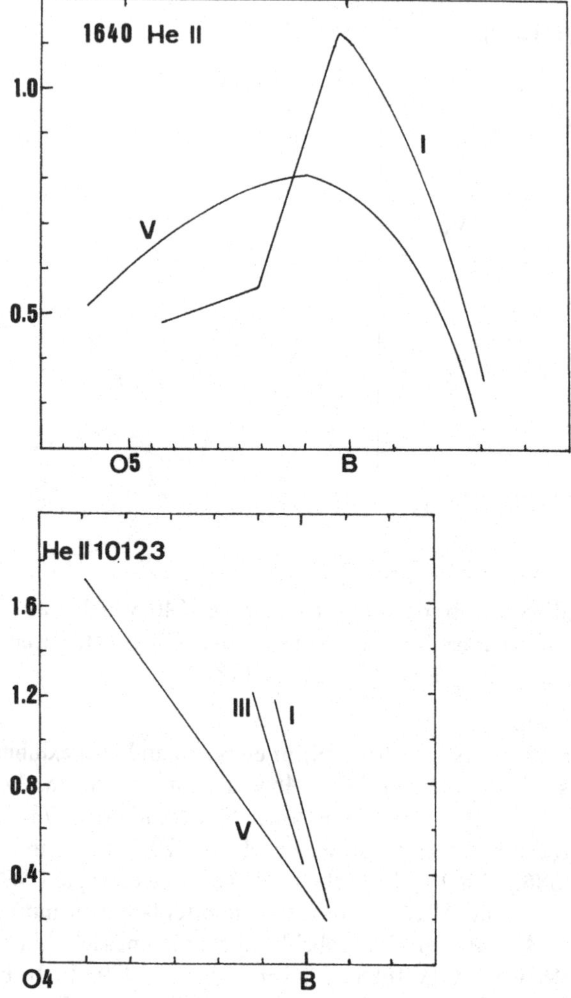

sometimes accompanied by other lines of the same species (Baratta *et al.* 1991, Freitas Pacheco and Costa 1992).

Behavior in non-normal stars

He is very strong in the so-called *He-strong stars*, also designated as *extreme helium stars* or *hydrogen-deficient stars*. In the extreme cases, He substitutes completely for the hydrogen. In less extreme cases, He lines coexist with hydrogen lines in the spectrum. Stars of such kind exist in both populations I and II. In population II the *sdO* (Baschek *et al.* 1982) and *sdB* (Baschek and Norris 1970) stars may be He-rich. In such cases, anomalies are present in the behavior of the different He series. In the He-rich sdO stars, there exist differences in the behavior of the lines of the ^{2}F and ^{2}G series. In these stars carbon and nitrogen are also enhanced (Rauch *et al.* 1991).

In sdB stars one finds anomalous relations between the line strengths of the singlet and triplet series of He I. For details see Jaschek and Jaschek (1987a) and for a theoretical explanation see Baschek and Norris (1970).

He is very strong in some of the *central stars of planetary nebulae* of the so-called H-deficient class (Mendez 1991). He is also very strong in hot *extreme helium stars* (Jeffery and Heber 1993) and in the spectra of *pre-degenerates* (Werner *et al.* 1991), although one pre-degenerate is known that has neither He nor H (Werner 1991).

He is weak in a number of stars that are called '*helium-weak*' and such stars are present in both populations I and II. In population I for instance all *Bp stars* of the subgroups Hg–Mn and Si are He weak. In the latter group *W*(4026) is weaker by a factor 2–6 (Deutsch 1956).

In the N-variety of the *CNO stars*, also called OBN stars by some authors (Schoenberner *et al.* 1988), He I lines are slightly stronger than in normal stars of the same spectral type.

He is weak in *HB stars*, with varying degrees of weakness (Heber 1991) ranging from moderate weakening to almost complete disappearance.

He lines are very strong in the so-called *DB stars* and for some years it was almost a dogma that besides He no other elements are visible in the spectra of these stars. However, recently the number of exceptions has become rather large – in many stars one sees Ca lines (Sion *et al.* 1988) and in 20% of all DB stars one also sees (faint) hydrogen lines (Shipman *et al.* 1987), see also Weidemann and Koester (1991).

He lines are very weak or absent ($W<0.1$) in *DZ stars* (Sion *et al.* 1990).

In *symbiotic stars*, both He I and He II lines are often seen in emission (see above).

In *novae* He I absorption lines are enhanced in the 'Orion spectrum' phase. Emission lines of He II are seen in the ultraviolet spectral region during the principal spectrum phase (Warner 1989). He I and He II emissions are also seen in nova remnants. In particular, 4686 of He II appears in many classical and recurrent novae but is weak or absent in dwarf novae (Warner 1989).

He lines are absent in the spectra of *supernovae* of Class Ia, they are very strong in those of class Ib and present in those of class II (Branch 1990).

Isotopes

He has three stable isotopes, namely He 3, 4 and 5, and two short-lived ones. In the solar system and also in the sun, only the first two are present and the He^3/He^4 ratio is very small (10^{-4}). The best way to investigate the presence of He^3 is to use the isotopic shifts. The most favorable case in the visual red region is He I 6678, where the shift is about 0.50 Å. For other He I lines like for instance 5875 the shift is much smaller (0.04). In the infrared, the He I 10830 line has an isotopic shift of 1.15 Å (Zirin 1968).

The first star in which He^3 was detected was an early type *Bp star* (Sargent and Jugaku 1961). Hartoog and Cowley (1979) studied a large number of early type stars and give a list of eight secure and four possible or probable stars with a large fraction of He in the form of He^3. Afterwards several *HB* and *sdO stars* were discovered in which He is predominantly in the form of He^3. However, this does not apply to all stars of these groups (Heber 1991).

Origin

He was either produced by cosmological nucleosynthesis at the beginning of the universe or later by hydrogen burning.

HYDROGEN H Z=1

This element was discovered by Cavendish in London in 1766. Its name alludes to the fact that water is generated when hydrogen burns in air.

Ionization energies

HI 13.6 eV.

Hydrogen is the simplest of the chemical elements. It is also the most abundant in the universe and is in many ways the most important chemical element for astrophysics.

Absorption lines

Table 1. *The wavelengths of the first lines of the first four series*

n	Excitation potential (eV)	$m = 1$ Lyman	$m = 2$ Balmer	$m = 3$ Paschen	$m = 4$ Brackett
2	10.2	1216			
3	12.0	1026	6563		
4	12.7	973	4861	18751	
5	13.0	950	4340	12818	40511
6	13.2	938	4101	10938	26251
7	13.3	931	3970	10049	21655
8	13.3	926	3889	9546	19445
Limit	13.5	912	3646	8203	14584

Note: Wavelengths as usual are in ångström units.

It is customary to designate the first lines by the Greek letters alpha, beta, gamma, etc. In the Balmer series the last letter used is epsilon. The following lines are numbered according to their upper level, so the next line after H epsilon is H 8. 'Limit' designates the wavelength for which $n = \infty$.

Table 2. *Equivalent widths of Balmer lines*

	4340(gamma)		4862(beta)		
Type	V	I	V	III	Ia
O 3	2.2	1.75(f)			
O 4	1.5				
O 6	1.8				
O 8	2.0				

Table 2. (*cont.*)

	4340(gamma)		4862(beta)		
Type	V	I	V	III	Ia
B 0	2.8	1.3(Ia)	2.5		0.8
B 2	5.3	3.0(ib)	4.1	2.7	1.0
B 3	6.9	1.6(Ia)			
B 4			5.2	5.0	1.2
B 5	6.7	1.7(Ia)			
B 6	7.8		6.5	5.8	1.5
B 7	8.1				
B 8	10.0	2.5(Ib)	9.0	7.2	1.8
B 9.5	9.9				
A 0	16.2		11.5	10.5	2.0
A 1	19.4	3.1(Ia)			
A 2	17.8	3.7(Ia)	13.0	10.5	2.0
A 3	15.9				
A 4			11.0	10.1	2.8
A 5	14.7		10.2	10.1	3.1
A 7	13.3		9.9	9.8	3.9
F 0	8.0	8.7(II)	8.8	8.7	4.5
F 2	5.5		6.8	6.8	4.8
F 5	5.5	7.3(Ib)	4.7	5.1	5.5
F 8		4.4(Ib)	3.2	4.8	4.0
G 0	3.0		2.3	2.3	2.5
G 2	2.6				
G 5	3.0				

Note: For K0III, $W(4340) = 1.0$. All values have been smoothed, because equivalent width measurements of Balmer lines have relatively large errors attached.
Source: Data are from Didelon (1982) and Jaschek and Jaschek (1987a).

Table 3. *Equivalent widths of the Paschen line P 7 (10048)*

Type	V	III	I
O 6	2.9		2.3 f
O 9	3.0	2.8	2.6 b
B 0	3.9	3.4	2.7 a
B 2	5.4	3.9	2.5 a
B 5	7.5	6.5	2.7 ab

Table 3. (*cont.*)

Type	V	III	I
B 8	8.5	6.8	2.6 a
A 0	10.6	9.5	4.8 b
A 2	11.4		
A 5	10.7		5.2
F 0		5.5	
F 2	(3.1)		
F 5	3.0		3.4 b
G 0	0.6	0.7	0.7 b

Source: Date are from Andrillat *et al.* (1993). For type I the luminosity class is given after the *W* value.

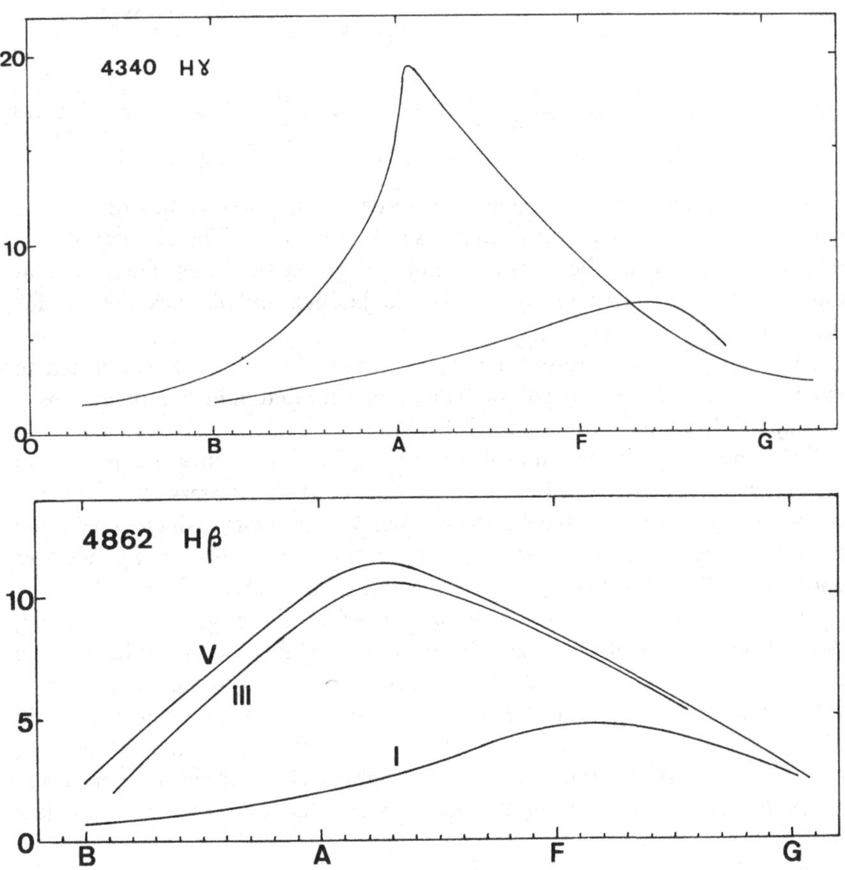

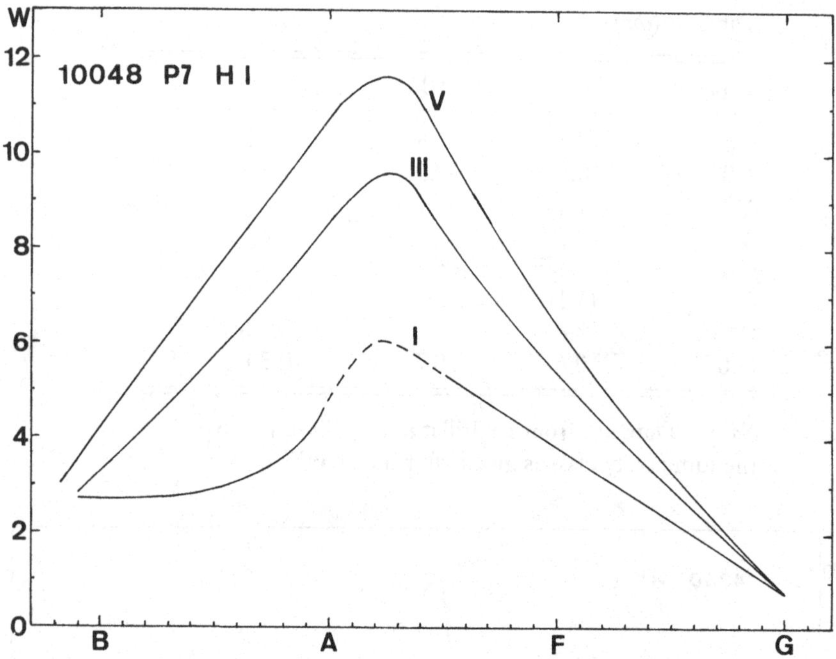

It can be seen that H lines appear in O-type stars, increase in strength to a maximum at about A 2 and decrease slowly thereafter. The strength of the hydrogen lines can also be measured photoelectrically by means of narrow band interference filters (see Golay (1974) and Jaschek and Jaschek (1987a) for details).

All hydrogen lines present a negative luminosity effect. This has led to extensive use of the strength of the Balmer lines for determination of luminosity (Golay 1974, Jaschek and Jaschek 1987a).

The line shape is determined by the Stark effect. Lines are narrow in supergiants and broad with large wings in dwarfs. With the Stark effect, the line profile varies with $x^{-2.5}$ (x is the distance from the line center), which explains the extremely large wings of the lines. For instance the H gamma line extends over about 6 Å in B 3Ia, 30 Å in B 3 III and 50 Å in B 3 V (Petrie 1952).

Toward the end of the series, the lines overlap. The degree of overlapping depends on both the intensity and the narrowness of the lines (provided that the resolving power of the spectrograph is adequate). At the maximum strength of the Balmer lines (around A 2) one can see up to $n=28$ in supergiants, $n=22$ in giants, about $n=18$ in dwarfs, $n=9$ in subdwarfs and about $n=7$ or 8 in degenerates. Since the number of the last visible Balmer line also depends very much on the plate factor of the spectrogram, the numbers given are just indicative.

In shell stars, for instance in 48 Lib, sometimes one can see up to $n=41$

(Merrill and Sanford 1944), although admittedly this is an extreme case. The narrowness of the lines indicates that the lines are formed in an extremely low-density medium. For this same reason one observes $n=35$ in the solar chromosphere (Mitchell 1947).

The convergence of the hydrogen lines towards the limit of the series produces a pronounced depression of the continuum, which extends well beyond the limit of the series. This discontinuity is also called a jump. Its size depends on the sharpness of the lines; it is therefore larger in supergiants than in dwarfs. The Balmer discontinuity has been used extensively in the Barbier–Chalonge classification system (see for instance Fehrenbach (1956)) and in broad band and intermediate band photometry, see Golay (1974).

Hydrogen lines in emission

In *Oe stars* Balmer and Paschen lines are in emission, H alpha having up to $W=13$ (Conti and Leep 1974, Andrillat *et al.* 1982).

In *Of stars* some Balmer lines are seen in faint emission (Conti 1974). In the infrared region of the later *WN stars* P 7 is frequently seen in emission (W up to 35 Å) (Conti *et al.* 1990).

In *Be stars* and in *B[e] stars* one most frequently observes hydrogen lines in emission. The emissions are usually stronger in the first lines of the series (Balmer, Paschen, etc.) than in the higher lines. All emissions are variable with time. Many publications have been devoted to studying the emissions and their variation with time, starting with the discovery of the first Be star by Secchi in 1878. For recent work see for instance Andrillat and Fehrenbach (1982) for H alpha, Hanuschik *et al.* (1988) and Slettebak *et al.* (1992) for H beta, gamma, etc., Andrillat *et al.* (1990) for P 12–20 and Andrillat *et al.* (1990) for P 7.

The H alpha emissions may be quite strong, with W up to 90 Å, but are usually of the order of about 30 Å.

As a rule, the Paschen emissions are weaker than those of the Balmer series. Few observations exist for the Brackett and Pfund series, but in general the Brackett emissions are weaker than the Paschen emissions and those of the Pfund series are weaker than the Brackett emissions (Sellgren and Smith 1992).

The Balmer (or Paschen) discontinuity (or jump) is seen in emission in some *Be stars*. This was discovered by Barbier and Chalonge (1941). For illustrations see Schild (1976).

Balmer, Paschen and Brackett lines in emission are also seen in *Herbig Ae–Be stars* (Harvey 1984), *compact infrared sources, high-luminosity early type objects* like P Cyg and S Dor (for a general survey see Jaschek (1991)) and in *symbiotic stars* (Schild *et al.* 1992). In the latter W(Br alpha) and W(Br gamma) can be as large as 740 and 41 Å respectively.

In *dKe or dMe* stars H alpha is seen in emission, and for this reason the star is classified with the suffix 'e'. dMe stars are most frequent towards later types and from dM4 onwards practically all stars are dMe. The emission is usually accompanied by emission in the other Balmer lines down to H delta and even to H 9.

Equivalent widths of H alpha are usually of the order of a few ångström units (W<7 Å). These emissions are at least partially of chromospheric origin (see Part Two, section 3.1) and are usually of variable strength. There exists a general tendency for the emission to be stronger in later stars. If the variability is strong and accompanied by other emissions, then the star is called a '*flare star*'. The other accompanying emissions are (other) chromospheric lines such as Ca II, as well a He I 5876 and 6678 and the Na I D lines. Furthermore, in these stars, Lyman alpha is usually also present in emission (Doyle *et al.* 1990). In the M-type flare stars the Lyman alpha is as strong as the H alpha emission. It must be added that Lyman alpha emission is a more general phenomenon than the Balmer line emissions and in fact one finds Lyman alpha emissions in stars as early as type F. A catalog of Lyman alpha emissions has been published by Landsman and Simon (1993).

Emission lines are prominent in *T Tau stars*. W(H alpha) for instance is always larger than 5 Å and may reach 100 Å. However, the higher Balmer lines do not always behave in a regular way, in the sense of a gradual decrease of intensity. All lines are variable with time and show a wide variety of line profiles (Sun *et al.* 1985, Basri 1987, Bertout 1989). Balmer line emissions are weak in *RR Lyr variables*, and the same happens in *Cepheids*. Emissions are stronger in *RV Tau* and *W Vir* stars and are very strong in *Miras* of both populations. For an illustration of the different emissions see Gillet (1988).

In *long-period variables*, emission lines are present over a large part of the cycle, except around the minimum light (Merrill 1952, Joy 1954). The maximum strength is reached after maximum light. The Balmer series can be seen up to n=14 at times when the star is on the descending branch of the light curve. Emission line intensities do not, however, follow the traditional pattern of a regular decrease of emission intensity, since H alpha and H beta are weaker than H gamma, delta and zeta (Merrill 1960).

Hydrogen emissions are present sometimes in *semiregular variables*, with variable strength from cycle to cycle (Querci and Querci 1989).

In *novae*, emissions are seen in the principal spectrum phase (Evans 1989). *Supernovae* may exhibit all series in emission (Arnett *et al.* 1989). Emission lines appear also in *nova remnants*, a few decades after outbursts (Warner 1989).

Balmer lines in emission are seen also in *R CrB stars* (Merrill 1951b).

Emissions are usually observed in *symbiotic stars*, in all of the Balmer, Paschen and Brackett series. In all cases the emission strengths are variable with time. For details see *IAU Coll.* **103** (1988) and Baratta *et al.* (1991). The Balmer jump is also found in emission in all *symbiotic stars* (Allen 1988).

Behavior in non-normal stars

Balmer lines are very weak or absent in *WR stars* (Andrillat and Vreux 1991). Hamann *et al.* (1991) find that, in hot WN stars, hydrogen is absent, whereas it is present in cooler WN stars.

In some *subdwarfs* of types O and B, hydrogen is weak when compared with helium. So for instance H gamma, which is much stronger than the 4387 line in normal stars, can become comparable to the latter or much weaker. In the most extreme cases, H is absent (see for instance Hunger *et al.* (1981)).

H lines are weak or absent in the so-called *extreme He stars* (see the section on helium) and in many *degenerates*. According to the usual classification scheme for degenerates (see for instance Jaschek and Jaschek (1987a)) about half of all degenerates do show hydrogen lines whereas the other half do not show hydrogen at all. Of these the most important subgroup is that of the DB (degenerates of type B), where only He lines appear. Shipman *et al.* (1987) found in a study of DB stars that at least 20% do exhibit traces of H lines.

H lines are also very weak or absent in the spectra of DZ stars ($W<0.1$ Å), see Sion *et al.* (1990).

H lines are very weak or absent in the spectra of one third of all *central stars of planetary nebulae* – in the so called H-deficient subgroup (Mendez 1991). In the other central stars, H lines are normal or strong and this constitutes the H-rich subgroup.

H lines are very weak or absent in the spectrum of *pre-degenerate stars* (Werner *et al.* 1990).

H line are absent in the spectra of *supernovae* of class I (and present in those of class II) (Branch 1990). Hill (1993) challenged this scheme for classification of supernovae on the grounds that it is an oversimplification based upon very few stars.

H lines are very weak in the spectra of *R CrB stars* (Cottrell and Lambert 1982).

In *C stars* hydrogen lines are weak, especially in C stars later than C 3. This contrasts with the strengths of the H lines in M-type stars of the same temperature (Yamashita 1967).

In stars in which the H content is abnormal, the Balmer discontinuity is abnormal too. This happens for instance in some A-type *supergiants* of the Magellanic Clouds, where both the Balmer jump and the hydrogen lines are too strong relative to the metallic lines (Humphreys *et al.* 1991).

H lines in stars with an intense magnetic field are split up into components. In normal stars with small magnetic fields such components are not seen because of the degeneracy of the levels. So for instance H alpha(6562) is split into three components located at 5870,7129 and 8450 in a field of about 2×10^8 G (Angel *et al.* 1985). *Degenerates* have polar fields up to 500 MG and *neutron stars* have fields in the range 10^{11}–10^{13} G.

Magnetic fields have been measured in some 30 degenerates of the DA and DC type (Bergeron *et al.* 1992). Strong magnetic fields are also present in *cataclysmic variables* (degenerates accreting matter from a companion star). For a recent summary see Chanmugam (1992) (see also the discussion on magnetic fields, p. 279).

Isotopes

H has two stable isotopes, H^1 and H^2 (deuterium), whereas a third one, H^3, is short-lived.

In the solar system the ratio H^2/H^1 is about 1.5×10^{-4}. Deuterium has never been observed in stars and only upper limits to the H^2/H^1 ratio can be provided. Ferlet *et al.* (1983) were able to derive an upper limit of 5×10^{-7} for an F 0 II star. Vidal-Madjar *et al.* (1988) derived 5×10^{-6} for a B 2 V star.

Origin

Hydrogen was produced by cosmological nucleosynthesis.

INDIUM In Z=49

This element was discovered in 1863 in Freiberg, Germany, by F. Reich and H. Richter. The name was chosen because in the spectrum of this element there appears a bright line of indigo color.

Ionization energies

InI 5.8 eV, InII 18.9 eV, InIII 28.0 eV.

Absorption lines of InI

The equivalent width of InI 4511(M.1 resonance line) in the sun is 0.004 (Lambert *et al.* 1969). The In lines are stronger in sunspot spectra. The element is also present in M 2 III stars (Davis 1947).

Emission lines of InI

InI 4511 appears in emission in *long-period variables* around maximum light (Merrill 1952, Joy 1954, Deutsch and Merrill 1959).

Behavior in non-normal stars

Cowley *et al.* (1974) found InII lines in the spectrum of one *Ap star* of the Cr–Eu–Sr subgroup.

In is enhanced in at least one *Ba star* (Lambert 1985).

Isotopes

In occurs in two stable forms, In^{113} and In^{115}. There exist also 32 short-lived isotopes and isomers. In the solar system the frequency of In^{113} is 4% and that of In^{115} is 96%.

Origin

In^{113} is produced by the r, the s or the p process. In^{115} is only produced by the r process or the s process.

IODINE I $Z=53$

This element was discovered by B. Courtois in Paris in 1811. Its name comes from the Greek *iodes* (violet).

Ionization energies

I II 10.4 eV, I III 19.1 eV, I IIII 33.2 eV.

No I line is observed in the sun. Ionized iodine was found in one *Ap star* of the Cr–Eu–Sr subgroup by Jaschek and Brandi (1972).

Isotopes

I occurs in the form of I^{127}. There exist a long-lived isotope, I^{129}, with a half life of 1.7×10^7 years and 22 short-lived isotopes and isomers. I^{129} in principle could be used for radioactive dating.

Origin

I is an r process element.

IRIDIUM Ir Z=77

This element was discovered by S. Tennant in London in 1803. The name comes from the Latin *iris* (rainbow), because iridium salts are highly colored.

Ionization energies
IrI 9.1 eV, IrII 17.4 eV.

Absorption lines of IrI
In the sun, IrI 3221(5) has W=0.032.

Behavior in non-normal stars
IrI was reported in one *Ap star* of the Cr–Eu–Sr group by Guthrie (1972).

Isotopes
Ir occurs in two stable forms, Ir^{191} and Ir^{193}. In the solar system Ir^{191} represents 37% and Ir^{193}, 63% of total iodine. There exist also 23 short-lived isotopes and isomers.

Origin
Both isotopes are only produced by the r process.

IRON Fe   Z=26

This element was known in antiquity. In Latin iron is called *ferrum*, therefore the abbreviation Fe.

Ionization energies

Fe I 7.9 eV, Fe II 16.2 eV, Fe III 30.6 eV, Fe IV 54.8 eV, Fe V 75.5 eV, Fe VI 100 eV, Fe VII 128.3 eV, Fe VIII 151.1 eV, Fe IX 235 eV, Fe X 262 eV.

Fe is an element whose lines dominate the spectra of all stars of type F and later. In the solar spectrum, iron lines account for about 30% of the lines, a percentage not surpassed by any other element. Because of this, a large part of the line blocking is due to Fe and therefore Fe was taken to represent all elements other than H and He (see also Part Two, section 2.1).

Absorption lines of Fe I

Table 1. *Equivalent widths of Fe I*

	4045(43)		5216(36)		
Group	V	Ib	V	III	Ib
B 9	0.067				
A 0	0.085	0.08			
A 1	0.101				
A 2	0.18	0.100,0.97(Ia)			
A 3		0.168(O)			
A 7	0.37				
F 0	0.43	0.41(Ia)			
F 2		0.36			
F 5	0.48	0.68	0.14	0.158	
F 8		1.00			0.265
G 0	1.010				0.316
G 2	1.175		0.14		0.275
G 5	1.79				0.288
G 8				0.26	
K 0	3.1			0.29	
K 2					0.288
K 3				0.324	0.275
K 5	3.53				0.490
M 0				0.72	
M 2					0.630
M 3				0.52	
M 4				0.91,1.10	

Table 2. *Equivalent widths of FeI 5269(15)*

Group	V	III	Ib
F 5	0.25		0.30
F 8			0.44
G 0	0.21		
G 2	0.44		
G 5			0.587
G 8		0.66	0.767
K 0		0.78,0.93	
K 3		1.08	
M 2.5		0.71	

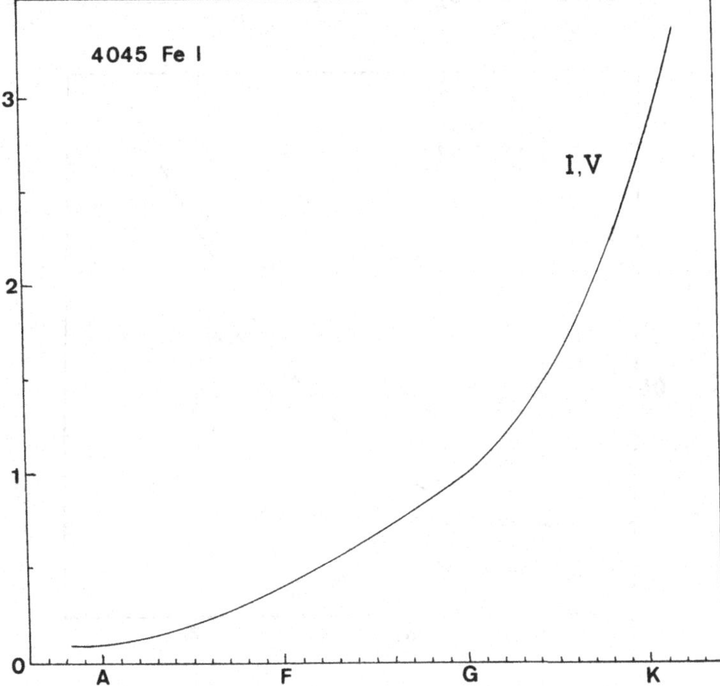

Fe I lines (for instance 4045, 5216 and 5269) appear in late B-type stars and grow rapidly toward later types. No luminosity effect is observed.

Among the most intense infrared lines one has 10216 (1247). In the sun, $W(10216)=0.094$.

High-excitation-potential Fe I lines (about 7 eV) were detected in the sun and in a late type giant at 2550–2500 cm^{-1} (Johansson *et al.* 1991).

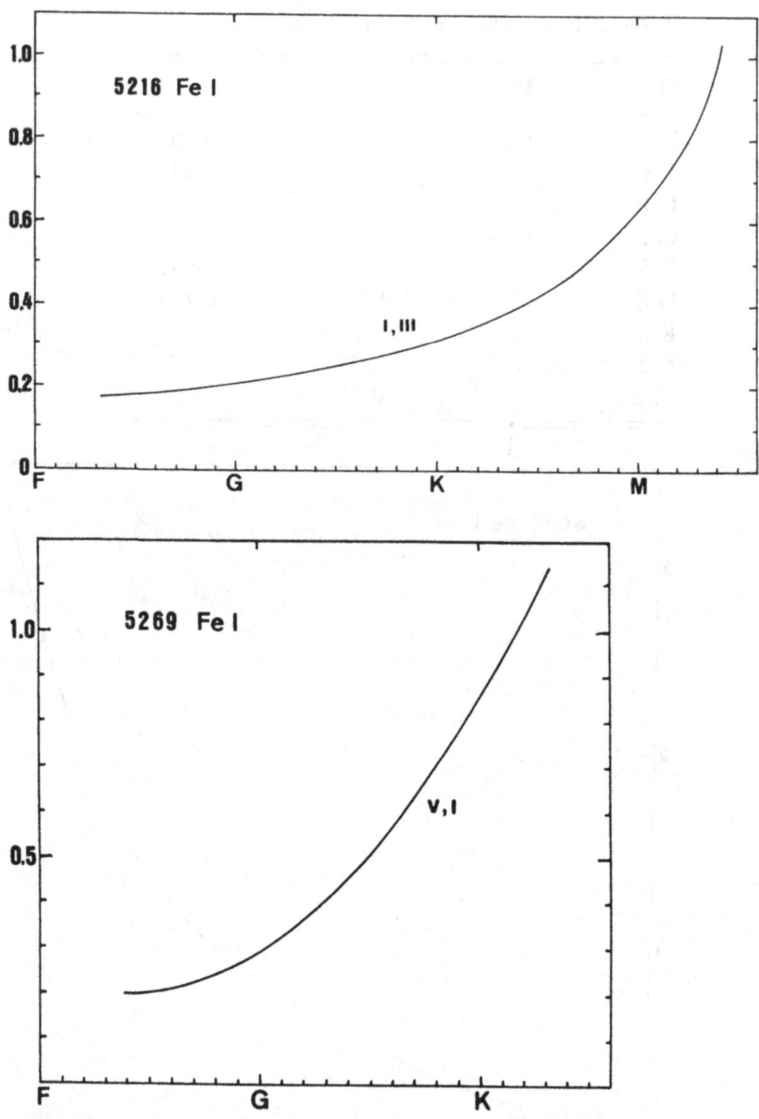

Emission lines of FeI

FeI emissions are found in the ultraviolet spectrum of K giants and M-type supergiants (van der Hucht *et al.* 1979).

In *T Tau stars* several multiples (for instance M.2) are seen in emission (Joy 1945). In addition the lines 4063 and 4132 (both of M.43) are often unexpectedly strong. This is due to fluorescence excited either by CaII or by H epsilon (Willson 1975).

In *long-period variables* many lines of Fe I are seen in emission, usually around maximum light; they disappear about at the minimum (Joy 1954).

Particularly strong are the 4202 and 4308 lines, both of M.42 (Querci 1986).

Several lines of Fe I are seen in emission in the *supernova* 1987A (Arnett *et al.* 1989).

Absorption lines of FeII

Table 3. *Equivalent widths of FeII*

	4233(27)			4515(37)	
Group	V	III	Ib	V	Ia
B 3	0.08			0.005	
B 5	0.035			0.015	0.036
B 6		0.120		0.018	
B 7	0.09				
B 8			0.100,0.110(Ia)		0.028
B 9	0.11				
A 0	0.07		0.33(Ia)		0.129
A 1				0.086	
A 2			0.23	0.094	0.223
A 3			0.590(0)		
A 7	no			0.130	
F 0		0.24(II)			0.537
F 2					0.407(Ib)
F 4				0.13	
F 5	0.20		0.383	0.16	0.40(Ib)
F 8			0.48	0.10	0.46(Ib)
G 0	0.17			0.08	
G 1				0.12	
G 2	0.097			0.08	
S				0.075	
G 5	0.19			0.115	
K 0				0.11	
K 2				0.09	

The 4233 line is present up to M 2 III (Davis 1947).

FeII (see for example 4233) lines appear in mid-B-type stars and strengthen toward A-type stars. They have a maximum for late F and decrease for G type. A positive luminosity effect is present, which is much used for spectral classification.

The Fe II lines 6149 and 6147 (M.74) are very sensitive to the Zeeman effect, but in different ways, because of their different splitting pattern and the operation

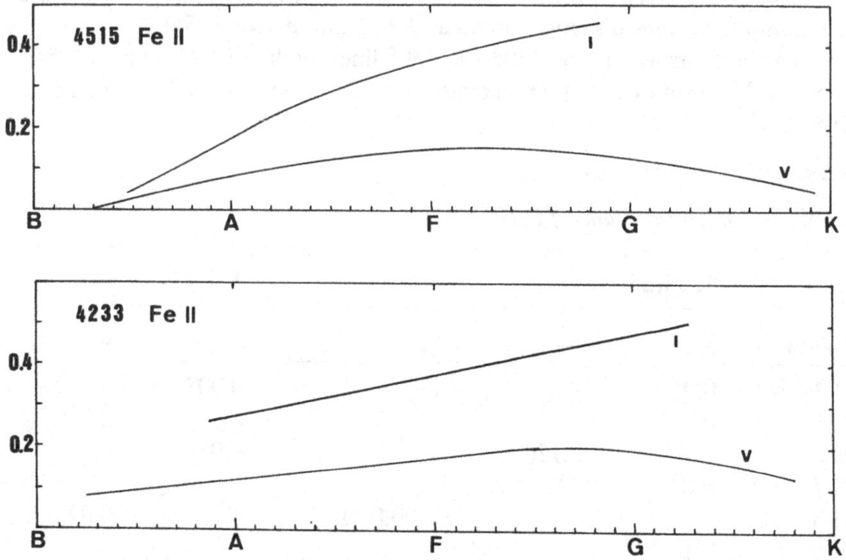

of the Paschen–Back effect (Mathys 1990). They have been used to measure the magnetic field in *Ap stars* (Mathys and Lanz 1992).

Emission lines of FeII

FeII emissions are ubiquitous and one finds them whenever a low-density medium is heated by shock waves, expanding atmospheres or radiation from hot spots or a binary companion. They are thus present in extended envelopes and in stellar chromospheres. Many details on FeII and [FeII] emissions can be found in the colloquium *Physics of Formation of FeII Lines Outside LTE* edited by Viotti and Friedjung (1988). FeII is seen in emission in the infrared in *Oe stars* (Andrillat *et al.* (1982). FeII and (FeII) emissions are present in variable OB stars (*S Dor type*) and *P Cyg stars* (Zickgraf 1988). (For the latter see also Stahl *et al.* (1991).) Often both types of objects are listed together under the title of *luminous blue variables*. For a summary of their properties, see Wolf (1993).

FeII lines (M.42) are seen in emission in early *Be stars* (Slettebak *et al.* 1992), but an inspection of the infrared region shows that FeII emissions (for instance 9997) are present in all Be stars, with an intensity depending upon the spectral type. This is probably due to fluorescence (Andrillat *et al.* 1990). *W*(9997) can reach up to 1 Å. [FeII] lines in emission are characteristic of *B[e]* stars (Andrillat and Swings 1976). These stars also show permitted FeII lines in emission. FeII emissions are also seen in *Herbig Ae–Be* stars (Talavera 1988).

Numerous (faint) emission lines of FeII are seen in the spectrum of the *chromosphere* of late type stars of types K and M (van der Hucht *et al.* 1979, Linsky *et al.* 1982, Judge and Jordan 1991). In one K giant these lines are responsible for more than 80% of all emission lines (Carpenter and Wing 1985).

In *symbiotic stars* one observes high-excitation ultraviolet fluorescence Fe II emissions (1360,1776,1869,1881,1884 and 1975) produced by the strong O VI, C IV and HI lines (Feibelman *et al.* 1991). FeII lines are also present. In the infrared the 9997 line is usually seen in emission (Baratta *et al.* 1991).

T Tau stars often exhibit in emission the lines of M.27, 28, 37 and 38, besides those of the chromospheric spectrum (Sun *et al.* 1985, Jordan 1988). In addition some [Fe II] lines are also seen in emission (Joy 1945).

In *semiregular variables* Fe II is always seen in emission, with time variations of the line strength (Querci and Querci 1989).

In *Mira-type variables*, Fe II emissions (for instance M.27 and 28) are present over a large part of the cycle, except for a few weeks after minimum (Joy 1954). Some forbidden Fe II lines (M.6F and 7F) are visible around the minimum in *long-period variables* of type S. Many Fe II lines appear in emission around maximum light: 4233,(M.27), 4583(37), 4924 and 5018(42) (Merrill 1952).

Fe II emissions are also found in *R CrB stars* and in binaries with a hot companion, like *cataclysmic variables*, dwarf novae, *symbiotic stars*, *VV Cep stars* (Viotti and Friedjung 1988) and in *novae* (Cassatella and Gonzalez Riestra 1988) (for the novae see also below) and in *supernova* 1987A (Arnett *et al.* 1989). [Fe II] emission lines are also present in *VV Cep stars* (Rossi *et al.* 1992) and in *supernova* 1987A (Jennings *et al.* 1993).

Absorption lines of Fe III

Table 4. *Equivalent widths of Fe III*

	4164(118)			4420(4)
Group	V	III	Ia	Ia
O 9	0.014			
B 0	0.070		0.118	
B 0.5	0.061	0.066		
B 1		0.055		0.064
B 2	0.025	0.052		0.138
B 3			0.040	0.118
B 5				0.071

Source: Some data are from Kilian and Nissen (1989).

Fe III (see for example 4164) appears in late O- and early B-type stars. For dwarfs the maximum lies around B 0. A positive luminosity effect exists. Fe III is prominent in the ultraviolet of early type stars, as is the case for Fe II.

Emission lines of Fe III

Lines of Fe III appear in emission in the ultraviolet spectrum of K and M giants and supergiants (van der Hucht *et al.* 1979). [FeIII] is seen often in *B[e] stars*

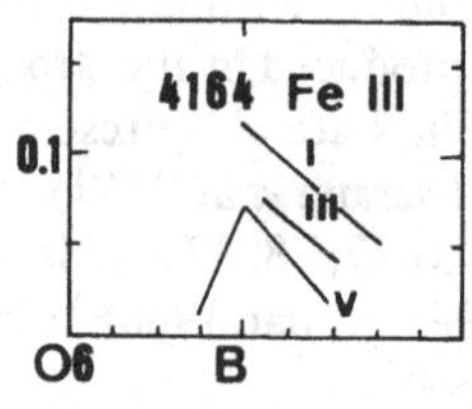

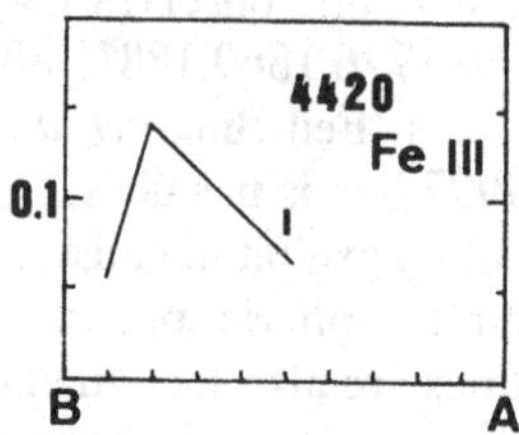

(Andrillat and Swings 1976) and [FeIII] lines 4658 and 5270 have been observed in some *novae* (Swings 1952).

Absorption lines of FeIV, Fe V and Fe VI

Many weak lines of FeIV, FeV and FeVI appear in the ultraviolet spectrum of one *O-type subdwarf* (Bruhweiler *et al.* 1981), the FeVI lines being the weakest. Nemry *et al.* (1991) found weak FeIV and FeV lines also in one *WR star* ($W \simeq 0.1$).

Emission lines of FeIV, Fe V and Fe VI

Emissions of forbidden FeV(3895, 3891) and FeVI (3664) have been observed in some *novae* and some *symbiotic stars* (Merrill 1948, Swings 1952).

Lines of Fe VII

Joy and Swings (1945) reported FeVII in one recurrent *nova*. The 6087 line of [FeVII] is always seen in emission in *symbiotic stars*. It should, however, be remarked that often one sees [FeII] and [FeVII] without the intermediate species.

Wallerstein *et al.* (1991) also report observation of the 3586 line of [FeVII] in emission in several *symbiotic stars*.

Forbidden lines of higher ionization stages

Forbidden lines of FeX, XI, XII, XIII, XIV and XV (7060) with ionization potentials between 233 and 390 eV are seen in the spectrum of the solar corona (Zirin 1988). (See also Part Two, section 3.2.)

[FeX] and [FeXIV] in emission have been observed in some *recurrent novae* (Joy and Swings 1945, Swings 1952).

Behavior in non-normal stars

Fe is weak in some of the *extreme helium stars* (Jeffery and Heber 1993).

Fe appears enhanced in *Bp stars* of Si type. For instance FeII 4233 is intensified by factors up to three, but there is a continuous transition to normal stars (Didelon 1986). Fe, on the other hand, may be very weak in some Hg–Mn stars (for instance 53 Tau), where even strong ultraviolet multiplets are missing. Nevertheless, in other stars of the Hg–Mn subgroup Fe lines are strong.

In *Ap stars* of the Cr–Eu–Sr subgroup there also exist cases of stars in which iron lines are strong (Cowley 1981). In some stars of this subgroup one finds iron simultaneously in three ionization stages, namely FeI, FeII and FeIII (Jaschek and Lopez García 1966).

Fe is slightly strengthened in *Am stars* (Burkhart and Coupry 1991).

Fe lines are weakened in *lambda Boo stars*. Typically the *W* values are smaller by factors of about two than in normal stars of the same temperature (Venn and Lambert 1990).

Fe lines are weakened in F-type *HB stars* by factors up to five in *W* values (Adelman and Hill 1987).

Fe I lines are present in the spectra of some *DZ degenerates* (Sion *et al.* 1990), with *W*(3820<5.5 Å).

Fe is also underabundant in *metal-weak stars*. It should be remarked that generally authors speak of 'metal abundances', rather than of iron abundances.

Fe is slightly weak in *subgiant CH stars* (Luck and Bond 1982).

Fe is underabundant in *globular cluster stars*, varying between −0.8 dex for 'metal-rich' globular clusters, to about −1.3 dex in 'intermediate clusters' and −2.3 dex or less in 'metal-poor' clusters.

The Fe/H ratio seems to be normal (solar) in the *SMC stars*. In the *LMC stars* real variations in Fe/H may exist between different stars (Luck and Lambert 1992).

Fe is slightly underabundant in *Ba stars* (Lambert 1985) and is probably normal in *S-type stars* (Lambert 1989).

Forbidden iron lines are visible in *novae*, especially in the 'nebular stage' phase. Their behavior varies from star to star. In some objects one sees lines from [Fe II] up to [Fe VII] whereas other stars have shown lines from [Fe XI] to [Fe VIII], with [Fe XI] appearing before [Fe VIII] (Warner 1989).

Forbidden iron lines are often seen in *symbiotic stars* – for instance from Fe II, III, IV, VI and VII (for instance Barba *et al.* (1992), Freitas Pacheco and Costa (1992)).

In *supernovae*, absorption lines of Fe II appear in all subtypes (Ia, Ib and II). Emission lines of both [Fe II] and [Fe III] are present in the nebular phase of supernovae of type Ia (Branch 1990).

Isotopes

Fe occurs in four stable isotopes, one long-lived and five short-lived ones. The stable isotopes are Fe 54, 56, 57 and 58, which occur in the solar system in 6%, 92%, 2% and 0.3% abundances. The long-lived isotope is Fe^{60} with a half life of 3×10^5 years.

Origin

Fe^{54} is produced by explosive nucleosynthesis, Fe^{56} and Fe^{57} also by this process or by a statistical equilibrium process; Fe^{58} can be produced by either helium burning or a nuclear statistical equilibrium process and also by carbon burning.

KRYPTON Kr $Z=36$

This element was discovered by W. Ramsay and M. W. Travers in 1898 in London. Its name comes from the Greek *kryptos* (hidden).

Ionization energies
KrI 14.0 eV, KrII 24.4 eV, KrIII 36.9 eV.

Absorption lines of KrI
No Kr line is seen in the solar spectrum.

Absorption lines of KrII
Following Bidelman's (1960) discovery of KrII in 3 Cen A, this element was observed in some other hot *Bp stars*. $W(4355)=0.025$ according to Hardorp (1966).

Emission lines of KrIII
Lines of [KrIII] are seen in the spectrum of at least one recurrent *nova* (Joy and Swings 1945).

Isotopes
Kr occurs in the form of six stable isotopes and 17 short-lived isotopes and isomers. The stable isotopes are Kr 78, 80, 82, 83, 84 and 86. In the solar system they represent 0.3%, 2%, 12%, 11%, 57% and 17% abundances respectively. Among the unstable isotopes, Kr^{81} has a half life of 2×10^5 years.

Origin
Kr 83, 84 and 86 can be produced by both the r and the s process, Kr^{82} is a pure s process product, Kr^{78} a pure p product and Kr^{80} can be produced by either the s or the p process.

LANTHANUM La Z=57

This element was discovered in 1839 by C. Losander in Stockholm, Sweden. The name comes from the Greek *lanthanein* (to lie hidden).

Ionization energies

La I 5.6 eV, La II 11.1 eV, La III 19.2 eV.

Absorption lines of La I

Table 1. *Equivalent widths of La I 6395(7)*

Group	III
K 5	0.022
M 0	0.028

Source: Data are from Gopka *et al.* (1990).

Absorption lines of La II

Table 2. *Equivalent widths of La II*

	4333(24)			6262(33)		
Group	V	III	Ib	V	III	Ib
F 0			0.049			
F 5	0.024		0.105			
F 8			0.290			
G 0	0.070					0.083
G 1						0.109
S	0.035			0.003		
G 2	0.040					0.103
G 5	0.090					0.085
G 6						0.168
G 8	0.024	0.050		0.011	0.040	0.173
K 0					0.020	
K 2					0.173	0.223
K 3					0.162	
K 5	0.100				0.141	0.177

Table 3. *Equivalent widths of La II 6320(19)*

Group	III	Ib
G 0		0.060
G 2		0.083
G 5		0.072
K 0	0.026–0.031	
K 2	0.032	0.158
K 3		0.148
K 5		0.129
M 2		0.126

La II lines (see for instance those at 4333 and 6262) appear in F-type stars and increase toward a maximum for K-type. A positive luminosity effect exists.

Emission lines of La II

Some La II lines like 6130(47) and 6627(61) have been observed in emission in one *Mira-type variable* (Grudzinska 1984). They are excited by fluorescence from H beta.

Behavior in non-normal stars

La II lines are normal or strong in the spectrum of *Ap stars* of the Cr–Eu–Sr subgroup (Adelman 1973b). Typically $W(4942)=0.035$ (Sadakane 1976). Cowley (1984) observed that La III is weak or absent in Ap stars. La II is observed exceptionally in a *Bp star* of the Si subgroup (Cowley 1984).

La II is strong in *Am stars*. Typically $W(4087)=0.150$ for late Am stars (Smith 1973, 1974).

La behaves in a manner parallel to that of Fe in *metal-weak* G and K *dwarfs* (Magain 1989), though Gilroy *et al.* (1988) find that it is overabundant with respect to iron (see also Part Two, section 2.2).

In *globular cluster stars* it behaves in a manner parallel to that of Fe (Wheeler *et al.* 1989).

As with other rare earths, La II is enhanced in *Ba stars* (Lambert 1985), which leads to large overabundances. Typically $W(4322)=0.19$ for a K 0 Ba star (Danziger 1965).

La II lines are enhanced in *S-type stars* (Bidelman 1953) by factors of the order of three; and to a lesser degree in *MS stars* (Smith *et al.* 1987).

La I lines are highly enhanced in the spectra of many *SC stars* (Wallerstein 1989, Kipper and Wallerstein 1990).

La lines are strong in the spectra of *C stars* (Keenan 1957), which was confirmed later by detailed analysis (Kilston 1975) for stars later than C 3 (Dominy 1984, Utsumi 1984).

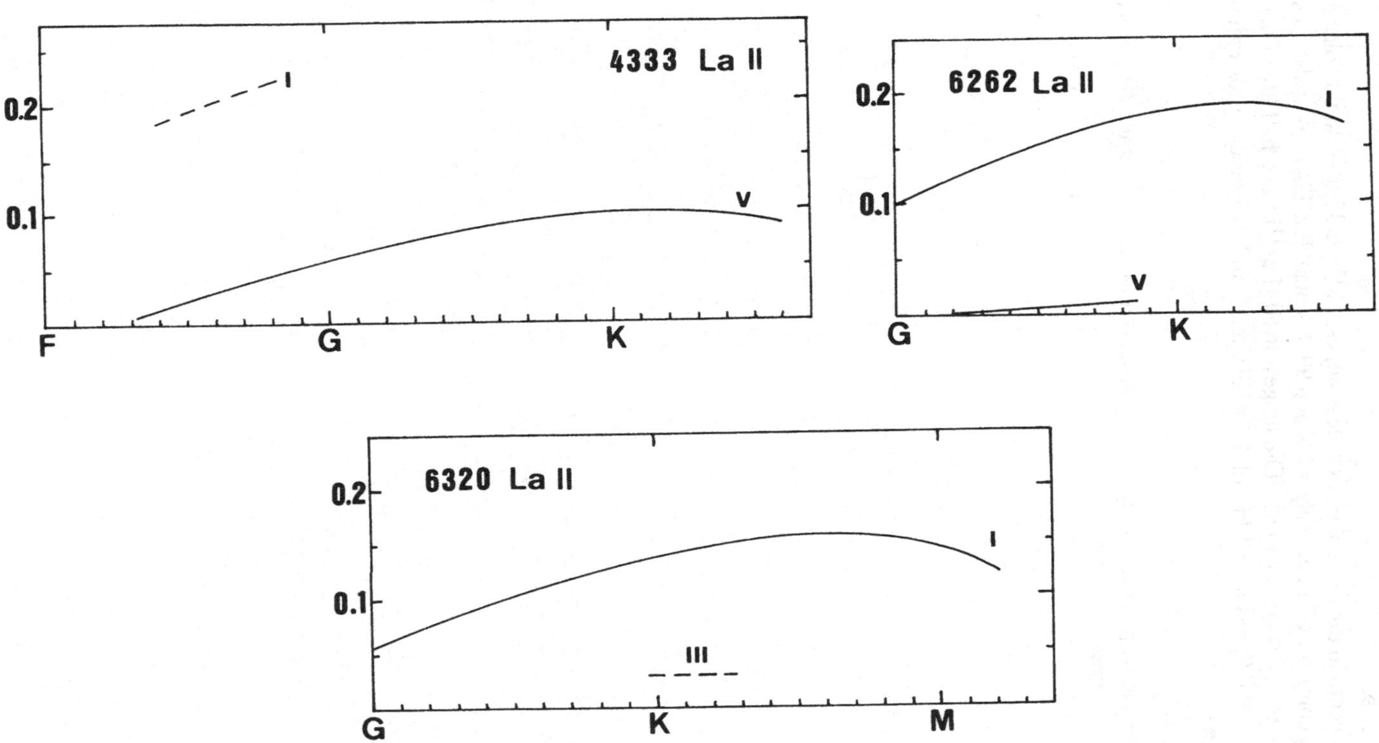
4333 La II
0.2
0.1
I
V
F
G
K
6262 La II
0.2
0.1
I
V
G
K
6320 La II
0.2
0.1
I
III
G
K
M

Isotopes

La occurs in the form of two stable isotopes, La^{138} and La^{139}. In the solar system they contribute respectively 0.1% and 99.9% of all La. There exist also 17 short-lived isotopes and isomers. The longest lived (La^{137}) has a half life of 6.7×10^4 years. La^{138} has a half life of 1.1×10^{11} years and can be used for radioactive dating.

Origin

La^{138} can be produced only by the p process and La^{139} by either the s process or the r process.

LEAD Pb Z=82

This element has been known since antiquity. Lead in Latin is *plumbum.*

Ionization energies
Pb I 7.4 eV, Pb II 15.0 eV, Pb III 31.9 eV.

Absorption lines of Pb I
In the sun the Pb I line 3683(1) has W=0.010. Fay *et al.* (1968) identified some lines of this element in M-type stars.

Emission lines of Pb I
The 3683 line is seen as a sharp strong emission line in the solar chromosphere (Pierce 1968).

Behavior in non-normal stars
Pb I was detected by Burbidge and Burbidge (1955) in one *Ap star* and confirmed by Guthrie (1972) and two Ap stars of the Cr–Eu–Sr subgroup.

Pb II 2203(2) was found by Severny and Lyubimkov (1985) in one Ap star of the Cr–Eu–Sr type.

Sadakane (1991) identified the 2203 line in the spectrum of one *Am star.*

Warner (1965) found that Pb is enhanced in *Ba stars.*

Isotopes
Lead occurs in the form of four stable isotopes, namely Pb 204, 206, 207 and 208. They contribute respectively 1%, 24%, 22% and 52% of the existing Pb. There exist also two unstable isotopes of long half life, namely Pb^{202} (3×10^5 years) and Pb^{205} (3×10^7 years) and 23 short-lived isotopes and isomers. The longer lived isotopes can be used for radioactive dating.

Origin
Pb 206, 207 and 208 are produced by either the r process or the s process. Pb^{204} can only be produced by the s process.

LITHIUM Li $Z=3$

This element was discovered in 1817 by J. Arfvedson in Stockholm. The name comes from the Greek *lithos* (stone).

Ionization energies
Li I 5.3 eV, Li II 75.6 eV.

Absorption lines of Li I

Table 1. *Equivalent widths of Li I 6707 (M.1 resonance line)*

Group	V	III
F 0	0.076	
F 4	0.080	
F 6	0.025	
F 8	0.047	
G 1	0.043	
S	0.004	
G 5	0.027	
K 0	0.025	0.028
K 2	0.015	
K 3		0.021
K 5	$<$0.040	

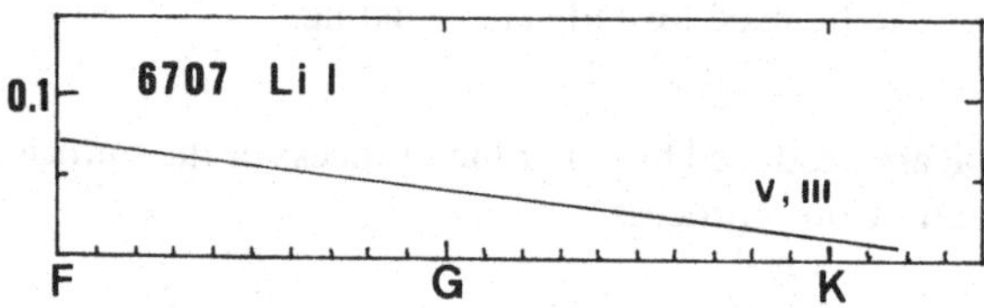

Li I (for instance 6707) appears in A-type stars and decreases toward K-type. No Li II line has been observed in stars.

Behavior in the stars
Li is normal or weak in late *Am stars* (Burkhart and Coupry 1991) and behaves erratically in *Ap stars* (Faraggiana *et al.* 1986).

In F-type stars in *open clusters* a curious gap appears around F 7. Before and after the gap (or dip), Li is abundant, whereas in the gap it almost disappears (Boesgaard and Tripicco 1986, Balachandran 1990). A detailed investigation by Lambert *et al.* (1991) of F-type stars shows that the gap may also depend upon the stellar population.

In G-, K- and M-type stars large scatter usually exists in Li line strengths, which is attributed to the evolutionary status of these stars, but other factors may also play a role. In one star Cayrel de Strobel and Cayrel (1989) find $W(6707)=0.2$ Å.

In *old stars* Li is still present, although rather weak (Rebolo 1991, Hobbs and Thorburn 1991). Spite (1992) finds no correlation between Li and Fe for *metal-weak stars* in the range $-3<\text{Fe/H}<-1.2$ dex. Thorburn (1992) has recently found some metal-weak stars with apparently no Li at all.

In giant stars Li lines are generally weak, except for a few stars in which Li is very strong, for instance the *S-type stars* and the so-called *weak G-band stars* (Vauclair 1991). Brown *et al.* (1989) analyzed over 600 G- and K-type giants and found strong Li lines (up to $W(6707)=0.45$) in about 1% of these stars. It is possible that such overabundances might be connected with anomalies in the C^{12}/C^{13} ratio. This was confirmed by Pilachowski *et al.* (1990).

G and K supergiants show a large scatter in Li strengths (Luck 1977).

In chromospherically active *RS CVn stars* Li is very strong, whereas in inactive stars it is weak or absent (Pallavicini *et al.* 1992). An investigation of *T Tau stars* showed that a large variety of Li strengths exists ($0.25<W<1.17$ Å) (Basri *et al.* 1991).

The '*Li stars*' of type *SC* or *CS* were analyzed by Catchpole and Feast (1976). In a sample of 380 S and C stars they found five Li stars. These Li-rich (also called 'super-lithium-rich') stars have $W(6707)$ between 3 and 8 Å. The latter value was measured in WZ Cas, which has been called the 'lithium star'. Keenan (1957) even found an object with $W=10$ Å. Such high values imply large overabundances and probably mean that Li is created *in situ* (Abia *et al.* 1991). It should, however, be mentioned that there exist also late C and SC stars in which Li is highly deficient (Kilston 1975, Wallerstein 1989, Kipper and Wallerstein 1990). Denn *et al.* (1991) measured Li I in many cool *carbon stars* and found no relation with the C^{12}/C^{13} ratio. They remarked that the Li abundance seems to be lower in these stars than in M-type giants.

Li-strong stars are also present in the *Magellanic Clouds* (Smith and Lambert 1990a, 1990b).

Li isotopes

Li has three stable isotopes, Li 5, 6 and 7. In the sun, one observes only Li^6 and Li^7 and their ratio is about 0.08. This ratio can be ascertained through the isotopic shift of the Li(6707) line, which is of the order of 0.2 Å. According to the discussion of Boesgaard (1976a), apparently Li^6 is not present in large quantities in stars. More recent data suggest a Li^6/Li^7 ratio of less than 0.1 (Ryan 1992).

Origin

Li^6 is produced by cosmic ray spallation. Li^7 arises from cosmological nucleosynthesis, cosmic ray spallation or hydrogen burning.

LUTETIUM Lu Z=71

This element was discovered independently by G. Urbain in Paris in 1907 and by von Welsbach. The name comes from the Latin name for Paris (*Lutetia*).

Ionization energies

Lu I 5.4 eV, Lu II 13.9 eV, Lu III 21.0 eV.

Absorption lines of Lu II

The equivalent width of Lu II 3397(4) in the sun is 0.028.

Behavior in non-normal stars

The element is present in some *Ap stars* of the Cr–Eu–Sr subgroup (see for instance Jaschek and Brandi (1972)), in the form of Lu I and Lu II. Poli *et al.* (1987) found Lu I in another Ap star.

Isotopes

Lu occurs in the form of two stable isotopes, Lu^{175} and Lu^{176}, which are present in the solar system with 97% and 3% respectively of all Lu. There exist 20 unstable isotopes and isomers. Lu^{176} has a half life of 3.7×10^{10} years and can be used for radioactive dating.

Origin

Lu^{175} can be produced by the r or the s process, whereas Lu^{176} is a pure s process product.

MAGNESIUM Mg $Z=12$

This element was first isolated by H. Davy in London in 1808. The name comes from the Greek city Magnesia, where a mineral known to the ancients was found.

Ionization energies

Mg I 7.6 eV, Mg II 15.0 eV, Mg III 80.0 eV, Mg IV 109.2 eV, Mg V 141.3 eV, Mg VI 186.5 eV, Mg VII 225 eV.

Absorption lines of Mg I

Table 1. *Equivalent widths of Mg I*

	5172(2)			5711(8)		
Group	V	III	Ib	V	III	Ib
B 9	0.12					
A 0	0.15					
A 7	0.25					
F 0	0.41					
F 4	0.48					
F 5	0.485		0.42	0.069		0.064
F 6	0.53					
F 8	0.64		0.615			0.10
G 0				0.12		0.15
G 1	0.78		0.534			
G 2	1.09,1.30		0.640			0.16
S	1.259					
G 5	1.23		0.770	0.11		0.18
G 8		1.29				0.16
K 0	2.08	1.95,1.73		0.17	0.19	0.14
K 2	2.08	1.45				0.22
K 3					0.17	0.27
K 5				0.20		0.22
M 0					0.15	
M 2						0.18

Mg I (see for instance 5172) appears in A-type stars and increases toward later types. A negative luminosity effect appears in the 5172 line from late F-type on. In contrast, a positive luminosity effect is present in the 5711 line.

The 5172 feature has such a large equivalent width that it can also be measured photoelectrically, see for instance Guinan and Smith (1984).

Fanelli *et al.* (1990) have analyzed the behavior of 2852 (UV M.1). The line

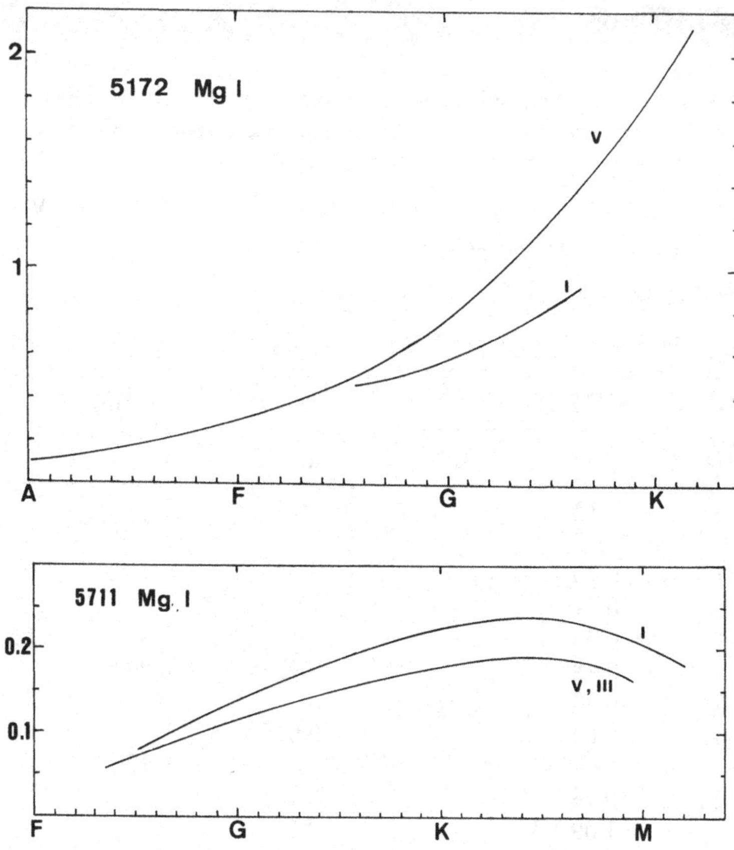

appears around A 0, increases slowly toward F 7 and then rises rapidly to a maximum at K 3. The line has a negative luminosity effect, like that of the line at 5172, after F 7.

Emission lines of Mg I

Mg I emissions appear in the chromospheric spectrum of K and M giants and supergiants (van der Hucht *et al.* 1979).

Some lines of Mg I appear in emission in *T Tau stars* (Joy 1945, Appenzeller *et al.* 1980).

In *long-period variables* of type S many Mg I lines appear in emission around the maximum – for instance 3829, 3832, 3838(3) and 5167, 5172 and 5183(2). Near the minimum the line 4571(1) is in emission in all long-period variables (Merrill 1952).

The 2852(1) line appears in emission in some M-type supergiants. Fix and Cobb (1987) found Mg I emission lines in the 1.6 μm region of an F-type supergiant that is an OH *maser*.

Absorption lines of Mg II

Table 2. *Equivalent widths of MgII*

	4481(4)			2800		
Group	V	III	Ia	V	III	Ib
O 7	0.044					
O 9	0.074					
B 0	0.104		0.150			
B 1	0.133		0.16	0.9		
B 3	0.19		0.404	1.3		
B 5	0.225	0.29	0.444			
B 6	0.343	0.395				
B 7				2.4		
B 8			0.43	4.0		
B 9	0.39					
A 0	0.4		0.59(Ia)			
A 1	0.41					
A 2	0.43,0.38		0.676,0.913(0)			
A 3			0.50(Ib),0.573(0)	7.0		
A 5					8.9	
A 7	0.50					
F 0	0.31		0.776			
F 1						6.9
F 2			0.468(Ib)	9.9		
F 4	0.36					
F 5	0.33,0.24		0.48(Ib)	8.4		
F 6	0.35					
F 8	0.29		0.44(Ib)			
G 0	0.20					
G 1	0.28					
G 2	0.125					
S	0.120					
G 5	0.22					

Note: The 2800 feature is a blend of 2790(3), 2795(1), 2798(3) and 2802(1).
Source: The 2800 line data are from Kondo *et al.* (1977). This line's behavior was also studied by Fanelli *et al.* (1990), but they do not provide equivalent widths.

MgII (see for instance 4481) appears in B-type stars, has a maximum at late A types and a positive luminosity effect. The strong ground level transitions at 2800 (2795 and 2802), which are analogous to the H and K lines of CaII, show a

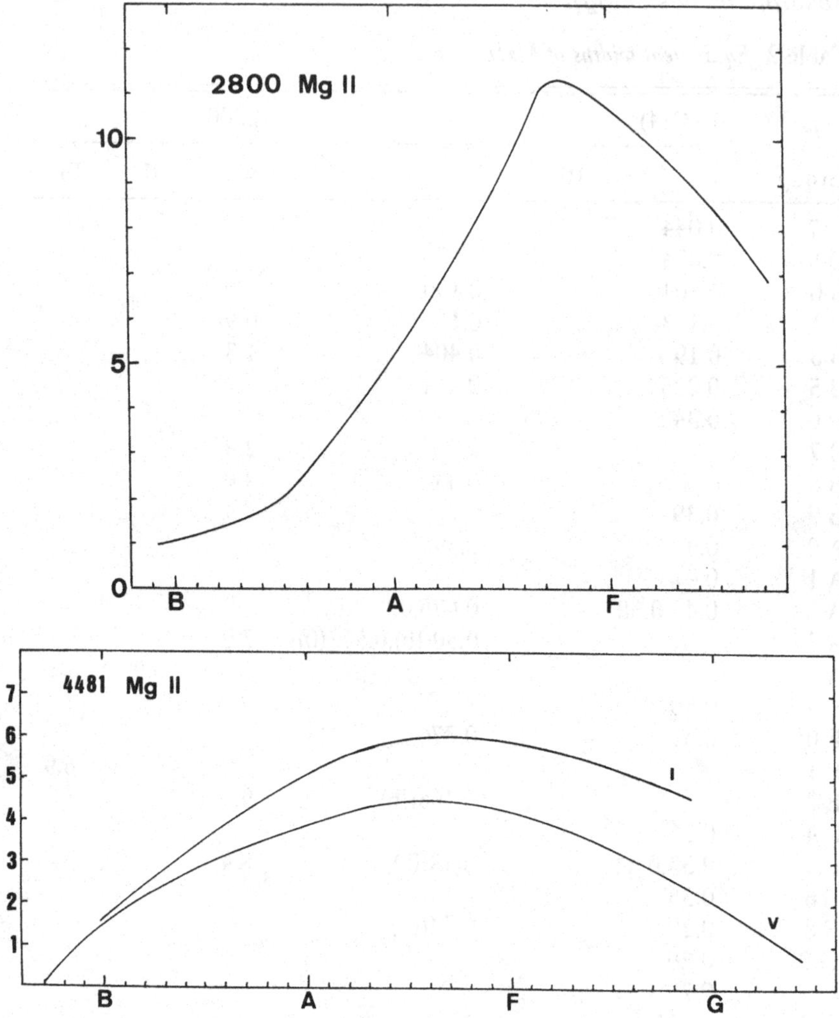

similar behavior, but the matter should be investigated in more detail with more stars. For the latter see also Gurzadyan (1975).

Mg II emissions

Emissions in the Mg II resonance doublet 2795 and 2803 are similar in appearance to those of the Ca II doublet, showing a centrally reversed emission core within the broad absorption line. (For illustrations of the line profiles see for instance Basri and Linsky (1979).) In general Mg II emissions appear in the same type of stars as Ca II emissions, namely late type stars of all luminosity classes from F-type onwards. They become very frequent from type K 3 onwards. In general the Mg II emissions are the strongest ones in the near ultraviolet region.

Their presence signals the existence of a stellar chromosphere. (See Part Two, section 3.1.) Their presence and strength are thus not correlated directly with metal weakness (Dupree *et al.* 1990) as was thought for some time. Several surveys of these emissions have been made, among others by Stencel *et al.* (1980) and Doherty (1985).

The behavior of the 2800 feature is similar to that of Ca II and one can derive a relation between the width of the emission feature (*d*) and the absolute magnitude of the star (Kondo *et al.* 1977). Elgoroy (1988) gives

$$M_v = -19.58 \log d + 41.36$$

This relation is analogous to the Wilson–Bappu relation for Ca II.

The Mg II lines 7877 and 7896 of M 8 appear in emission in some early type peculiar objects, like *Be, B[e]* and *Herbig Ae–Be stars* (Jaschek *et al.* 1993). In early *Be stars* one sees also 9217–9243(1) in emission (Johnson 1977).

Mg II in emission is often observed in *T Tau stars* (Appenzeller *et al.* 1980) and always in *semiregular variables* (Querci and Querci 1989). Profiles of the Mg II emissions have also been used to map the bright plages (bright emission regions) in *RS CVn stars*. For a summary see Rodono (1992).

Behavior in non-normal stars

Mg is overabundant by 0.6 dex in the spectrum of at least one *extreme He star* (Schoenberner and Drilling 1984).

The 4481(4) line of Mg II is weakened in *shell spectra* (Struve and Wurm 1938).

Bidelman (1966) called attention to one early *Ap star* characterized by strong Si II and Mg II lines. Jaschek and Jaschek (1974) later listed a small subgroup of stars that share this characteristic and can be called *Mg-strong stars*. For a detailed study of one star of the group see Naftilan (1977).

Mg II tends to be weak in Ap stars of the Cr–Eu–Sr subgroup (Adelman 1973b). Typically *W*(4481)=0.30 (Sadakane 1976).

Mg tends also to be weak in *Am stars*. Smith (1973, 1974) found that *W* values in Am stars are about 70% those of normal stars.

Mg II lines are generally weak in *lambda Boo stars* (Holweger and Stuerenburg 1991). Typically *W*(4481)=0.100 for A0V. Abt (1984) has used the Mg weakness as a criterion to select candidates for new lambda Boo stars.

Mg lines are weakened in *F-type HB stars* with regard to normal stars of the same colors, by factors up to ten in *W* (Adelman and Hill 1987). Mg I(3832, 3838) is seen in *DZ degenerates* (Sion *et al.* 1990), with *W* up to 8 Å.

Mg seems to be slightly overabundant with respect to iron in *metal-weak stars* by factors of the order of two (Gratton and Sneden 1987, Magain 1989, Francois 1991). This is also true for the most metal-deficient stars, with Fe/H about −4 dex (Molaro and Bonifacio 1990). In *globular cluster stars* Wheeler *et al.* (1989) found that Mg is underabundant with respect to Fe, but Smith and Wirth (1991) remarked that the behavior of Mg is somewhat erratic, if stars of different CN strength are examined.

In *novae*, MgII appears both in absorption and in emission. Emissions are usually seen in the ultraviolet spectral region. Mg appears again at the nebular stage phase in the form of [MgV] and [MgVII] (Payne-Gaposchkin 1957, Warner 1989).

Strong MgII lines appear in the spectra of *supernovae* of types Ia and II. Strong emissions of [MgI] are present in the nebular stage of supernovae of type Ib (Branch 1990).

Mg Isotopes

Magnesium has three stable isotopes, Mg 24, 25 and 26, which occur in the solar system with respectively 79%, 10% and 11% abundance. There also exist five unstable isotopes.

Since the bands of $Mg^{24}H$, $Mg^{25}H$ and $Mg^{26}H$ lie close together, they can be used for an analysis of the comparative abundance of the three Mg isotopes. Such an analysis was performed for two *Ba stars* by Tomkin and Lambert (1979) and the isotope ratios found are compatible with the solar values. The analysis was extended by Malaney and Lambert (1988) to more Ba stars, with the same result (i.e. solar ratios). Smith and Lambert (1988) investigated in a similar way the *MS* and *S stars* and also found no obvious anomalies in the isotopic ratio. McWilliam and Lambert (1988) concluded from an analysis of *old disk stars* that the isotope ratio is about solar. It is possible that, in *metal-weak stars*, the isotopes Mg^{25} and Mg^{26} diminish with respect to Mg^{24}.

Origin

Mg^{24} can be produced either by hot hydrogen burning or by explosive nucleosynthesis. Both Mg^{25} and Mg^{26} can be produced by neon burning, carbon burning or explosive nucleosynthesis.

MANGANESE Mn $Z=25$

This element was discovered by J. Gahn in 1774 in Stockholm. The name probably comes from *magnes* (magnet) because of the magnetic properties of the manganese ore pyrolusite.

Ionization energies
Mn I 7.4 eV, Mn II 15.6 eV, Mn III 33.7 eV.

Absorption lines of Mn I

Table 1. *Equivalent widths of Mn I*

	4030 (two resonance lines)		6022(27)		
Group	V	Ib	V	III	Ib
B 9	0.013				
A 1	0.027				
A 2	0.04				
A 7	0.23				
F 2		0.40			
F 4	0.36		0.045		
F 5			0.083		0.068
F 6	0.41		0.045		
F 8	0.36		0.060		0.090
G 0	0.34		0.130		0.165
G 1	0.56				0.139
G 2			0.11		0.149
S	0.326		0.096		
G 5	0.45		0.15		0.166
G 8			0.14(IV)	0.16	0.203
K 0			0.16	0.165	
K 2			0.16		0.238
K 3				0.245	0.323
K 5			0.22		0.205
M 0			0.16		
M 2					0.200

Mn I (for instance 4030) lines appear in A-type stars and increase in strength toward later types. For later types a positive luminosity effect exists (see the line at 6022).

Emission lines of Mn I

The 4032 blend appeared in emission during one flare of an *RS CVn star* (Doyle 1991), probably produced by fluorescence.

The blend at 4032 is in emission at post-maximum phases in *long-period variables* (Merrill 1947). Around minimum light one finds instead lines of [Mn I] (Querci 1986).

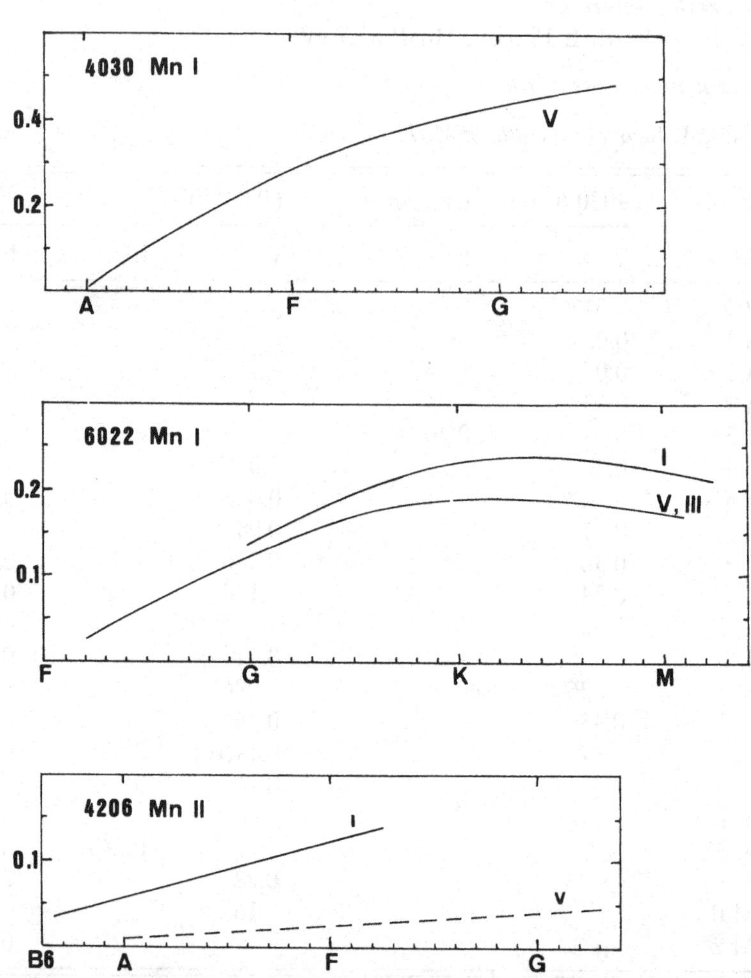

Absorption lines of Mn II

Table 2. *Equivalent widths of Mn II*

	3482(3)		4206(7)	
Group	V	Ib	V	Ia
B 8				0.045
A 0	0.05	0.16		0.040
A 1			0.011	
A 3		0.342(0)		
F 0				0.120
F 2				0.141(Ib)
S	0.153		0.041	

Mn II (for instance 4206) lines appear in late B-type stars and grow in intensity toward later types. A positive luminosity effect exists.

Emission lines of Mn II

Mn II emissions are seen in one typical *B[e] star* (Swings 1973). Mn II emission lines appear in the spectra of *long-period variables* around maximum light, in *T Tau stars* (Appenzeller *et al.* 1980) and in some *VV Cep stars*.

Behavior in non-normal stars

Mn lines are very strong in the spectra of the so-called Hg–Mn subgroup of *Bp stars*. $W(4030)=0.040$ and $W(4206)=0.100$ (Kodaira and Takada 1978).

Gratton (1989) has analyzed the behavior of Mn in *metal-weak stars* and found that in general Mn is underabundant with respect to Fe, by a factor of two except in stars where the metal weakness is small, i.e. $Fe/H>-1$ dex. This implies that Mn cannot be taken as a typical metal. According to Wheeler *et al.* (1989) Mn is also slightly underabundant in *globular cluster stars*. Peterson *et al.* (1990) found that Mn is underabundant in an extreme metal-weak globular cluster giant by a factor of ten, suggesting that the deficiency of Mn may change with increasing metal deficiency.

Mn seems to be normal in *S-type stars* (Fujita *et al.* 1963).

Isotopes

Mn occurs in the form of Mn^{55}. It has ten unstable isotopes and isomers, of which the longest lived is Mn^{53}, with a half life of 2×10^6 years.

Origin

Mn can be produced either by explosive nucleosynthesis or by the nuclear statistical equilibrium process.

MERCURY Hg $Z=80$

This element was known in antiquity. In Latin it was called *hydragyrum* (liquid silver).

Ionization energies

Hg I 10.4 eV, Hg II 18.8 eV, Hg III 34.2 eV.

Absorption lines of Hg I

The resonance line at 1849(2) is present as a weak absorption line in the sun (McAlliatrer 1960).

Behavior in non-normal stars

Hg II was first identified by Bidelman (1962a) to be the element responsible for a strong line at 3984 in the spectra of many *Bp stars* of the Mn subgroup. Because in general Mn is accompanied by Hg, the group is usually called the Hg–Mn subgroup instead of the former designation of Mn subgroup. In the Bp stars $0.050<W<0.300$ Å, whereas the line is absent in normal stars. The identification of Hg II was strengthened by the identification of Hg I lines (4046, 4358 and 5460 from M.l) by Bidelman (1966) in the *Bp and Ap stars*. Adelman (1988) measured $W(4358)=0.003$ in a Hg–Mn star.

A proposed identification of Hg III is very controversial. For a discussion, see Dworetsky (1985).

Isotopes

Hg has seven stable isotopes, namely Hg 196, 198, 199, 200, 201, 202 and 204. In the solar system their frequencies are respectively 0.2%, 10%, 17%, 23%, 13% 30% and 7%. There exist 26 unstable isotopes and isomers. Bidelman (1962a) called attention to the fact that the wavelength of the 3984 line does not correspond exactly to the laboratory wavelength, which fact he attributed to the different isotopic composition in different stars. What varies is the proportion of isotopes 198, 200, 202 and 204. Hg^{204} seems to predominate in late *Ap stars*, according to Dworetsky (1985). Leckrone *et al.* (1991a, 1991b) have recently shown that, in one Ap star of the Hg–Mn subgroup, practically all the Hg is in the form of Hg^{204}.

Origin

Hg 199, 200, 201 and 202 can be produced by either the r process or the s process. Hg^{196} is a pure p process, Hg^{198} a pure s process and Hg^{204} a pure r process product.

MOLYBDENUM Mo $Z=42$

This element was isolated in 1781 by P. Hjelm in Uppsala, Sweden. The name alludes to *molybdos*, which in Greek means lead, an element for which it was mistaken.

Ionization energies
Mo I 7.1 eV, Mo II 16.1 eV, Mo III 27.2 eV.

Absorption lines of Mo I

Table 1. *Equivalent widths of Mo I*

	5506(4)		5570(4)		
Group	V	III	V	III	Ib
F 5					0.016
G 2	0.004				
S	0.005		0.006		
G 9				0.022	
K 0		0.044		0.034	
K 2		0.069			0.100
K 3				0.088	
K 5					0.120
M 2					0.151

Mo I has many faint lines in the solar spectrum and that of later type stars. It increases in strength toward cooler stars (see 5570) and shows a positive luminosity effect.

Absorption lines of Mo II
The equivalent width of Mo II 3320(6) in the sun is 0.009. Mo II lines are also present in the ultraviolet at 2038(32) and 2045(22) (Sadakane 1991).

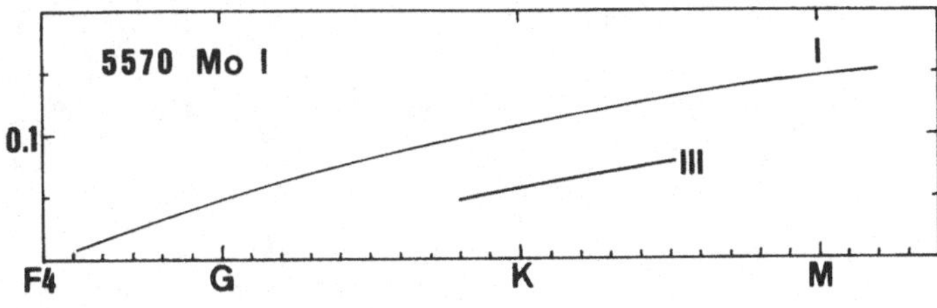

Behavior in non-normal stars

Jaschek and Brandi (1972) and Cowley *et al.* (1974) detected Mo II in two *Ap stars* of the Cr–Eu–Sr subgroup. Adelman (1973b) signaled the presence of Mo in more stars of this subgroup.

Mo I lines are enhanced in the spectra of *Ba stars*, which leads to overabundances of one order of magnitude (Lambert 1985, Smith 1984). Typical *W* values are twice as large as in normal giants (Danziger 1965).

Mo I lines are enhanced in *S-type stars*, which leads to overabundances of one order of magnitude (Smith and Wallerstein 1983) and in *SC stars* (Kipper and Wallerstein 1990).

Isotopes

Mo has seven stable isotopes and 14 short-lived isotopes and isomers. The stable isotopes are Mo 92, 94, 95, 96, 97, 98 and 100, which occur in the solar system with frequencies 15%, 9%, 16%, 17%, 10%, 24% and 9% respectively.

Origin

Mo^{92} and Mo^{94} can only be produced by the p process; Mo 95, 97 and 98 are produced by either the r or the s process. Mo^{96} is a pure s process and Mo^{100} a pure r process product.

NEODYMIUM Nd Z=60

This element was isolated by A. von Welsbach in Vienna in 1885. The name comes from the Greek *neos didymos* (new twin).

Ionization energies
NdI 5.5 eV, NdII 10.7 eV, NdIII 22.1 eV, NdIV 40.4 eV.

Absorption lines of NdI
No lines of NdI have been observed in the sun. Gopka *et al.* (1990) observed a weak line of NdI 5676 in K 5III (W=0.016) and M 0III stars.

Absorption lines of NdII

Table 1. *Equivalent widths of NdII*

	4061(10)		5431(80)		
Group	V	Ib	V	III	Ib
A 1	0.005	0.035			
F 0	0.075				
G 0	0.072				0.017
G 1					0.040
G 2			0.002		0.043
G 5	0.089				0.047
G 6					0.067
G 8					0.070
K 0				0.018	
K 2				0.013	0.081
K 3					0.068
K 5					0.085,0.120

NdII lines (for instance 4061 and 5431) appear in A-type spectra and strengthen toward later types. A positive luminosity effect is present (see for instance the line at 5431).

According to Smith and Lambert (1985), NdII is also visible beyond K 5. They quote that the line 7513 grows from 0.020 for K5III to 0.055 for M6III.

Behavior in non-normal stars
NdII lines are strong in the spectra of *Ap stars* of the Cr–Eu–Sr subgroup (Adelman 1973b). A typical value of W(4061) is 0.027 (Sadakane 1976). Cowley (1976) and Aikman *et al.* (1979) observed lines of NdIII in the spectra of several Ap stars with strong NdII lines. NdII and NdIII have also been observed in the spectrum of one *Bp star* of the Si subgroup (Cowley and Crosswhite 1978).

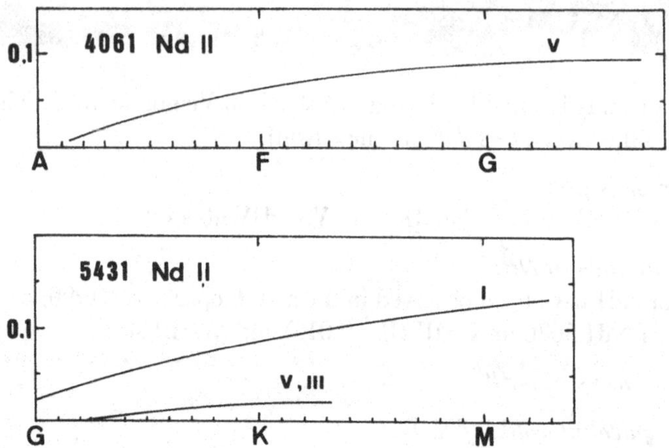

Cowley (1984) calls attention to some Ap stars in which Nd and Sm are conspicuously absent, whereas other rare earths are strong.

Nd II lines are also strong in *Am stars*. A typical value for late Am stars is $W(4061)=0.150$ (Smith 1973, 1974).

Nd II lines are enhanced in the spectra of *Ba stars*, which corresponds to large overabundances (Lambert 1985). Typical W values are twice as large as in normal giants of the same temperature (Danziger 1965). Nd II lines are also enhanced in the *subgiant CH stars* (Luck and Bond 1991). In *S-type stars* Nd II lines are prominent (Bidelman 1953), having W values larger by a factor of 2–3 than in normal giants. This leads to large overabundances (Smith and Lambert 1986, Smith *et al.* 1987). In *MS stars* the enhancements are less extreme.

Nd is strengthened in *C stars* later than C 3 (Utsumi 1984).

Nd II lines have been observed in G- and K-type *metal-weak dwarfs* by Gilroy *et al.* (1988). These authors found Nd to be overabundant with respect to iron (see also Part Two, section 2.2).

Isotopes

Nd occurs in the form of seven stable and nine short-lived isotopes and isomers. The stable ones are Nd 142, 143, 144, 145, 146, 148 and 150. In the solar system their respective contributions are 27%, 12%, 24%, 8%, 17%, 6% and 6%. For the use of Nd as a nuclear chronometer see Learner *et al.* (1991).

Origin

Nd 143, 144, 145 and 146 can be produced by either the r process or the s process. Nd^{142} is a pure s process product and Nd^{148} and Nd^{150} are both pure r process products.

NEON Ne Z=10

This element was isolated by W. Ramsay and M. Travers in 1898 in London. The name alludes to the Greek *neos* (new).

Ionization energies

Ne I 21.6 eV, Ne II 40.9 eV, Ne III 63.5 eV, Ne IV 97.1 eV, Ne V 126.2 eV, Ne VI 157.9 eV, Ne VII 207.3 eV.

Absorption lines of Ne I

Table 1. *Equivalent widths of Ne I 6402(1)*

Group	V	Ia
B 3	0.084	0.170
B 5		0.158

Ne I lines appear in B-type stars and are prominent in supergiants.

Emission lines of Ne I

Several multiplets of Ne I (for instance 1, 3, 5 and 6) are present in the spectrum of one R CrB star (Dahari and Osterbrock 1984).

Absorption lines of Ne II

Table 2. *Equivalent widths of Ne II*

	3482(6)	3766(1)		4219(52)		
Group	V	V	Ia	V	III	Ia
O 9				0.025		
O 9.5	0.039					
B 0	0.042	0.054	0.100	0.046	0.035	0.033
B 0.5		0.053		0.027		
B 2					0.015	

Source: Some values for 4219 are from Kilian and Nissen (1989).

Ne II lines appear in late O-type and early B-type stars, being slightly stronger in supergiants.

Emission lines of Ne V

The lines at 1136 and 1145 are seen in emission in the solar ultraviolet spectrum.

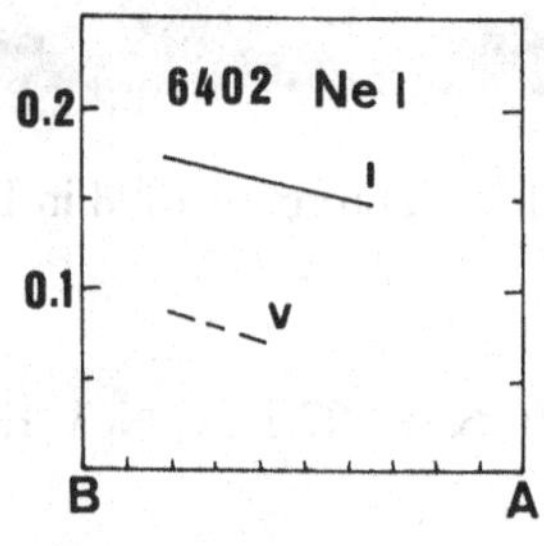

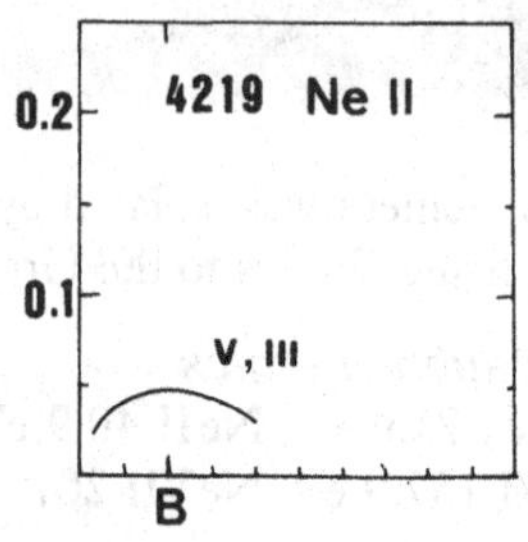

Emission lines of Ne VI

Four NeVI lines between 997 and 1010 are seen in emission in the solar ultraviolet spectrum (Feldman and Doschek 1991).

Behavior in non-normal stars

Ne lines are very strong in *extreme helium stars* (Jeffery and Heber 1993). NeI lines are very strong in the spectrum of the A-type *He-strong star* upsilon Sgr (Greenstein 1943).

[NeIII], [NeIV] and [NeV] lines are sometimes present in the nebular phase of the spectral evolution of *novae* (Joy and Swings 1945, Warner 1989). Smits (1991) has called attention to a group of novae that have large overabundances of Ne, as well as of at least one of the elements Na, Al, Mg and Si. Andreae (1993) calls this group O–Ne–Mg stars and finds that the overabundance of Ne reaches one order of magnitude.

In *symbiotic stars* lines of [NeIII] and [NeIV] have been observed (Merrill 1950, Freitas Pacheco and Costa 1992). In several stars of this group Wallerstein *et al.* (1991) report also the presence of [NeV].

Isotopes

Ne has three stable isotopes, Ne 20, 21 and 22, and five unstable isotopes. In the solar system the stable isotopes contribute 90.5%, 0.3% and 9.2% respectively.

Origin

Ne^{20} can be produced by carbon burning. Ne^{21} can also arise from this process or from explosive nucleosynthesis. Ne^{22} is produced by either helium or nitrogen burning.

NICKEL Ni   Z=28

This element was discovered by A. Cronstadt in Stockholm, Sweden in 1751. The name alludes to the mineral called by German miners *Kupfernickel* (false copper).

Ionization energies

Ni I 7.6 eV, Ni II 18.2 eV, Ni III 35.2 eV, Ni IV 54.9 eV, Ni V 75.4 eV, Ni VI 107.8 eV, Ni VII 132.7 eV.

Absorption lines of Ni I

Table 1. *Equivalent widths of Ni I*

	4470(86)		4714(98)	
Group	V	III	V	Ib
A 0	p			
A 2			0.02	
A 7			0.08	
F 0			0.13	0.085(II)
F 2				0.112
F 4			0.12	
F 5			0.155	0.16
F 6			0.17	
F 8			0.17	0.26
G 0	0.088		0.11	
G 1			0.23	
G 2	0.085		0.21	
S	0.069			
G 5	0.107		0.26	
G 8	0.117(IV)			
K 0			0.31	
K 2		0.105	0.34	

Table 2. *Equivalent widths of Ni I*

	5435(70)			5587(70)		
Group	V	III	Ib	V	III	Ib
F 5	0.022		0.022			
F 8			0.051			
G 0			0.125	0.126		

Table 2. (*cont.*)

	5435(70)			5587(70)		
Group	V	III	Ib	V	III	Ib
G 2	0.041		0.128			0.135
S				0.049		
G5			0.128			0.162
G 8			0.151,0.214			0.245
K 0		0.110			0.095	
K 2			0.173		0.120	0.251
K 3			0.190			0.323
K 5			0.192			0.380
M 0		0.125			0.125	
M 2			0.196			
M 2.5		0.132			0.132	

Ni I (see for example 4714) appears in A-type stars and grows monotonically toward later types. The luminosity effect is positive (see also the line at 5435).

In the infrared one of the most intense lines is 7555(187). In the sun, W(7555)=0.107.

Emission lines of Ni I

Lines of M.30 and 33 are seen in emission in *T Tau stars* (Joy 1945) and in at least one *supernova* (Arnett *et al.* 1989). Jennings *et al.* (1993) also observed [Ni I] at 11.3 μm in the *supernova* 1987A.

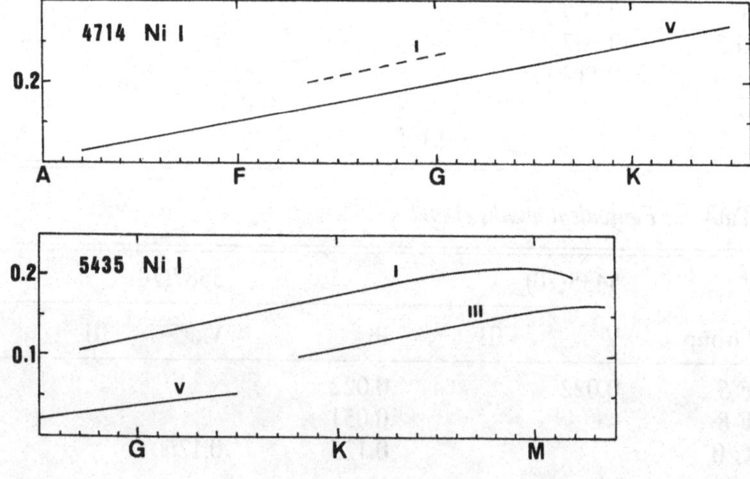

Absorption lines of Ni II

Table 3. *Equivalent widths of Ni II*

	4067(11)		4362(9)	
Group	V	I	V	Ib
B 5	0.014	0.22		
B 7	0.022			
B 9	0.051			
B 9.5	0.029,0.036		0.005	
A 0		0.100(Ia)		
A 1			0.016,0.038(IV)	
A 2			0.024	0.043(Ia)
A 3				0.072(Ia)
F 2				0.190
F 5			0.015	0.055
F 8				0.072
S			0.021	

Ni II is seen from late B-type on and disappears in G-type stars. A positive luminosity effect exists.

Emission lines of Ni II

Lines of [Ni II] appear in emission in one *VV Cep star* (Rossi *et al.* 1992), in the *luminous blue variable* eta Car (Thackeray 1953) and in the supernova 1987A (Arnett *et al.* 1989) Lines of [Ni II] are also seen in one typical *B[e]* star (Swings 1973).

Absorption lines of Ni IV and Ni V

Ni IV and Ni V have been found by Dean and Bruhweiler (1985) in the spectra of some O-type stars.

Forbidden lines of higher ionization stages

Lines of Ni XII (4231), XIII, XV(6701, 8024) and XVI (3601) with ionization energies between 318 and 455 eV are seen in the spectrum of the solar corona (Zirin 1988) and in at least one *nova* (Joy and Swings 1945).

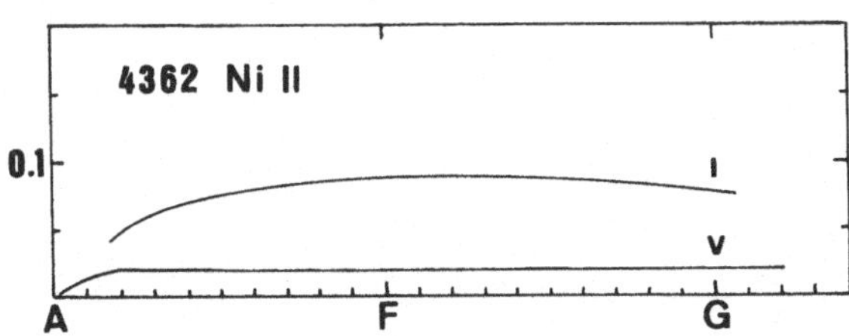

Behavior in non-normal stars

Ni seems to be weak in Ap stars of the Hg–Mn subgroup (Takada-Hidai 1991).

Sneden *et al.* (1991) found that Ni behaves in a manner similar to that of Fe in *metal-weak disk* and *halo stars*. Wheeler *et al.* (1989) found the same behavior for *globular cluster stars*. Ni can thus be regarded as a typical representative of the metals.

Isotopes

Ni has five stable isotopes, Ni 58, 60, 61, 62 and 64, which in the solar system occur with 68%, 26%, 1%, 4% and 1% abundances respectively. Besides these isotopes there also exist four unstable isotopes, the longest lived, Ni^{59}, having a half life of 8×10^4 years.

Origin

Ni 58, 61 and 62 can be produced by explosive nucleosynthesis and by nuclear statistical equilibrium. Ni^{61} is produced by these processes and by carbon burning. Ni^{62} is produced by explosive nucleosynthesis, by statistical equilibrium and by oxygen burning. Ni^{60} is produced only by the statistical equilibrium process and Ni^{64} only by explosive nucleosynthesis.

NIOBIUM Nb Z=41

This element was discovered by C. Hatchett in 1801 in London. Its first name was columbium, which alludes to Columbus, because the element was discovered in minerals brought from America. In 1846 H. Rose, a German chemist, proved that columbium is not identical to tantalum, as was believed at that time, and he proposed the name niobium, because Niobe was a daughter of Tantalus in Greek mythology. The name niobium was finally adopted in 1950.

Ionization energy
Nb I 6.9 eV, Nb II 14.3 eV, Nb III 25.0 eV.

Absorption lines of Nb I

Table 1. *Equivalent widths of Nb I 4152(1)*

Group	V	III
S	0.008	
K 5	0.074	
M 5		0.181

Absorption lines of Nb II

Table 2. *Equivalent widths of Nb II*

	3620(4)	4579(8)
Group	V	III
S	0.015	
K 0		0.098

In view of the small number of data, one can only conclude that both Nb I and Nb II are present in late type stars.

Behavior in non-normal stars
Jaschek and Brandi (1972) detected Nb I in one *Ap star* of the Cr–Eu–Sr subgroup. Cowley *et al.* (1974) detected Nb II in another star of the same subgroup.

Nb I lines are strengthened in *Ba stars*, which leads to overabundances of one order of magnitude (Lambert 1985, Smith 1984). Nb I lines are also strengthened in *subgiant CH stars* (Krishnaswamy and Sneden 1985).

Merrill (1947) detected Nb I in one *S-type star* and Davis (1984) added some more stars with Nb I. This detection was confirmed by Wallerstein and Dominy

(1988), who found a variety of Nb strengths, although in general it seems weaker in *S-type* than in *M-type giants* of the same temperature. Lambert (1989) found that Nb apparently goes with Tc; but there exists at least one exception to this rule. See also Vanture *et al.* (1991).

Isotopes

Nb has one stable isotope, Nb^{93}, and 23 short-lived isotopes and isomers. The longest lived is Nb^{94} with a half life of 2×10^4 years, which can be used for radioactive dating.

Origin

Nb is only produced by the s process.

NITROGEN N Z=7

This element was discovered by D. Rutherford in 1772 in Edinburgh. Its name comes from *nitron* (nitre) and *gene* (which forms).

Ionization energies

NI 14.5 eV, NII 29.6 eV, NIII 47.4 eV, NIV 77.5 eV, NV 97.9 eV, NVI 552 eV, NVII 667 eV.

Absorption lines of NI

Table 1. *Equivalent widths of NI*

	8683(1)		8711(1)	
Group	V	III	V	Ib
B 5 V	0.005			
B 8 V	0.028			
A 0	0.070			0.25
A 1				0.30
A 2 V	0.056	0.056		
A 5				0.35
A 3 IV	0.036			
F 0 V	0.038		0.05	0.33
S	0.008			

Source: Values for 8683 from Roby and Lambert (1990) and Sadakane and Okyudo (1989). Values for 8711 are from Jaschek and Jaschek (unpublished).

The most useful lines are those of the infrared region. NI appears around mid-B-type stars and is seen in A and F stars with a maximum in A-type stars (see for instance the line at 8683). A strong positive luminosity effect exists (see for instance the line at 8711). The resonance line of NI (UV 1) is at 1200.

Emission lines of NI

NI (M.1 and 8) are often seen in emission in early *Be stars*. The *W* values are less than 1 Å. In *Be shell stars* the lines are generally in absorption (Andrillat *et al.* 1990).

The ultraviolet lines 1134(2) and 1167–70(6) are seen in strong emission in the solar spectrum.

Absorption lines of NII

Table 2. *Equivalent widths of NII 4630(5)*

Group	V	Ia
B 0	0.03	0.067
B 1	0.05	0.135
B 2	0.05	0.537
B 3	0.039	0.355
B 5	0.010	0.199

NII lines (see for instance the line at 4630) are seen in early type stars with a maximum at B 2. A strong positive luminosity effect exists. The same behavior is shared by the NII lines 5667, 5676 and 5711(3).

Emission lines of NII

The [NII] line is present in emission in *P Cyg* (Stahl *et al.* 1991). Swings (1973) observed many [NII] lines in the spectrum of one typical *B[e] star*. The [NII] lines 6548 and 6584 have been observed in emission in two peculiar G-type supergiants (Sheffer and Lambert 1992). The [NII] line 6584 is prominent in the post-nova phase of *novae* (Warner 1989) and in *symbiotic stars* (Freitas Pacheco and Costa 1992). In *luminous blue variables* several faint [NII] emissions have been observed (Wolf 1993).

The ultraviolet lines 916(2) and 1085(1) appear in emission in the solar spectrum (Feldman and Doschek 1991).

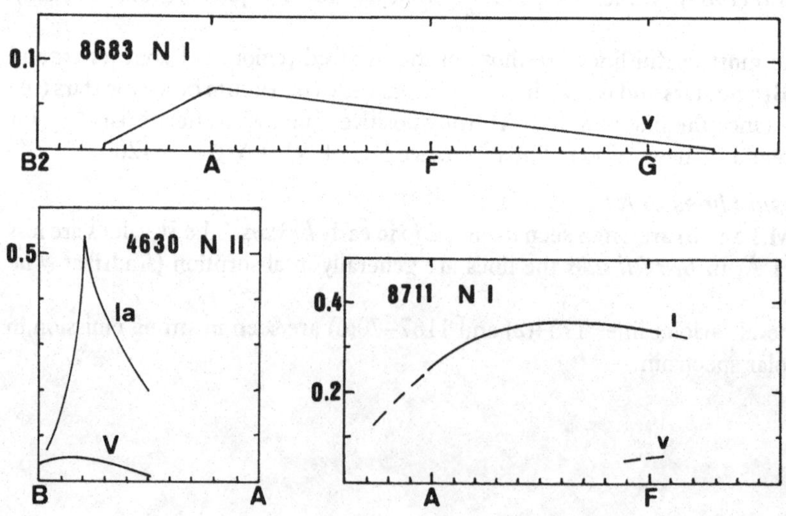

Absorption lines of N III

Table 3. *Equivalent widths of N III 4514(3)*

Group	V	III	I
O 6	0.055		
O 8	0.100		
O 9	0.115		
O 9.5	0.075		
B 0	0.048		0.038,0.075(Ia)
B 0.5	0.026		
B 2		0.006	

N III lines are seen in O- and early B-type stars, with a maximum at O 9. A strong positive luminosity effect exists. The ultraviolet line 1758 has a maximum around B 0 and disappears toward B 5 (Prinja 1990).

Emission lines of N III

For the emissions of N III in O-type stars see the discussion of behavior in non-normal stars.

N III appears in emission in some *symbiotic stars*. The ultraviolet lines of M.1, 989–91 appear in emission in the solar spectrum (Feldman and Doschek 1991).

Absorption lines of N IV

Table 4. *Equivalent widths of N IV 3478(1)*

Group	V	I
O 6	0.045	
O 7	0.070	
O 8		0.300
O 9	0.032	

Source: The only supergiant is plotted as a cross in the figure.

N IV lines are seen in O-type stars. A strong luminosity effect exists. N IV 6381 reaches its maximum strength at O 4–O 5 in supergiants. The ultraviolet N IV line 1718(7) has a maximum at O 4 (Heck *et al.* 1984).

Absorption lines of N V

NV (resonance line at 1238 and 1243(1)) is present in all stars hotter than type B 0 for the main sequence, B 1 for the giants and B 2 for the supergiants (Abbott *et al.* 1982, Dean and Bruhweiler 1985). The line often has a shell component.

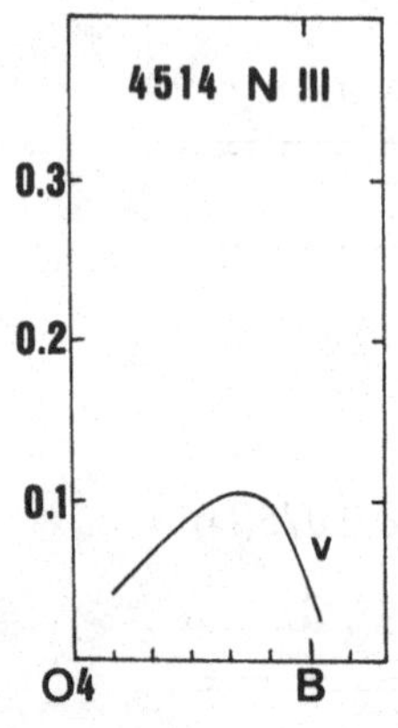

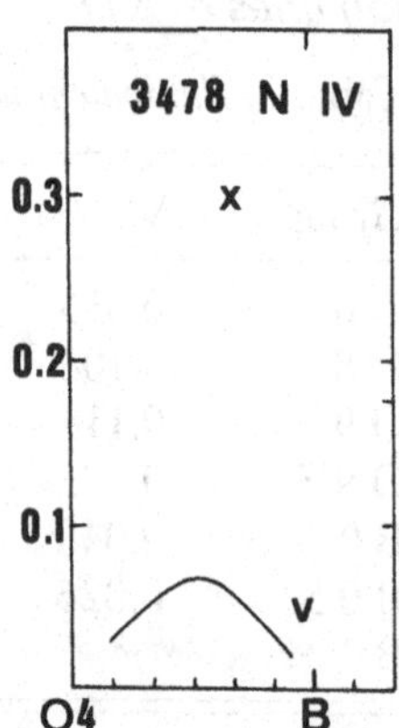

The line structure is interpreted as signifying mass loss (Howarth and Prinja 1989). At O 4V $W(1240) \simeq 15$, at B 1 Ib, W=2 Å.

In *Be stars* the 1240 feature is visible up to B 1–2V (Marlborough 1982).

Emission lines of N V

For O-type stars and B-type supergiants, see above.

N V is present in at least one *pre-degenerate star* (Werner *et al.* 1991). The 1238–1242 lines, when seen in late type stars, indicate the existence of a transition region (see Part Two, section 3.1) characteristic of 'coronal stars'. For instance the feature is seen in *T Tau stars*, together with emission lines of N IV (Appenzeller *et al.* 1980).

The nitrogen sequence in WR stars

Wolf–Rayet stars with strong nitrogen features are called WN stars. Nitrogen appears in the form of N III, N IV and N V, with the latter being more prominent in early subtypes (WN 3–WN 6). The most important feature is a strong and broad emission in the region 4600–4640, formed by a blend of N V (4604 and 4620) and N III (4634, 4640 and 4642) lines. In earlier stars N V predominates, with a subsequent blue shift of the average wavelength of the feature, whereas in later stars one finds mostly N III. The combined feature has an equivalent width between 10 and 100 Å and half-widths of up to 65 Å. Another important emission feature is N IV at 3478–83, with equivalent widths similar to those of the 4600–40 feature. N IV is present in practically all WN stars, with a maximum around WN 6. The N IV emission line 7115 has equivalent widths up to 220 Å (Conti *et al.* 1990).

Other emission features present in WN stars include He II 4686 ($10<W<500$) and He I 5876 ($4<W<60$ Å). For more details see He I and He II.

Another strong emission that appears in most WN stars is that of C IV at 5801–12, whose W may reach 1000 Å.

In the infrared spectral range the same species appear in emission, with the addition of H I. One thus finds He II 10124, 8237, 6891 and 6683, the first of the four lines being the most intense (*W* values up to 400 Å) whereas the others are weaker and may be absent. In addition one finds in emission He I (7065 and 6678), with 6678 more intense than 7065. These He I lines are, however, not present in all WN stars. The most interesting emission in that of H I, P 7 (10049), with *W* up to 50 Å. Usually H lines are not seen in the blue spectral region, or at best they are seen as weak absorption features, so that the P 7 emissions give the most direct information that H is present in WN stars. However, the line is not present in all WN stars (Conti *et al.* 1990).

In the ultraviolet one also finds the same atomic species in emission, namely N V, N IV, N III and N II, C IV and C III, He II and in addition O V (in the earliest subtypes) and O IV (in the later subtypes) plus Si IV. For a detailed account see Willis *et al.* (1986).

Behavior in non-normal stars

The lines 4640, 4634 and 4642 of M.2 of N III appear in emission in supergiants with $W<2$ Å. Usually, but not always, the N III emission is accompanied by He II 4686, in emission (see He) and by N IV 4058. When N III and He II both appear in emission, the star is called *Of.* When N III is in emission and He II in weak emission or in absorption, the star is called O(f) and, if N III is in emission whereas He II is in strong absorption, the star is called O((f)). For more details see Jaschek and Jaschek (1987a).

N is about normal in *sdO and sdB stars* (Heber 1987) but it is strong in He-rich *subdwarfs* (Husfeld *et al.* 1989). In *HB stars*, it is about normal (Heber 1991).

N II and N III lines are highly enhanced in the spectra of the N variety of the so-called *CNO stars* (which some authors call OBN), where one finds *W* values larger by factors of 2–4 than in normal stars of the same spectral type (Schoenberner *et al.* 1988). In the ultraviolet spectrum of these objects one also sees the N V lines 1239(1) and 1243(5) prominently (Walborn and Panek 1985). In contrast these lines are very weak in the *CNO stars* with enhanced carbon (Walborn and Panek, 1985).

N I has been investigated in early peculiar stars by Roby and Lambert (1990). They find that N I is slightly weak in *Am* and *Bp* stars of the Si subgroup, and much weaker in Hg–Mn and Cr–Eu–Sr stars. In the latter the *W* values of the N I lines are weakened by factors of the order of four or more with respect to normal stars. N I lines are about normal in *lambda Boo stars* (Bascheck *et al.* 1984).

Strong N I lines are present in the spectrum of the A-type *He-strong star* upsilon Sgr (Greenstein 1943).

N seems to behave in a manner parallel to that of Fe in stars of the *Magellanic Clouds* (Barbuy *et al.* 1981, Spite and Spite 1990).

N abundances in late type stars can only be derived through the behavior of the CN molecule – but of course one needs also to analyze other molecules to cope with the C abundance, such as CO, NH and C_2. From such analysis it turns out that N is about normal in *O-rich giants* (M stars and *S stars*) and perhaps slightly underabundant in *C stars* (Lambert *et al.* 1986). There exist, however, some *S stars* that are strong in N (Lambert 1989).

N behaves in a manner parallel to that of Fe in *metal-weak stars* (Carbon *et al.* 1987, Wheeler *et al.* 1989). Curiously, there exist a few *metal-weak dwarfs* in which N is overabundant (with respect to iron). The percentage of stars with such anomalies is of the order of 4%. Spite and Spite (1986) have shown that, in these stars, Li behaves normally as in other halo dwarfs. Very recently also *metal-weak giants* have been found in which nitrogen is either abnormally strong or weak with respect to iron (Anthony-Twarog *et al.* 1992).

Nitrogen is strong in the *weak G-band stars* (Cottrell and Norris 1978) and in *R CrB stars*. However, in the latter a large scatter exists among the different stars (Cottrell and Lambert 1982).

In *globular cluster giants* it has been shown that, at equal Fe/H, there exist variations in CN strength from one cluster to another (Langer *et al.* 1992). Although most of this is attributed to variations in C, variations in the N strength may also exist.

[N II] lines (like that at 5755) are seen in emission in the principal spectrum phase of *novae*, together with emission lines of N III, N IV and N V in the ultraviolet spectral region. During the Orion spectrum phase, absorption lines of N III (4097 and 4103) and N V (4603 and 4629) are present. At mid-phase of the Orion spectrum there appears a strong 4640 emission, produced by a blend of N III and N II lines, which persists well into the nebular stage phase. Besides this emission feature one observes also emissions of N II (4995, 5680 and 5950), N III (4097, 4103 and 3484) and N IV (4058). This exceptional enhancement of the N lines is called 'nitrogen flaring'.

At the post-novae phase, [N II] 6584 is seen in emission (Warner 1989, Payne-Gaposchkin 1957). In general N is overabundant by two orders of magnitude in novae (Andreae 1993).

Isotopes

N has two stable isotopes, namely N^{14} and N^{15}, and six unstable ones. In the solar system the two stable isotopes occur with frequencies of 99.6% (N^{14}) and 0.4% (N^{15}).

Querci and Querci (1970) identified some lines of C^{12}/N^{15} in an *early C star*, which implies a ratio N^{14}/N^{15} of several thousands, whereas in the solar system the ratio is about 270.

Origin

N^{14} and N^{15} are both produced by hydrogen burning and N^{15} also by explosive hydrogen burning.

OSMIUM Os Z=76

This element was discovered by S. Tennant in London in 1803. The name comes from the Greek *osme* (smell).

Ionization energies

Os I 8.7 eV, Os II 16.6 eV, Os III 24.9 eV.

Absorption lines of Os I

The equivalent width of Os I 3232(3) in the sun is 0.018. Os I is present in M 2 III stars (Davis 1947).

Behavior in non-normal stars

Os I and Os II lines were identified by Guthrie (1969) in one *Ap star* of the Cr–Eu–Sr subgroup *W*(4608, Os II)=0.038 and *W*(4421, Os I)=0.040. Brandi and Jaschek (1970) and Cowley (1987) later identified this element in other Ap stars.

Isotopes

Os has seven stable isotopes, Os 184, 186, 187, 188, 189, 190 and 192. These occur in the solar system with frequencies 0.02%, 2%, 2%, 13%, 16%, 26% and 41% respectively. There also exist 12 unstable isotopes and isomers.

Origin

Os 189, 190 and 192 are pure r process products. Os^{188} can be produced by both the r process and the s process. Os^{186} and Os^{187} are pure s process products and Os^{184} is produced only by the p process.

OXYGEN O Z=8

This element was discovered independently by J. Priestley in Leeds, England in 1774 and by C. Scheele in Uppsala, Sweden in 1771. The name comes from the Greek *oxy genes* (acid forming).

Ionization energies

OI 13.6 eV, OII 35.1 eV, OIII 54.9 eV, OIV 77.4 eV, OV 113.9 eV, OVI 138.1 eV, OVII 739 eV, OVIII 871 eV.

Absorption lines of OI

Table 1. *Equivalent widths of OI 7774(1)*

Group	V	III	Ib	Ia
B 0	(0.10)			
B 2	0.35			
B 5	0.56			(1.6)
A 0	0.75	0.75	1.40	2.00
A 5	0.80	0.90	1.60	2.20
F 0	0.71	0.95	1.48	2.25
F 5	0.46	(0.80)	1.20	
G 0	(0.60)			1.98†
S	0.19			
G 8		0.14†		
K 0		0.040*		
K 2		0.024*		

Source: Data are from Jaschek *et al.* (1992). The data are averages and refer to the whole blend 7772–7774–7775. The * data are from Arellano Ferro *et al.* (1991). The † data refer to the 7772 line alone.

OI (7772) appears in B-type stars, has a maximum around A 5 and declines toward later stars. It has a pronounced positive luminosity effect.

The luminosity effect was first noticed by Keenan and Hynek (1950) and has since then been used frequently for calibration purposes. For instance Arellano Ferro *et al.* (1991) have measured equivalent widths of OI 7774 in F and G stars to obtain a quadratic relationship between *W*(7774) and absolute magnitude. Faraggiana *et al.* (1988) have measured equivalent widths for early type stars.

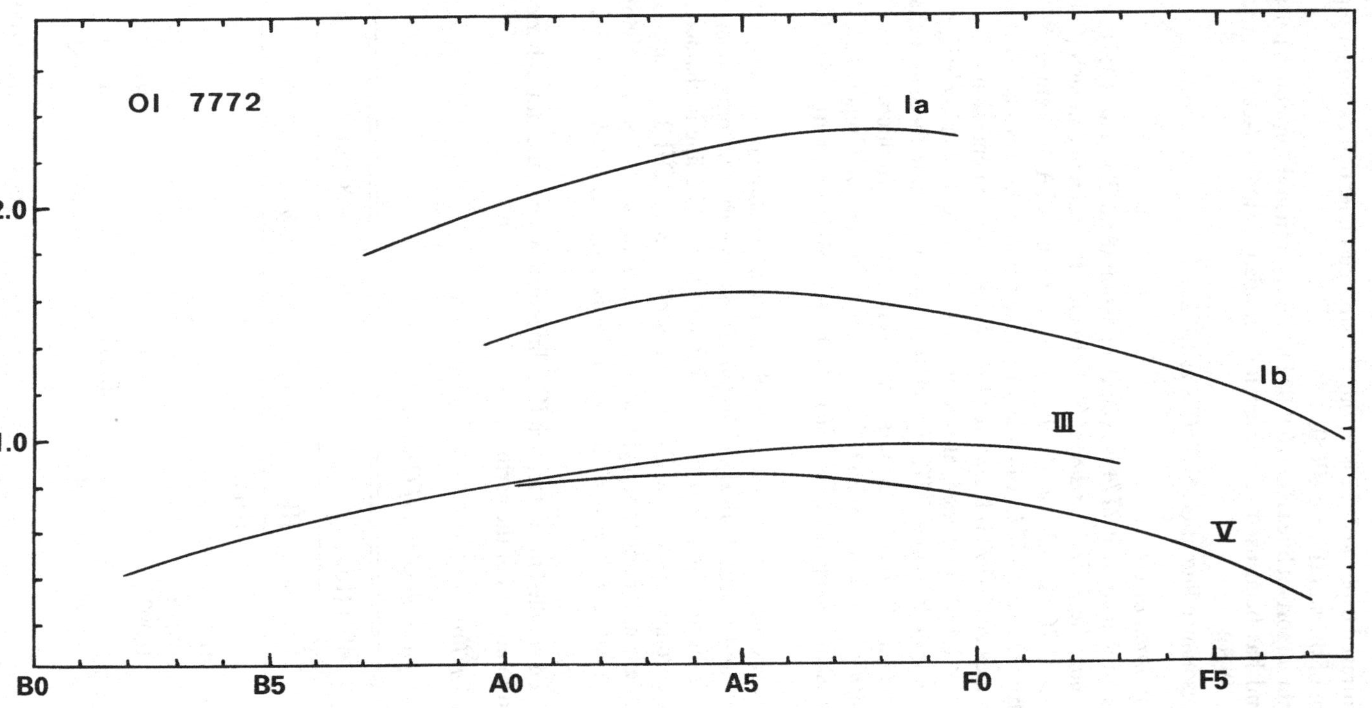
OI 7772
Ia
Ib
III
V
2.0
1.0
B0
B5
A0
A5
F0
F5

The infrared triplet at 7772 is among the strongest lines of the chromospheric spectrum of the sun (Pierce 1968).

Another important line is (O I) 6300. The line is seen in absorption in late type giants and has been used very often for abundance studies (see for instance Barbuy (1988)).

The resonance line of O I (UV 2) is at 1302.

Emission lines of O I

In *Oe stars* the feature at 7774 is seen in emission (Andrillat *et al.* 1982).

In *Be stars*, the 7772(1) and 8446(4) lines are often seen in emission (Polidan and Peters 1976, Andrillat *et al.* 1990) with $W(8446) < 3$ Å. Sometimes 8446 is much stronger than 7772, which fact is attributable to fluorescence of Lyman beta. Jaschek *et al.* (1993) found that O I 7772 is always in emission in Be stars ($W(7772) < 1.4$ Å). The line is also seen in very strong emission in *Herbig Be–Ae* and in *B[e]* stars. The lines 7772 and 8446 do not always show the same behavior (Praderie *et al.* 1987). In B[e] stars one observes [O I] in addition.

In *compact infrared sources* 8446 is usually in emission (McGregor *et al.* 1988) and the same happens in *symbiotic stars* (Kenyon and Fernandez-Castro 1987).

The ultraviolet 1152(6) line is visible in the solar chromospheric spectrum (Feldman and Doschek 1991). This line is also observed in other cool stars, where it indicates the existence of a chromosphere. The line is characteristic of 'non-coronal' stars. See Part Two, section 3.1. [O I] lines appear in emission around minimum light in *long-period variables* (Querci 1986), in *planetary nebulae, T Tau stars, novae, supernovae, gaseous nebulae* and *symbiotic stars*.

Attention is called to the fact that [O I] lines also occur in the atmospheric airglow and in auroras of the earth.

Absorption lines of O II

Table 2. *Equivalent widths of O II*

	4076(10)			4189(36)	
Group	V	III	Ia	V	Ia
O 8		0.064			
O 9	0.077				
O 9.5	0.096				
B 0	0.092		0.116		0.076
B 1			0.270		0.110
B 2	0.060	0.170(II)			0.063
B 3	0.028		0.199		
B 5			0.114(Ib)	0.005	0.041(Ib)

Table 3. *Equivalent widths of O II*

	4415(5)		4649(1)	
Group	V	Ia	V	Ia
O 7			0.085	
O 9	0.035			
O 9.5	0.075		0.145	
B 0	0.096	0.112	0.130	0.600
B 0.5	0.140(IV)			
B 1		0.415		
B 2		0.130–0.275		
B 3	0.024(IV)	0.098	0.030	0.086
B 5		0.027		0.111

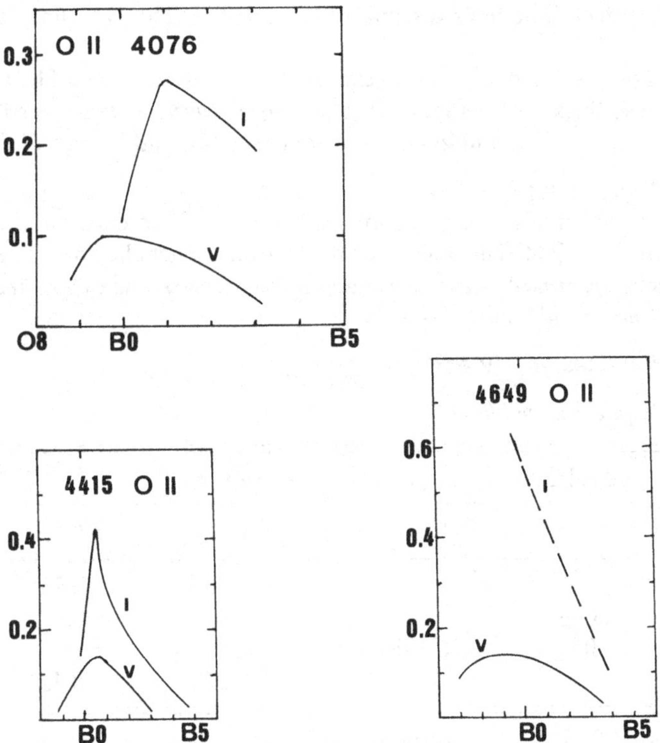

O II (for instance 4076) appears in late O-type stars, has a maximum at B 1 and disappears around B 3. A strong positive luminosity effect exists (see also 4649 and 4415).

Forbidden lines of O II

Forbidden lines of O II appear in emission in *B[e] stars*. For discussion of a typical object, see Swings (1973).

T Tau stars often exhibit 3726 and 3729 in emission (Sun *et al.* 1985).

Absorption lines of O III

Table 4. *Equivalent widths of O III 5592(5)*

Group	V	I
O 7	0.235	
O 9	0.14	
B 0	0.0542	0.16

The only supergiant is plotted as a cross in the figure.

O III lines (5592) are strongest for types O 6–O 7 and show a positive luminosity effect. The lines disappear at about B 0. The O III line 3759 has W=0.30 for O 7.

O III lines (M.3 and 15) can be excited by fluorescence from He II Lyman alpha. These lines are prominent in emission in *symbiotic stars* (Merrill 1950, Wallerstein *et al.* 1991) and in one recurrent *nova* (Joy and Swings 1945).

Forbidden lines of O III

T Tau stars sometimes exhibit the lines 5007 and 4959 in emission (Sun *et al.* 1985). These two [O III] lines are usually the strongest nebular lines. They occur in objects like *planetary nebulae, novae* and *gaseous nebulae*, and in post-maximum spectra of *novae* and *symbiotic stars*.

Absorption lines of O IV and O V

Table 5. *Equivalent widths of O IV*

	3411(2)		1339	
Group	V	I	V	I
O 5			0.55	0.75
O 7	0.045		0.45	0.62
O 8	0.035	0.195		
O 9			0.25	0.45

Source: Data are from Prinja (1990) and are average values derived from MK standards.

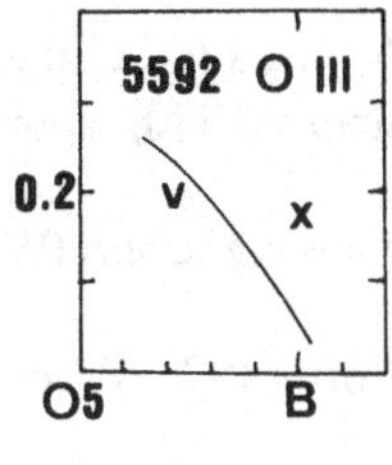

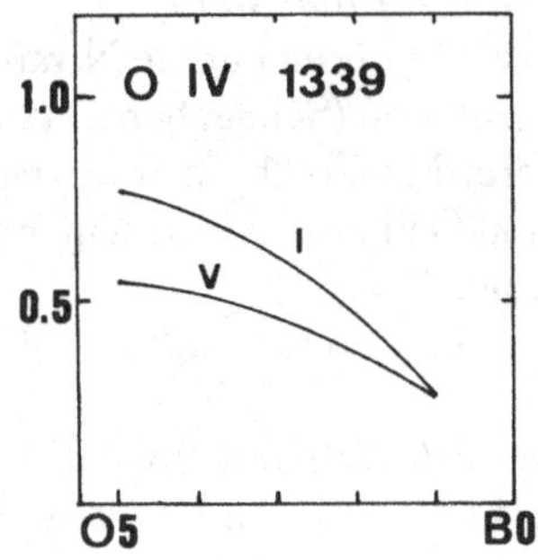

O IV is present in O-type stars (Dean and Bruhweiler 1985). The ultraviolet line 1371(7) of O V decreases from O 3 to O 7, where it disappears (Heck *et al.* 1984).

Emission lines of O V

The presence of O V 1371(7) indicates the existence of a transition region, characteristic of *coronal stars* (see Part Two, section 3.1).

One O V emission line at 1.554 μm is observed weakly in *WC stars* (Eenens *et al.* 1991).

Absorption lines of O VI

The blend at 1035 (1032–1037) (UV M.1) is present in early O-type stars, whereas in *Be stars* it remains visible up to O 9.5 (Marlborough 1982).

Emission lines of O VI

O VI lines in emission (centered on broad absorption lines) are characteristic of *pre-degenerates* (Motch *et al.* 1993). The resonance lines 1032 and 1037 are seen in very strong emission in the solar ultraviolet spectrum (Feldman and Doschek 1991). Their presence in stars (for instance in *symbiotic stars*) can be inferred from the presence of strong fluorescence ultraviolet Fe II emission lines (Feibelman *et al.* 1991).

Emission lines of O VIII

Possibly the 6068 line is present in emission in *pre-degenerates*, according to Motch *et al.* (1993).

The Wolf–Rayet oxygen sequence

Barlow and Hummer (1982) have introduced a small subgroup of the *WR sequence* characterized by a very strong emission of O VI 3811 and O IV at 3400 (a blend of 3381 and 3426–M.2 and 3). Eenens and Williams (1991) have recently studied one of these objects and also found O V in emission. Nugis and Niedzielski (1990) suggest that the lines of O VI may be present in many WN stars, but that these lines are masked by blends of other lines.

Behavior in non-normal stars

O II lines are about normal in the N variety of the *CNO stars* (also called OBN stars by some authors) (Schoenberner *et al.* 1988). Walborn (1980) reported that O III is reinforced in CNO stars of N type.

Both O V and O VI lines are present in the spectra of *pre-degenerates* (Werner *et al.* 1991, Werner 1991).

O is weak in some *sdO and sdB stars* (Baschek and Norris 1970, Baschek *et al.* 1982).

Gerbaldi *et al.* (1989) and Roby and Lambert (1990) have investigated the behavior of O I lines in *Bp* and *Ap stars*. They find that O I is somewhat weak in stars of the Hg–Mn and Si subgroups, whereas in stars of the Cr–Eu–Sr subgroup it is very weak. O I lines are somewhat weak (Roby and Lambert 1990) in *Am stars*. In *lambda Boo stars*, O is about normal (Baschek *et al.* 1984).

O seems to be slightly weak in *galactic supergiants* by a factor of two, but it is not certain that this underabundance is real (Luck and Lambert 1992).

O seems to be underabundant in *blue stragglers* of an old open cluster (Mathys 1991).

O is overabundant with respect to iron in metal-weak (late) giants by factors of the order of two, both in *halo stars* and in *globular cluster giants* (Brown *et al.* 1991, Gonzalez and Wallerstein 1992) (see also Part Two, section 2.1). It has to be remarked, however, that these determinations are delicate because they are based either upon the analysis of forbidden O lines or upon molecules (TiO, CO). In the latter case one has to assume something about the behavior of the other elements that enter into the molecule. This has produced discrepancies between the results of different authors, which have not been completely cleared up (Barbuy 1992). For the use of CO see for instance Tsuji (1991) and for that of forbidden oxygen lines, Spite and Spite (1991).

Oxygen-rich giants are usually surrounded by circumstellar shells, which are rich in oxygen, characterized by silicate emission features peaking at 10 and 29 μm.

O is strengthened in *R CrB stars* (Cottrell and Lambert 1982) and seems to have a normal abundance in *C stars* (Lambert *et al.* 1986), in *Ba stars* (Lambert 1985) and in *S* and *MS stars*.

O is probably deficient within *Magellanic Cloud supergiants* (Lennon *et al.* 1991) but some authors find that it behaves in a manner parallel to that of iron (Barbuy *et al.* 1981, Spite and Spite 1990). Luck and Lambert (1992) find that this is true only for the Small Magellanic Cloud, whereas in the Large Magellanic Cloud the O/Fe ratio may differ from star to star.

O I absorption lines are strong in the principal spectrum phase of *novae*, whereas O II lines are strong in the Orion phase. Lines of [O I] (6300 and 6363) and [O III] appear in emission in the same phase, accompanied in the ultraviolet region by emission lines of O III, O IV and O V. Sometimes an 'O I flash' appears before the Orion phase, during which the O I lines rival in intensity the Balmer lines. Toward the end of the Orion phase there appears the [O III] flash. Both

[O I] and [O III] lines are present in emission in the nebular stage (Payne-Gaposchkin 1957, Warner 1989). O is overabundant by one order of magnitude in the spectra of the C–O nova subgroup (Andreae 1993).

O I appears in absorption in the spectra of *supernovae* of type Ib and in the nebular stages of types Ib and II (Branch 1990). The intensification of O is sometimes so great that a special type of supernova, called Ic, has been created (Hill 1993).

Oxygen isotopes

Oxygen has three stable isotopes: O 16, 17 and 18. In the solar system their respective frequencies of occurrence are 99.7%, 0.04% and 0.2%. There exist also five unstable isotopes.

Harris *et al.* (1988) studied the isotope ratio in five K-type giants, using the CO molecules formed with the different oxygen isotopes in the 5 μm region. The result is that the O^{16}/O^{17} ratio is about 400 and the O^{16}/O^{18} ratio about 500. The solar system values are about 2600 and 500 respectively.

Dominy (1984) has searched for isotope anomalies in early *C-type stars* and found no deviation from solar system values.

Harris *et al.* (1987) re-analyzed the early C stars, using the CO bands at 4289 cm^{-1}. They found O^{16}/O^{17} ratios between 550 and 4100 and O^{16}/O^{18} ratios between 700 and 2400.

J stars have been studied by the same authors, who found O^{16}/O^{17} ratios between 350 and 850.

Harris *et al.* (1985a) have studied these ratios in *Ba stars* and found O^{16}/O^{17} ratios btween 100 and 500 and O^{16}/O^{18} ratios between 60 and 550.

In yet another study, Harris *et al.* (1985b) investigated some *MS* and *S stars*. They found O^{16}/O^{17} ratios between 500 and 3000 and O^{16}/O^{18} ratios between 850 and 4600.

It seems clear that it is difficult to draw general conclusions from these values, which are scattered rather widely.

For completeness it should be added that isotopic ratios can also be determined radioastronomically from a variety of molecules – for instance CO and CS (Kahane *et al.* 1992).

Origin

O^{16} is produced by He burning, O^{17} by hydrogen burning or by explosive hydrogen burning and O^{18} by He or N burning.

PALLADIUM Pd Z=46

This element was discovered in 1803 by W. Wollaston in London. The name alludes to the asteroid Pallas discovered in 1802 by Olbers. Pallas Athene is the name of the goddess of wisdom in Greek mythology.

Ionization energy

Pd I 8.3 eV, Pd II 19.4 eV, Pd III 32.9 eV.

Absorption lines of Pd I

The equivalent width of Pd I 3405(2) in the sun is 0.036.

Behavior in non-normal stars

Pd was first discovered by Bidelman (1966) in one *Ap star*. Pd II was found by Adelman (1974) in one *Bp star* of the Si-4200 subgroup and by Adelman *et al.* (1979) in another Ap star of the Cr–Eu subgroup.

Isotopes

Pd occurs in the form of six stable isotopes, namely Pd 102 104, 105, 106, 108 and 110, which occur in the solar system with frequencies 1%, 11%, 22%, 27%, 27% and 12% respectively. There exist also 15 short-lived isotopes and isomers. The longest lived is Pd^{107} with a half life of 7×10^6 years, which can be used for radioactive dating.

Origin

Pd 105, 106 and 108 are produced by both the r process and the s process. Pd^{110} is a pure r process product, Pd^{104} a pure s process product and Pd^{102} are pure p process product.

PHOSPHORUS P $Z=15$

This element was discovered in 1669 by H. Brandt in Hamburg, Germany. The name *phosphorus* in Greek means bringer of light, which was also the name given to the planet Venus.

Ionization energies

P I 10.5 eV, P II 19.7 eV, P III 30.2 eV.

Absorption lines of P I

The equivalent width of P I 9796(2) in the sun is 0.012. According to Underhill (1977) it is present in an A 2 Ia star.

Absorption lines of P II

The P II line 4127(16) is present in B 5 Ia stars and ultraviolet lines are also visible in B 7 and B 9V stars (Artru *et al.* 1989).

Absorption lines of P III

The equivalent width of P III 4280(1) in a B 8V star is W=0.005. P III has its ultraviolet resonance line at 1334 (UV M.1).

Behavior in non-normal stars

P II is very enhanced in some *early type He-strong stars*, like 3 Cen A, where it was discovered by Bidelman (1960). According to Smith (1981) the line 4127 has W=0.028. P II and P III are also highly enhanced in many but not all *Bp stars* of the Hg–Mn type (Adelman 1987). According to Kodaira and Takada (1978) one has W(4127)=0.050 and W(4080)=0.040. The phosphorus enhancement has led to the introduction of a subgroup of Bp stars, the so-called '*P-strong stars*' (Jaschek and Jaschek 1974). In some, but not all, *HB stars*, P is enhanced (Heber 1991).

Isotopes

P has one stable isotope, P^{31}, and six short-lived isotopes.

Origin

P can be produced by both Ne burning and explosive nucleosynthesis.

PLATINUM Pt Z=78

The element was known in pre-Columbian America. Ulloa discovered this element in 1735. The name is derived from the Spanish *plata* (silver).

Ionization energies
PtI 9.0 eV, PtII 18.6 eV, PtIII 29.0 eV.

Absorption lines of PtI
In the sun the equivalent width of PtI 3065(2) is 0.083.

Behavior in non-normal stars
PtI was detected by Brandi and Jaschek (1970) in several *Ap stars* of the Cr–Eu–Sr subgroup.

PtII lines were detected by Dworetsky (1969) in some *Bp stars* of the Hg–Mn subgroup. A typical value is $W(4046)=0.054$. PtII was also detected by Cowley (1987) and by Fuhrmann (1989) through the ultimate line 1777(13) in several Ap stars of the Si and Cr–Eu–Sr subgroup.

Isotopes
Pt has six stable isotopes: Pt 190, 192, 194, 195, 196 and 198, which occur in the solar system with frequencies of 0.01%, 0.8%, 33%, 34%, 25% and 7%. There exist also 26 unstable isotopes and isomers.

Origin
Pt 194, 195, 196 and 198 can only be produced by the r process, Pt^{192} only by the s process and Pt^{190} by the p process.

PLUTONIUM Pu $Z=94$

This element was discovered by G. T. Seaborg, A. C. Wahl and J. W. Kennedy in 1940 in Berkeley, USA. Its name alludes to the planet Pluto. The name of Pluto corresponds to the Roman god of the underworld.

Ionization energies
Pu I 6.1 eV.

Lines of this element were reported by Jaschek and Brandi (1972) in one *Ap star*. No later confirmation of this identification has been made.

Isotopes
The longest lived isotope is Pu^{244} with a half life of 7.5×10^7 years (Burbidge *et al.* 1957) and could be used for radioactive dating.

Origin
The element is produced by the r process.

POTASSIUM K Z=19

This element was discovered by H. Davy in London in 1807. The name comes from the English word *potash* (pot ashes); in Latin the element is called *kalium*.

Ionization energies

KI 4.3 eV, KII 31.6 eV, KIII 50.3 eV, KIV 67.0 eV, KV 90.9 eV.

Absorption lines of KI

Table 1. *Equivalent widths of KI 7699(1)*

Group	V	III	Ia
G 2	0.15		0.18
S	0.154		
K 2		0.21	
M 2			0.45

The KI resonance lines are 7664 and 7699(1), but 7664 falls near a strong atmospheric line of O_2 and is therefore difficult to observe. The 7699 line shows a complicated luminosity effect, in the sense that it is strong in dwarfs and supergiants and weaker in giants (Keenan 1957). In subdwarfs KI is even stronger than in dwarfs. The interpretation of the observations is complicated because the lines lie in the region of TiO bands. If TiO bands are missing – for instance in *C stars* – the K lines become apparently much stronger. The same happens if the TiO bands are weaker, as occurs in supergiants (because for equal spectral type they have a higher temperature than dwarfs). A further complication of the interpretation resides in the fact that, in high-luminosity objects, part of the KI resonance lines may be due to interstellar or circumstellar absorption. For the latter case see for instance Sanner (1976), who also provides line profiles of the resonance lines for several stars.

The K lines are also useful as a classification criterion for very late (M 8–9) type stars (Solf 1978, Kirkpatrick *et al.* 1991). Kirkpatrick *et al.* (1993) have used also the lines at 1.169, 1.177, 1.243 and 1.252 μm (M.5 and 6).

Emission lines of KI

KI lines (M.1) were seen in emission in one *S-type Mira variable* (Bretz 1966, Wallerstein 1992) and those from multiplet 3 by Deutsch and Merrill (1959).

Forbidden lines

Forbidden lines of KIV and KV were observed in one *recurrent nova* (Joy and Swings 1945).

Isotopes

Two stable isotopes exist, namely K^{39} and K^{41}, plus one long-lived isotope, K^{40}, with a half life of 1.3×10^9 years. These three isotopes occur in the solar system with frequencies 93%, 7% and 0.01% respectively. There also exist seven short-lived isotopes and isomers. K^{40} could be used for radioactive dating.

Origin

K 39, 40 and 41 are all produced by explosive nucleosynthesis, but K^{40} can also be produced by the s process and by Ne burning.

PRASEODYMIUM Pr Z=59

This element was discovered by A. von Welsbach in Vienna in 1885. The name comes from the Greek *prasios didymos* (the green twin).

Ionization energies
PrI 5.4 eV, PrII 10.6 eV, PrIII 21.6 eV, PrIV 39.0 eV.

Absorption lines of PrI
No PrI lines are visible in the sun.

Absorption lines of PrII

Table 1. *Equivalent widths of PrII*

	5219(35)		5259(35)		
Group	V	Ib	V	III	Ib
F 5		0.012			
G 0					0.058
G 1		0.031			
S	0.002		0.003		
G 2		0.035			0.058
G 5		0.031			0.069
G 8		0.035			
K 0				0.062	
K 2		0.062			0.105
K 3					0.123
K 5		0.059			0.105

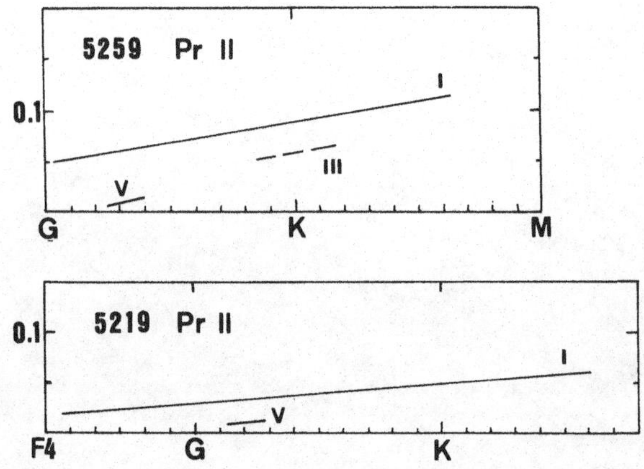

Pr II (for example 5259) is present in F-type supergiants and grows toward later types. A positive luminosity effect is present.

Behavior in non-normal stars

Pr II lines are strong in the spectra of *Ap stars* of the Cr–Eu–Sr subgroup (Adelman 1973b). A typical value of *W*(4222) is 0.035 (Sadakane 1976). In normal stars of the same temperature, the 4222(4) line is invisible (Smith 1972, 1973). The presence of Pr III was signaled by Bidelman (1966) in one Ap star. Later Aikman *et al.* (1979) detected Pr III in some of the Ap stars with strong Pr II lines. A detailed study by Mathys and Cowley (1992) in the red spectral region showed that Pr III is very strong in many Ap stars, without a clear relation to the peculiarity subgroups (*W*(6160)=0.1 Å).

Pr II is also strengthened in *Am stars*. For instance, $W(4222) \simeq 0.050$ for late Am stars.

In *Ba stars* Pr I and Pr II lines are enhanced (as are those of other rare earth elements), which leads to large overabundances (Lambert 1985). Typical values of *W* are twice as large as in normal giants of the same temperature (Danziger 1965). Because of the similar pattern of rare earths in Ba and in Ap stars, it is worth recalling the observation of Cowley and Dempsey (1984) that, whereas Pr is generally strong (or very strong) in Ba stars, there do exist some Ap stars in which Pr is weak.

Pr II lines are fairly strong in the spectrum of one *S-type star* (Bidelman 1953).

Pr II lines have been observed in *metal-weak G- and K-type dwarfs* by Gilroy *et al.* (1988). These authors find Pr to be overabundant with respect to iron (see also the discussion on rare earths).

Isotopes

Pr has one stable isotope, Pr^{141}, and 14 short-lived ones.

Origin

Pr can be produced by both the r process and the s process.

PROMETHIUM Pm Z=61

This element was discovered by J. A. Marinsky, L. E. Glendenin and C. D. Coryell at Oak Ridge, USA. Its name comes from the hero Prometheus, who stole fire from the Gods in the Greek legend.

Ionization energies

PmI 5.5 eV, PmII 10.9 eV.

Pm in stars

No Pm lines are observed in the sun. Pm was discovered in the *Ap star* HR 465 by Aller and Cowley (1970). The presence of this element was not confirmed by Wolff and Morrison (1972), but Poli *et al.* (1987) found it to be present in another Ap star.

Aslanov *et al.* (1975) showed that the element is present in at least one Ap star, but its intensity varies over the spectroscopic variability cycle of the star. The strongest line of PmII, that at 3877, has $W \leqslant 0.016$ Å.

Isotopes

Pm has 14 unstable isotopes and isomers. The longest lived, Pm^{145}, has a half life of about 18 years, implying essentially that, if it is found in cosmic sources, then it has been made *in situ*.

Origin

The element is produced by the r process.

RHENIUM Re $Z=75$

This element was discovered by W. Noddack, I. Tacke and O. Berg in 1925 in Berlin. Its name comes from the Rhine river, called *Rhenium* in Latin.

Ionization energies

Re I 7.9 eV, Re II 13.1 eV, Re III 26.0 eV.

Re is not seen in the solar spectrum.

Re I was discovered by Jaschek and Brandi (1972) in one *Ap star* of the Cr–Eu–Sr subgroup and in another one by Guthrie (1972).

Isotopes

Re has two stable isotopes, Re^{185} and Re^{187}, which occur with frequencies 37% and 63% respectively in the solar system. There exist also 18 short-lived isotopes and isomers. Re^{187} has a half life of 4.5×10^{10} years and can be used for radioactive dating.

Origin

Re^{185} can be produced by the r process and the s process and Re^{187} only by the r process.

RHODIUM Rh Z=45

This element was discovered in 1803 by W. Wollaston in London. The name comes from the Greek *rhodon* (rose).

Ionization energies
Rh I 7.5 eV, Rh II 18.1 eV, Rh III 31.0 eV.

Absorption lines of Rh I
The Rh I lines 3452(2) and 3188(4) in the sun both have an equivalent width of 0.013. Gopka and Komarov (1990) measured Rh in one K 5 III star.

Behavior in non-normal stars
Jaschek and Brandi (1972) detected Rh I in one *Ap star* of the Cr–Eu–Sr subgroup.

Isotopes
Rh has one stable isotope, Rh^{103}, and 19 short-lived isotopes and isomers.

Origin
Rh is produced by both the r process and the s process.

RUBIDIUM Rb $Z=37$

This element was discovered by R. Bunsen and G. Kirchhoff in Heidelberg, Germany in 1861. The name comes from the Latin *rubidium* (deep red).

Ionization energies
Rb I 4.3 eV, Rb II 27.3 eV.

Absorption lines of Rb I

Table 1. *Equivalent widths of Rb I 7800(1)*

Group	V	III
G 2	0.005	
S	0.005	
K 2		0.039

This element is prominent in sunspot spectra.

The resonance lines of Rb I are 7800 and 7948 from M.1. These lines become intense in M 9 and M 10 dwarfs (Solf 1978).

Emission lines of Rb I
Rb I 7800 was detected in emission in one *S-type star* (Wallerstein 1992).

Behavior in non-normal stars
Rb is overabundant with respect to iron in *Ba stars* (Lambert 1985).*

Isotopes
Rb has two stable isotopes, Rb^{85} and Rb^{87}, which occur in the solar system with frequencies 72% and 28% respectively. There exist also 18 short-lived isotopes and isomers.

Lambert and Luck (1976) analyzed the resonance lines of Rb I 7800(1) and 7947(1) in one *K-type giant* and found the isotope ratio to have the solar value. The same result was found by Malaney and Lambert (1988) from an analysis of the 7800 line in two *Ba stars*. From a preliminary analysis, Lambert (1991) suggested that the same is true for a sample of *M*-, *MS*- and *S*-type stars.

Rb^{87}, with a half life of 4.9×10^9 years, can be used for radioactive dating.

Origin
Rb^{87} is a pure s process product, whereas Rb^{85} is produced by both the r process and the s process.

* Note added in proof. According to Gratton and Sneden (1994) *AA* **287**, 927, Rb is proportional to Fe in metal-weak stars.

RUTHENIUM Ru Z=44

This element was discovered by A. Sniadecki in Vilno (then in Russia) in 1808 and was rediscovered by Osann in 1828 and by Klaus in 1844. Its name comes from *Ruthenia* (Russia).

Ionization energies
Ru I 7.4 eV, Ru II 16.8 eV, Ru III 28.5 eV.

Absorption lines of Ru I

Table 1. *Equivalent width of Ru I 4081(7)*

Group	V
S	0.005
K 5	0.044

Absorption lines of Ru II
In the sun, Ru II 3177(2) has W=0.014.

Both Ru I and Ru II are present in the spectra of late type stars. Orlov and Shavrina (1990) have reported observations in G- and K-type giants.

Behavior in non-normal stars
Ru I was detected by Jaschek and Brandi (1972) in at least one *Ap star* of the Cr–Eu–Sr subgroup. Leckrone *et al.* (1991b) detected Ru II in one star of the Hg–Mn subgroup.

Ru is perhaps enhanced in *Ba stars* (Warner 1965, Lambert 1985).

Ru is enhanced in *subgiant CH stars* (Krishnaswamy and Sneden 1985) and in *S-type stars* (Smith and Wallerstein 1983), leading to an overabundance by one order of magnitude.

Isotopes
Ru has seven stable isotopes, namely Ru 96, 98, 99, 100, 101, 102 and 104, which occur in the solar system with respective frequencies 5%, 2%, 13%, 13%, 17%, 31% and 19%. There exist also nine short-lived isotopes.

Origin
Ru^{96} and Ru^{98} are pure p process products, Ru^{100} is produced by the s process and Ru^{104} by the r process. Ru 99, 101 and 102 can be produced by both the r process and the s process.

SAMARIUM Sm *Z*=62

This element was discovered in 1879 in Paris by P. Lecoq de Boisbaudran. Its name alludes to the mineral Samarskite, named after a Russian mine official, Colonel Samarski.

Ionization energies
Sm I 5.6 eV, Sm II 11.0 eV, Sm III 23.4 eV, Sm IV 41.4 eV.

Absorption lines of Sm II

Table 1. *Equivalent widths of Sm II*

	4566(32)		4719(3)		6731(59)
Group	V	III	V	III	Ib
F 5	0.003		0.003		
G 0					0.061
G 2			0.007		
K 0		0.062		0.026	
K 1					0.116
K 2				0.046	
K 5	0.026				
M 2					0.100(Ia)

Sm II lines are visible from F 5 onwards and strengthen monotonically. Gopka *et al.* (1990) have observed Sm II lines in giants K 5 and M 0.

Behavior in non-normal stars
Sm II lines are observed in the spectra of many *Ap stars* of the Cr–Eu–Sr subgroup. Aikman *et al.* (1979) also observed Sm III lines in the spectra of some stars of this subgroup. Sm II and Sm III are also present in one *Bp star* of the Si subgroup (Cowley and Crosswhite 1978).

Sm II lines are enhanced in the spectra of *Am stars* (Smith 1973, 1974). The *W* values are larger by factors of 2–3 than in normal stars of the same temperature.

Sm lines are enhanced in some *Ba stars*, which leads to large overabundances (Lambert 1985). However, this element is not observed in all Ba stars. Typically *W*(4566)=0.081 for a K 0 III Ba star (Danziger 1965).

Sm lines are enhanced in many *C stars* cooler than C 3 (Utsumi 1966, 1984).

Numerous Sm II lines have been observed in at least one *S-type star* (Bidelman 1953). Sm is enhanced in *SC stars* (Kipper and Wallerstein 1990).

Sm II lines have been observed by Gilroy *et al.* (1988) in the spectra of *metal-*

weak G and K dwarfs. These authors find the element to be overabundant with respect to iron (see also Part Two, section 2.2).

Sm is enhanced in *Magellanic Cloud stars* (Luck and Lambert 1992).

Isotopes

Sm has seven stable isotopes and isomers, Sm 144, 147, 148, 149, 150, 152 and 154, which occur in the solar system with frequencies of 3%, 15%, 11%, 14%, 7%, 27% and 23% respectively. There exist also ten short-lived isotopes, among which is Sm^{146} with a half life of 7×10^{7} years. Sm^{146} can be used for radioactive dating.

Origin

Sm^{144} is a pure p process product, Sm^{148} and Sm^{150} are pure s process products and Sm^{154} is a pure r process product. Sm 147, 149 and 152 can be produced by both the r process and the s process.

SCANDIUM Sc $Z=21$

This element was discovered by L. Nilson in Uppsala, Sweden in 1879. Its name comes from *Scandia* (Scandinavia).

Ionization energies
Sc I 6.5 eV, Sc II 12.8 eV, Sc III 24.8 eV, Sc IV 73.5 eV, Sc V 91.7 eV, Sc VI 111.1 eV, Sc VII 138 eV.

Absorption lines of Sc I

Table 1. *Equivalent widths of Sc I*

	5671 (a12-resonance line)			5343(4)	
Group	V	III	Ib	III	Ia
F 5	0.010		0.014		
F 8			0.021		
G 0			0.027		
G 2	0.015		0.028		
S	0.014				
G 5			0.063		
G 8		0.059	0.126		
K 0		0.079			
K 2		0.112	0.190		
K 3		0.181	0.219		
K 5	0.153		0.245		
M 0		0.186		0.039	
M 2					0.044
M 2.5		0.208			
M 3				0.030	
M 4				0.042	

Sc I (see for instance 5671) appears in mid-F-type stars and grows monotonically toward later types. From late G-type onwards, a positive luminosity effect is visible for supergiants.

Emission lines of Sc I
The line at 3907(8) is seen in emission in *long-period variables*. It is probably excited by flurorescence (Joy 1954).

Absorption lines of ScII

Table 2. *Equivalent widths of ScII*

	4246(7)			5031(23)	
Group	V	III	Ib	V	Ib
B 9	0.042				
A 0	0.062		0.05		
A 1	0.06				
A 2	0.09		0.114(Ia)		
A 7	0.20				
F 0	0.24		0.575(Ia)		
F 2			0.500		
F 4	0.20			0.050	
F 5	0.20,0.21		0.39		
F 6				0.052	
F 8	0.17		0.48	0.056	0.422
G 0	0.15,0.19				
G 1	0.23			0.100	
G 2	0.16			0.060	
S	0.171				
G 5	0.20				
G 9	0.16(IV)			0.012(IV)	
K 0	0.176	0.200			
K 5	0.21			0.240	

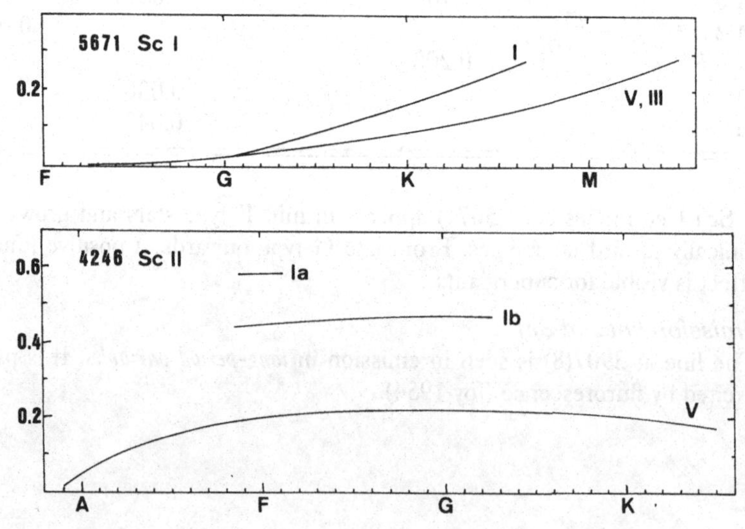

Table 3. *Equivalent widths of ScII 5657(29)*

Group	V	III	Ib
G 0			0.26
G 2	0.07		0.23
S	0.064		
G 5			0.19
G 8			0.24
K 0		0.099,0.123(III)	
K 2		0.13	0.20
K 3		0.12	0.22
K 5			0.19
M 0		0.11	
M 2.5		0.12	

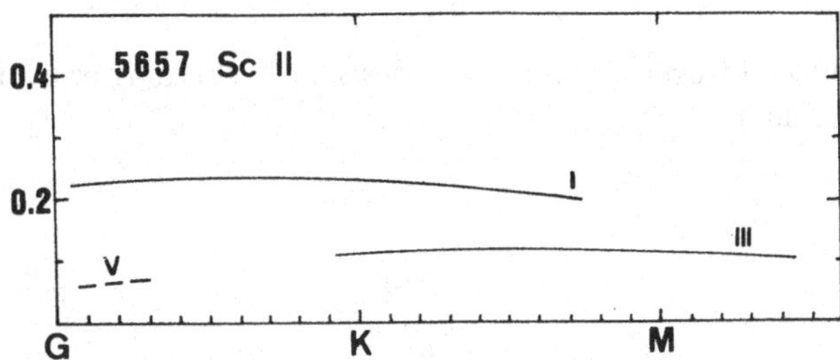

Sc II lines (see for instance 4246) appear in late B-type, grow toward F-type and remain practically constant thereafter. A strong positive luminosity effect is present. The line at 5657 exhibits the same behavior. Sc II also has two rather strong red lines at 6211(2) and 6306(2), which characterize late type stars (Keenan 1957).

Emission lines of Sc II

In *T Tau stars* (Joy 1945) the line at 4246 as well as other Sc II lines appear in emission.

Emission lines of higher ionization stages

The line 4823 of [Sc VII] was observed in at least one *recurrent nova* by Joy and Swings (1945).

Behavior in non-normal stars

Sc II lines tend to be strong in *Bp stars* of the Hg–Mn subgroup. *W*(4246) averages about 0.040 (Kodaira and Takada 1978). In contrast, Sc II lines tend to be weak in *Ap* stars of the Cr–Eu–Sr subgroup (Adelman 1973b). They are also weak in *Am* stars, which fact was discovered by Bidelman and further explored by

Conti (1965). The average *W* values are smaller by about a factor of two. It must be added, however, that there exist Am stars in which Sc is of normal strength and even some stars in which Sc is strong (Cowley 1991). In *delta Del* stars the Sc II lines are weak (Kurtz 1976). However, Sc seems to behave normally in stars intermediate between Am stars and A-type giants (Berthet 1990).

Sc II lines are weakened by factors up to ten in F-type *HB stars* with regard to normal stars of the same colors (Adelman and Hill 1987). Adelman and Philip (1992a) remark that this applies to HB stars in general. Sc lines are probably variable in the *blue stragglers* of an old open cluster (Mathys 1991).

Sc seems to behave in a manner parallel to that of Fe in *metal-weak dwarfs* (Magain 1989) and in *globular cluster stars* (Wheeler *et al.* 1989). In *extreme halo dwarfs* Zhao and Magain (1990) found Sc to be overabundant with respect to Fe.

Strong Sc II absorption lines appear in the spectra of *supernovae* of class II (Branch 1990).

Isotopes

Sc has one stable isotope, Sc^{45}, and 14 unstable isotopes and isomers.

Origin

Sc can be produced by explosive nucleosynthesis, by Ne burning or by nuclear statistical equilibrium.

SELENIUM Se $Z=34$

This element was discovered by J. Berzelius in 1817 in Stockholm. Its name alludes to the moon, *Selene* in Greek.

Ionization energies

Se I 9.75 eV, Se II 21.2 eV, Se III 30.8 eV.

No Se I or Se II lines have been observed in the sun.

Se in stars

Bidelman (1966) announced the probable detection of Se II in one *Ap star* of the Cr–Eu–Sr subgroup.

Isotopes

Se occurs in the form of six stable isotopes, Se 74, 76, 77, 78, 80 and 82, which occur in the solar system with frequencies 1%, 9%, 8%, 23%, 50% and 9% respectively. There also exist 14 short-lived isotopes and isomers, among them Se^{79} with a half life of 7×10^4 years.

Origin

Se^{74} can only be produced by the p process, Se^{76} by the s process and the p process. Se 77, 78 and 80 are produced by the r process and the s process. Se^{82} is a pure r process product.

SILICON Si Z=14

This element was discovered by J. Berzelius in 1824 in Stockholm, Sweden. The name comes from the Latin word for flint (*silici*).

Ionization energies

Si I 8.1 eV, Si II 16.3 eV, Si III 33.5 eV, Si IV 45.1 eV, Si V 166.8 eV, Si VI 205 eV, Si VII 246 eV, Si VIII 303 eV, Si IX 351 eV.

Absorption lines of Si I

Table 1. *Equivalent widths of Si I 5948(16)*

Group	V	III	Ib
F 0	0.09		
F 4	0.10		
F 5	0.115		0.104
F 6	0.10		
F 8	0.10		0.142(Ia)
G 0	0.085		0.15
G 1	0.12		
G 2	0.11		0.17
S	0.088		
G 5	0.14		0.15,0.16
G 8	0.09	0.12	0.13,0.15
K 0	0.13	0.12	
K 2	0.09	0.09	0.15
K 3		0.13	0.18
K 5	0.082,0.06		0.10
M 0		0.08	

Si I (for instance 5948) appears in A-type stars and is present up to M-type stars, with a flat maximum near G 5. A positive luminosity effect exists for supergiants, from F-type onwards.

Si I lines have been identified in the ultraviolet spectrum of one A 0 V star (Rogerson 1989). The Si I resonance line (UV M.1) is at 2516 Å.

Emission lines of Si I

In *T Tau stars* (Joy 1945) the line at 3905(3) appears in emission.

In *long-period variables*, two Si I lines, 3905 and 4103(2), appear in emission around maximum light (Merrill 1952, 1960). The line 3905 is also visible in at least one *nova* (Joy and Swings 1945).

Absorption lines of SiII

Table 2. *Equivalent widths of SiII*

	3856(1)			4128(3)		
Group	V	III	Ia	V	III	I
B 1						0.052(Ib)
B 2				0.1		0.100(Ia)
B 3	0.091		0.175	0.057		0.182(Ia)
B 5	0.11		0.179	0.096		0.158(Ia)
B 6	0.144	0.16			0.156	
B 7	0.17			0.146		
B 8			0.32	0.188(IV)		0.182(Ia)
B 9	0.15			0.110		
B 9.5				0.122		
A 0	0.15			0.15		0.21(Ib)
A 1				0.115		
A 2			0.432	0.125		0.345(Ia)
A 3	0.3		0.34(Ib)			0.20(Ib)
A 7				0.15		
F 0			0.776			0.363(Ia)
F 5				0.14		0.29(Ib)

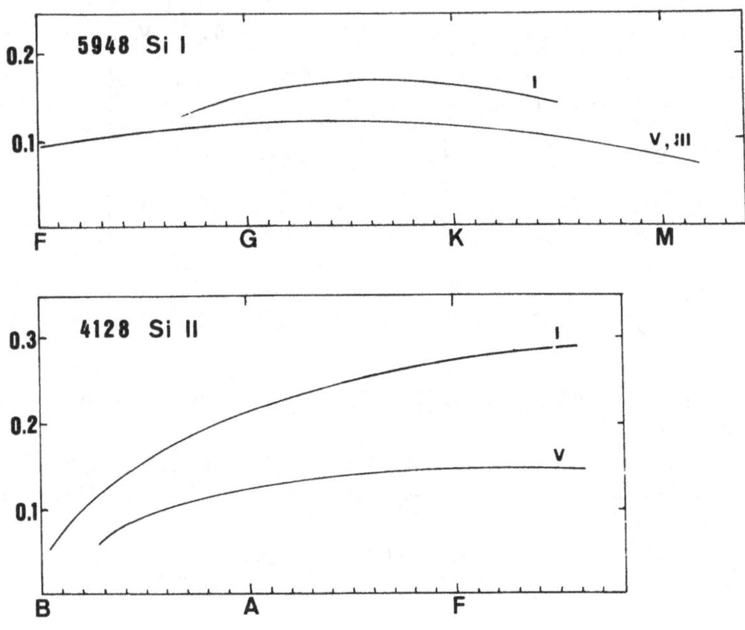

Table 3. *Equivalent widths of Si II 6347(2)*

Group	V	III	Ib
B 5			0.410(Ia)
B 6	0.195	0.344	
F 5	0.13		0.33
F 8			0.34
G 2	0.05		
S	0.054		
G 8		0.044	
K 0		0.035	
K 3		0.008	

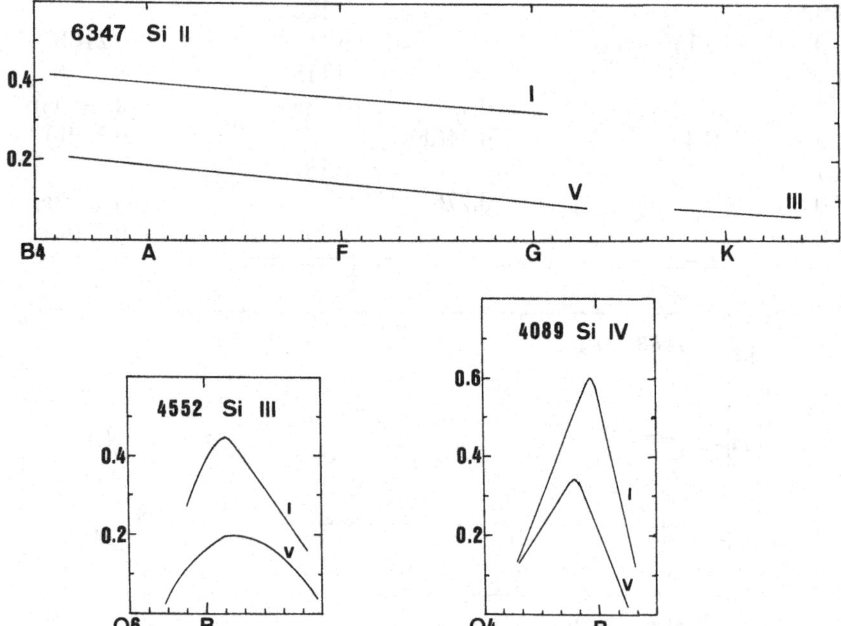

In stars of luminosity class O, *W*(3856)=0.755 and 0.404 for A 2 and A 3 respectively.

Si II (for instance the lines at 4128 and 3856) increases from early B-type to F-type and declines toward K-type stars (see the line at 6347). A positive luminosity effect is present.

The Si II resonance line is at 1808 (UV M.1).

Si II has a number of lines from high excitation levels, like 3955, 3992 and 4200, which are prominent in the spectra of the Si subgroup of *Bp stars* (see the discussion on behavior in non-normal stars).

Emission lines of Si II

Emission lines of Si II were observed by Beals (1951) in *P Cyg* and by Appenzeller *et al.* (1980) in *T Tau stars*. Emissions are also seen from one *nova* (Joy and Swings 1945).

The ultraviolet Si II line 1817(1), if seen in emission in late type stars, is indicative of a stellar chromosphere. The line is visible in late F-type stars, strengthens in G- and K- and disappears in early M-type stars. (See Part Two, section 3.1.)

Absorption lines of Si III

Table 4. *Equivalent widths of Si III 4552(2)*

Group	V	III	I
O 8	0.035		
O 9	0.120		
O 9.5	0.120		
B 0	0.132	0.280	0.345(Ia)
B 1	0.2		0.45(Ia)
B 3	0.15	0.250	0.123(Ib) 0.346(Ia)
B 5	0.025		0.180(Ib) 0.172(Ia)
B 6	0.074	0.155	

Si III (for instance the line at 4552) is present in late O-type and early B-type stars, with a maximum at about B 1. At B 5 the line is no longer visible. A positive luminosity effect is present.

The red line 5740(4) of Si III is visible around B 1(B 0.5 to B 1.5) and has a positive luminosity effect (Walborn 1980). The strong ultraviolet Si III lines around 1300 (of M.4) are present in types O 3 to O 9, where they disappear (Heck *et al.* 1984). Attention must be paid to the fact that after B 0 the 1300 feature is associated with Si II.

The resonance line of Si III is at 1206 (UV M.2).

Emission lines of Si III

Emission lines of Si III were observed by Beals (1951) in *P Cyg* and by Joy and Swings (1945) in one *nova.*

The presence of the Si III line at 1895(1) in the ultraviolet spectra of late type stars is indicative of a transition region. It is seen in G- and K-type stars and disappears in early M-type stars. (See Part Two, section 3.1.)

Absorption lines of SiIV

Table 5. *Equivalent widths of SiIV*

	4089(1)			1400		
Group	V	III	Ia	V	III	I
O 6	0.150		0.125(f)	1.5	4.5	14
O 8	0.300		0.400(f)	1.5	5	11
O 9	0.330	0.270				
O 9.5	0.340		0.63			
B 0	0.314		0.61,0.560	4	5.5	8.5
B 0.5	0.268					
B 1			0.20	8.5	5.5	7
B 2	0.0	0.17	0.15	4.5	5	6
B 4				1.5	2	4

Source: The data for 1400 were taken from Sekiguchi and Anderson (1987) and are mean values. The 1400 feature is a blend of 1394 (1) and 1403 (1).

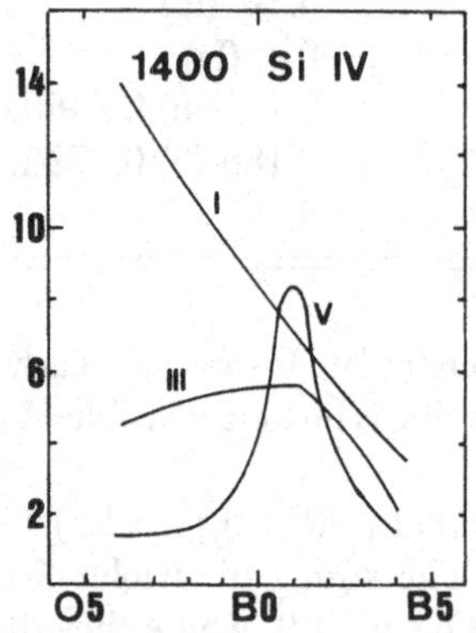

SiIV (for instance the line at 4089) is present in early O-type stars. It has a maximum at O 9 and disappears at B 2. A positive luminosity effect exists.

The ultraviolet SiIV 1400 feature can be used between O 5 and B 5 stars for classification purposes (Walborn and Panek 1984, Heck *et al.* 1984). In high-luminosity stars 1400 shows a P-Cygni profile, which is not seen in dwarfs. The P Cyg profile is usually interpreted as indicating mass outflow (Howarth and Prinja 1989). In *Be stars* the line is visible up B 8V stars (Marlborough 1982).

Emission lines of SiIV

SiIV emissions were seen in one *nova* by Joy and Swings (1945). If the ultraviolet SiIV feature at 1400 is seen in emission in late type spectra, then it indicates the presence of a stellar transition region. The line is visible in G and K dwarfs up to K 2. (See Part Two, section 3.1.)

Behavior in non-normal stars

Si is strong in hot *extreme helium stars* (Jeffery and Heber 1993).

Si is usually weak in *sdO* and *sdB* stars, with an indication that it is much weaker in the former than in the latter (Baschek and Norris 1970, Baschek *et al.* 1982, Lamontagne *et al.* 1985).

In *Bp* stars of the Si subgroup, the Si II lines (for instance that at 3856) are enhanced by factors up to three (Didelon 1986), but there exists a continuous transition between Si stars and normal stars. In certain stars of this subgroup several strong lines appear (4200, 3955 and 3992). The attribution of these lines to Si II was made independently by Bidelman (1962b) and Jaschek and Jaschek (1962). Such stars were called formerly 'λ 4200 stars' and are now called more correctly 'Si-λ 4200 stars'. These stars are the hottest objects of the Bp–Si subgroup.

In *Ap stars* of the Cr–Eu–Sr subgroup the element tends to be weak (Adelman 1973b). Typically $W(4128)=0.10$ (Sadakane 1976). Si is also weak in the Hg–Mn subgroup.

Si is about normal in *Am* stars (Burkhart and Coupry 1991).

Si lines are weakened in F-type *HB stars* with regard to normal stars of the same colors by factors up to ten in W (Adelman and Hill 1987).

Si seems to be overabundant by factors of the order of two with respect to iron in *metal-weak stars*, both in dwarfs and in giants (Gratton and Sneden 1987). Spite (1992) remarks that Si (and S) behave like magnesium in these stars. An overabundance of Si with respect to Fe by a factor of about two is also found in *globular cluster stars* (Wheeler *et al.* 1989, Francois 1991) (see Part Two, section 2.1).

Si II and Si III lines are seen in emission in the ultraviolet spectral region of *novae* during the 'principal spectrum' phase. Sometimes one sees also [Si VII] and [Si IX] lines during the nebular phase (Warner 1989). Si is overabundant by a factor of about 50 in the spectra of the novae of the O–Ne–Mg subgroup (Andreae 1993).

Si II emission lines are also observed in the spectra of *supernovae* of type Ia (Branch 1990).

A strong absorption blend of Si II at 6355(2) is characteristic of the *supernova* spectra of type Ia near light maximum. It is accompanied by other strong features of Si II.

Isotopes

Si has three stable isotopes, namely Si 28, 29 and 30, which in the solar system occur with frequencies 92%, 5% and 3% respectively. There also exist five unstable isotopes.

Si isotopes have been examined by Lambert *et al.* (1987) in four red giants, using the SiO molecular bands at 2495 cm^{-1}. These authors find for the

Si^{28}/Si^{29} ratio the solar value (which is about 20) for two stars. For an *MS star* and one M star they find ratios of 35 and 40. Si^{30} appears to be underabundant by a factor of two in all four stars. Instead of these bands, one can also use radio transitions of the SiO molecule.

The ratio Si^{29}/Si^{30} has also been derived from observations of the SiS molecular lines of the circumstellar envelope of the *C-type object* IRC + 10216 in the radio domain. The ratio found does not differ significantly from the solar system ratio.

Origin

All three Si isotopes can be produced by explosive nucleosynthesis. Si^{28} can also be produced by oxygen burning and Si^{29} and Si^{30} by Ne burning.

SILVER Ag $Z=47$

This element was known to ancient civilizations. The Latin name for silver is *argentum*.

Ionization energies
AgI 7.6 eV, AgII 21.5 eV, AgIII 34.8 eV.

Absorption lines of AgI
In the sun, the equivalent width of AgI 3281 (M.1-resonance line) is 0.044.

Behavior in non-normal stars
Ag was first detected in stars by Merrill (1947), who found it in one *S-type star*.

Jaschek and Brandi (1972) detected AgI in one *Ap star* of the Cr–Eu–Sr subgroup.

Isotopes
There exist two stable isotopes of this element, namely Ag^{107} and Ag^{109}, which occur in the solar system with frequencies 52% and 48% respectively. There exist also 25 short-lived isotopes and isomers.

Origin
Ag^{107} and Ag^{109} are produced by both the r process and the s process.

SODIUM Na $Z=11$

This element was discovered by H. Davy in 1807 in London, England. The name comes from the English *soda*, in Latin *natrium*.

Ionization energies

Na I 5.1 eV, Na II 47.3 eV, Na III 71.6 eV, Na IV 98.9 eV, Na V 138.4 eV, Na VI 172.1 eV.

Absorption lines of Na I

Table 1. *Equivalent widths of Na I*

	5889 (M.1 resonance line)			5688(6)		
Group	V	III	Ib	V	III	Ib
B 5			0.321(Ia)			
B 9	0.115					
A 0	0.20		0.26			
A 7	0.27					
F 0	0.35					
F 4	0.45					
F 5	0.45		0.73			0.23
F 6	0.51					
F 8	0.56		0.80			
G 0	0.80					0.24
G 1	0.78					
G 2	0.80					0.22
S	0.752					
G 5	0.80					0.26
G 8		1.26		0.14(IV)		
K 0	1.66	1.32		0.25	0.19	
K 2	1.74		0.985			0.32
K 3		2.75			0.25	0.35
K 5	4.00					0.35
M 0		2.94			0.26	
M 2			1.33(Ia)			0.40
M 3		2.74				
M 4		2.96				
M 5		4.4				

The blend of the lines 5889 and 5895 is often called 'D line' (following a designation given by Fraunhofer); in particular, 5889=D1 and 5895=D2.

Table 2. *Equivalent widths of the doublet 8183 = 8194*

Type	V	III
F 6	0.32	
F 8	0.73	0.64
G 0	0.79	
G 2	0.62	
G 5	0.65	0.70
G 8	0.71	0.57
K 0	0.81	0.60
K 3	1.14	0.90
K 5	1.38	0.78
M 0	1.86	0.74
M 1	2.18	0.64
M 4	2.60	

Source: Data are from Zhou Xu (1991).

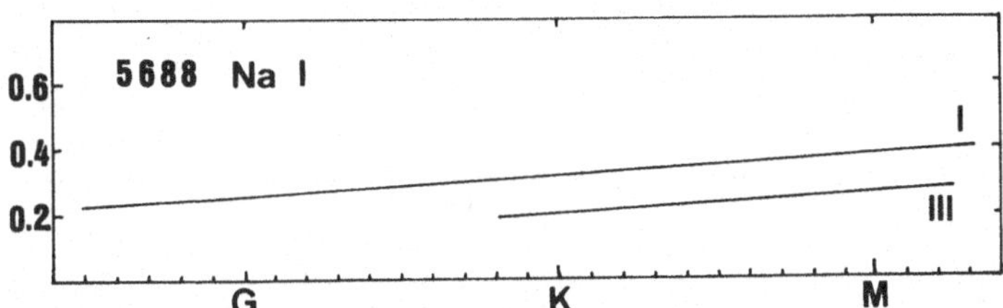

Na I (for instance the line at 5889) appears in B-type stars and increases monotonically toward later types. Before G-type, a positive luminosity effect is seen. After G-type, the luminosity effect reverses and is negative. (This was already known to Luyten (1923).) Zhou Xu (1991) found a similar behavior of the 8183–8194 doublet. This anomaly is not present in the line at 5688.

The intensity of the Na I 5889 and 5895 lines passes through a minimum at M 5, because of the superposed absorption of TiO. A similar phenomenon happens for C stars at C 5 to C 6, although in this case the veiling absorption is due to CN and C_2. In S-type stars there is also a band of ZrO superimposed, but since ZrO is less opaque than TiO, the Na I lines are stronger in S-type than in M-type stars (Keenan 1957).

Attention should be paid to the fact that Na I 5889 and 5895 lines (M.1) often have interstellar component(s).

Na I emission lines

One observes the 2.206–2.209 μm doublet in emission in some *high-luminosity stars* (McGregor *et al.* 1988) and in *F-type supergiants* (Lambert *et al.* 1981). The

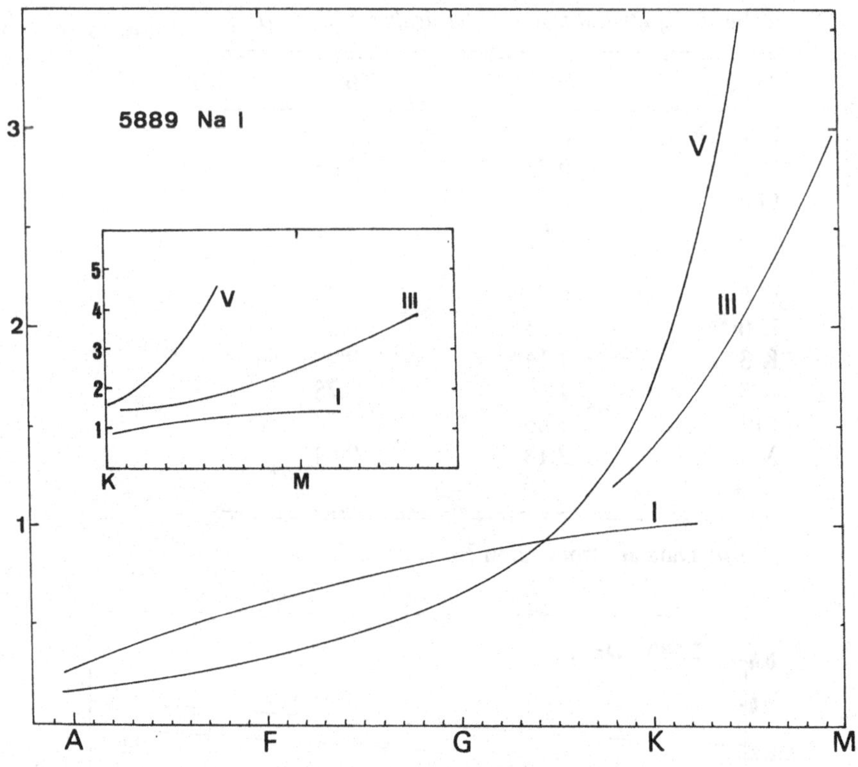

5889 and 5895 lines are seen in emission in at least one *B[e] star* (Merrill 1951a).

The Na I D lines are in emission in *dMe stars* (Pettersen 1989).

Na I lines are frequently in emission in *T Tau stars*, exhibiting complex profiles. Usually the emission is strong in stars that also exhibit Ca II emissions. These lines have been used extensively to study the physical state of the envelopes (Mundt 1984).

Behavior in non-normal stars

In *lambda Boo stars*, Na I lines are weaker than in normal stars of the same temperature. Typically *W*(5889)=0.080 for A 0V (Venn and Lambert 1990).

Na is normal in *Am stars* (Smith 1973, 1974).

Na seems to be overabundant by a factor of three in four giants of the Hyades *open cluster* (Arimoto and Cayrel de Strobel 1988) and similar overabundances were found in *metal-rich stars*. Since the lines used are strong ones, the results are to be taken with caution.

Na is probably underabundant with respect to Fe in the most *metal-weak stars*, Fe/H=−4 dex (Molaro and Bonifacio 1990). In moderately metal-weak stars Na may be normal with respect to Fe (Wheeler *et al.* 1989). In *globular cluster stars*

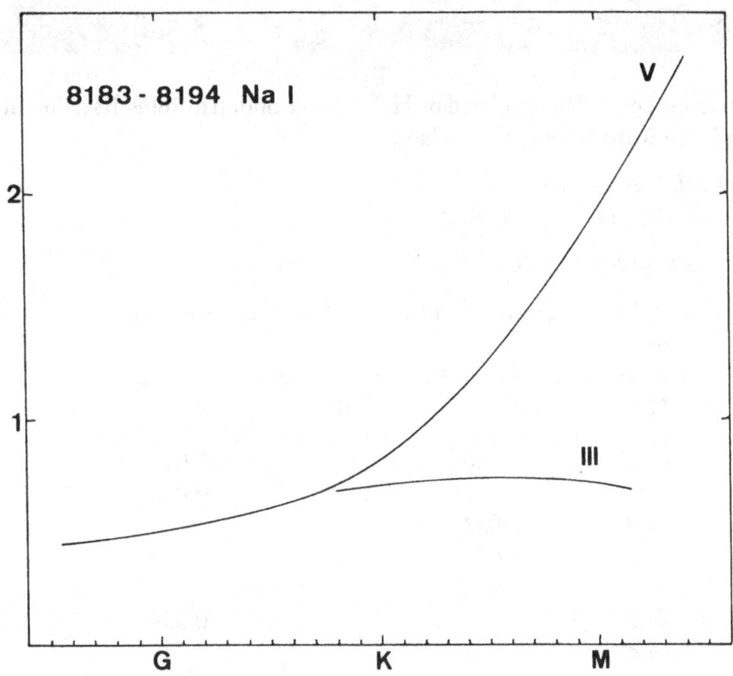

it probably behaves in a manner parallel to that of iron. However, Gonzalez and Wallerstein (1992) find Na to be enhanced in one *globular cluster supergiant*. Smith and Wirth (1991) find an erratic behavior of Na in globular cluster stars having different CN strengths, and Francois (1991) finds a general overabundance of Na. It is evident that it is best to consider the Na abundance as being an open problem.

Na I (5889) is very strong in the spectra of *C stars* (Fujita *et al.* 1963), which probably implies an overabundance of Na. In some very cool variables Keenan (1957) finds $W(5859+5895)$ up to 50 Å.

Na lines are very weak in the spectra of *CH stars* (Yamashita 1967). [Na V] and [Na VI] lines are sometimes seen in the spectra of *novae* during the nebular stage (Joy and Swings 1945, Warner 1989).

Strong emission lines of Na I are visible in the spectra of *supernovae* of types Ib and II (Branch 1990).

Isotopes

The only stable isotope is Na^{23}. There also exist six short-lived isotopes.

Origin

Na can be produced by carbon burning, neon burning or explosive nucleosynthesis.

STRONTIUM Sr Z=38

This element was first isolated by H. Davy in London in 1808. Its name alludes to the Latin word *Strontian* (Scotland).

Ionization energies
Sr I 5.7 eV, Sr II 11.0 eV, Sr III 43.6 eV.

Absorption lines of Sr I

Table 1. *Equivalent widths of Sr I 4607(2) (resonance line)*

Group	V	III	Ib
F 5	0.020		0.017
F 8			0.042
G 0	0.032		
G 2	0.033		0.040
S	0.036		
G 5			0.038
G 8			0.075
K 0		0.087	
K 2		0.094	0.122
K 5	0.161,0.170		

Table 2. *Equivalent widths of Sr I 6550(12)*

Group	V	III	Ib
S	0.005		
G 2			0.021,0.040
G 5			0.038,0.046
G 6			0.111
G 8			0.075
K 0		0.018	
K 2			0.076
K 3		0.040	0.107
K 5			0.083

Sr I (see the line at 4067, which is a resonance line) appears in F-type stars and increases toward later types. No luminosity effect is perceptible.

Absorption lines of SrII

Table 3. *Equivalent widths of SrII 4077 (one resonance line)*

Group	V	I
B 9	0.057	
A 0	0.06	
A 1	0.126	
A 2	0.24	0.141(Ia)
A 7	0.27	
F 0	0.33	0.32(II) 0.630(Ia)
F 4	0.38	
F 5	0.39,0.34	0.72,0.68(Ib)
F 6	0.41	
F 8	0.41	0.71(Ib)
G 0	0.35	
G 1	0.56	
G 2	0.33	
S	0.428	
G 5	0.4	
K 5	0.71,0.42	

Table 4. *Equivalent widths of SrII 10914(2)*

Group	V	III
G 2	0.125	
S	0.098	
K 2		0.200

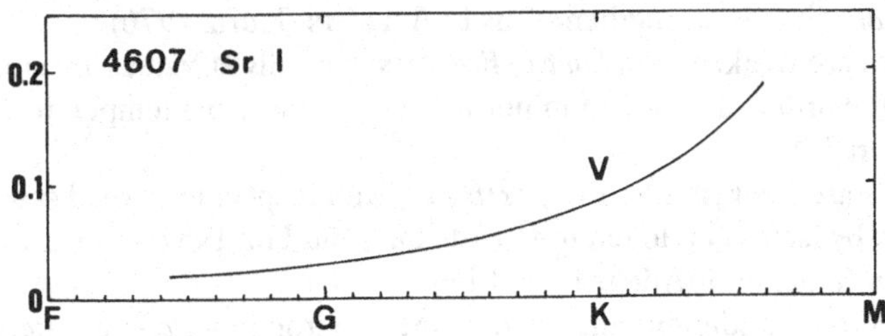

SrII (see for instance the line at 4077) appears in late B-type stars and increases toward later spectral types. It has a strong positive luminosity effect, which has frequently been used as a luminosity criterion.

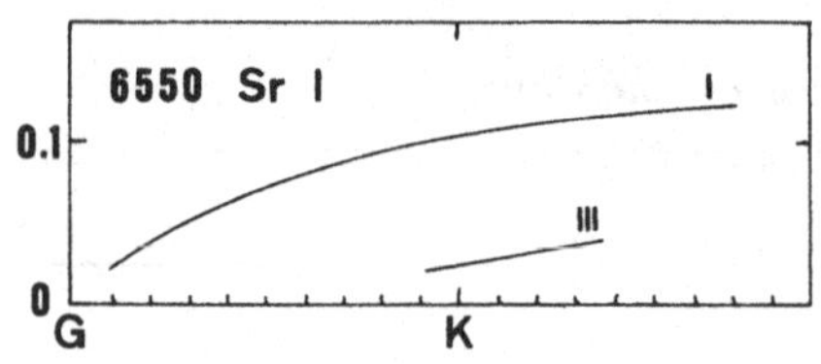

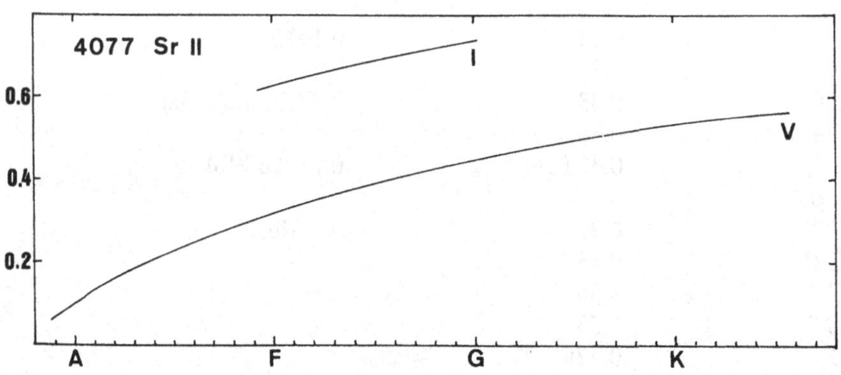

Emission lines of SrII

Lines of M.1 appear in emission in *T Tau stars* (Joy 1945) and in *long-period variables* after maximum light (Querci 1986).

Behavior in non-normal stars

SrII lines are strong in *Bp stars* of the Hg–Mn subgroup. The average *W*(4077)=0.090 (Kodaira and Takada 1978). SrII lines are also strong in the *Ap stars* of the Cr–Eu–Sr subgroup (Adelman 1973b). According to Sadakane (1976), typically *W*(4077)=0.34.

SrII lines are strong in *Am stars*; the *W* values are larger by a factor of about 1.5–2.0 than in normal stars of the same temperature (Smith 1973, 1974). In *delta Del stars*, SrII is strengthened, as in Am stars (Kurtz 1976).

SrII lines are weakened in *lambda Boo* stars: typically *W* values are smaller by factors of the order of two than in normal stars of the same temperature (Venn and Lambert 1990).

SrII lines are weakened in *F-type HB stars* with respect to normal stars of the same color, by factors up to ten in *W* (Adelman and Hill 1987). The same holds for *HB stars* in general (Adelman and Philip 1992a).

Sr seems to be underabundant with respect to iron in *metal-weak stars* (Luck and Bond 1985). The same is valid for the most metal-weak stars with Fe/H=10^{-4} (Molaro and Bonifacio 1990).

Sr lines are prominent in *Ba stars*, which leads to an overabundance of an order of magnitude (Lambert 1985, Smith 1984). Typically *W*(4607)=0.220 for a K

0 III Ba star (Danziger 1965). Sr lines are strengthened in *subgiant CH stars* (Luck and Bond 1982, Krishnaswamy and Sneden 1985).

Sr lines are enhanced in *S-type stars* (Smith and Lambert 1986) and in *C stars* cooler than C 3 (Dominy 1984).

Sr seems to be normal in the *Magellanic Cloud stars* (Luck and Lambert 1992).

Isotopes

Sr has four stable isotopes, namely Sr 84, 86, 87 and 88, which occur in the solar system with frequencies 0.6%, 10%, 7% and 83% respectively. There also exist 14 short-lived isotopes and isomers.

Origin

Sr^{84} is only produced by the p process, Sr^{86} and Sr^{87} by the s process. Sr^{88} can be produced by either the s process or the r process.

SULFUR S Z=16

This element was known to the ancients. The Latin word is *sulphurum*.

Ionization energies

S I 10.4 eV, S II 23.3 eV, S III 34.8 eV, S IV 47.3 eV, S V 72.7 eV, S VI 88.0 eV, S VII 281 eV, S VIII 328 eV.

Absorption lines of S I

Table 1. *Equivalent widths of S I*

	6046(10)		6052(10)		
Group	V	III	V	III	Ib
F 0	0.04				
F 4	0.022				
F 5					0.077
F 6	0.024				
F 8	0.018				
G 0	0.030				0.055
G 1	0.020				
G 2			0.009		
K 0		0.017		0.020	
K 2					
K 3		0.052		0.040	
K 4					0.088
M 2					0.115

Note that the resonance line of S I is at 8695(6). In the sun *W*(8695)=0.034.

Table 2. *Equivalent widths of S I 8695*

Group	V	III
A 0	0.035	
A 2	0.049	0.067
S	0.033	
K 0		0.025
K 2		0.009

Source: Some values are from Sadakane and Okyudo (1989).

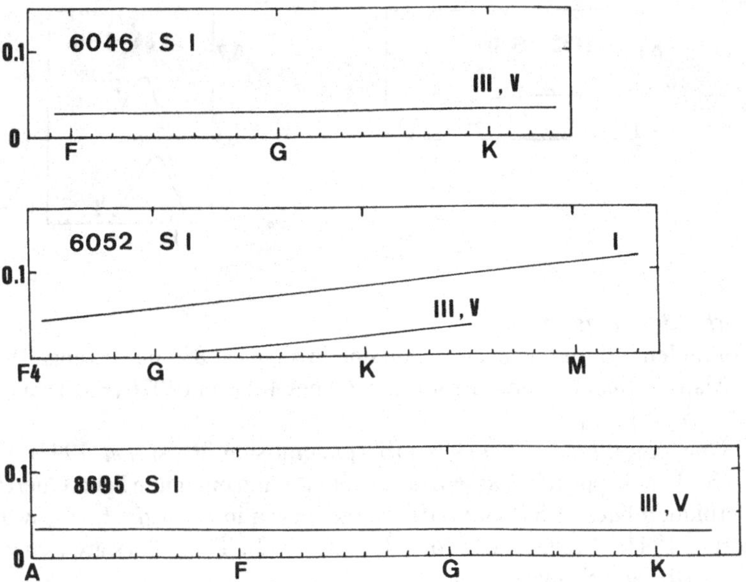

S I is present in A- to M-type stars without noticeable change in intensity. A positive luminosity effect exists for supergiants. The ultraviolet resonance line at 1807(2) seems not to have been studied.

Emission lines of S I

The lines of (UV M.9) excited by pumping of O I are present in the spectra of cool giant and supergiant stars (Joras 1986).

Absorption lines of S II

Table 3. *Equivalent widths of S II*

	4162(44,65)	
Group	V	Ia
B 0	0.002	
B 3		0.046
B 5	0.027	0.048
B 7	0.021	
B 9.5	0.005	
A 0		0.040

S II (for instance 4162) appears weakly in early B-type stars, decreasing its intensity toward A-type stars.

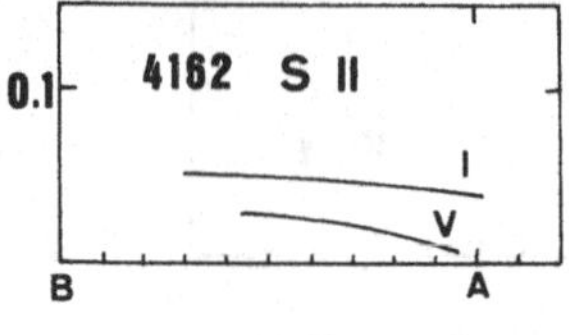

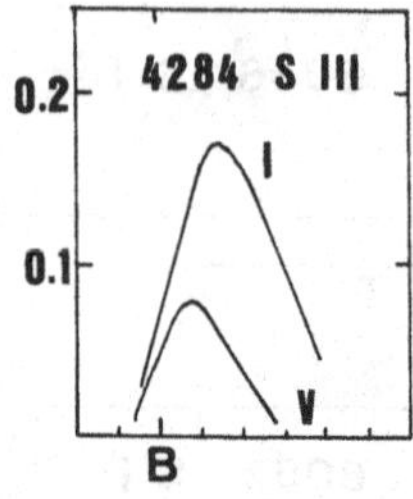

Forbidden lines of S II

Forbidden S II lines appear in emission in a typical *B[e] star* (Swings 1973).

Many *T Tau stars* show the forbidden lines 4068 and 4076 (FM 1) in emission (Sun *et al.* 1985).

One typical *VV Cep star* shows [S II] in emission (Rossi *et al.* 1992).

[S II] lines appear in *long-period variables* around minimum light (Querci 1986). Forbidden lines of S II and S III are often seen in *symbiotic objects* (Swings and Struve 1941a, Freitas Pacheco and Costa 1992). These lines are also visible in *novae* (Joy and Swings 1945).

Absorption lines of S III

Table 4. *Equivalent widths of S III*

	4253(4)		4284(4)		
Group	V	I	V	III	Ia
O 9			0.018		
B 0	0.073	0.138(Ia)	0.032		0.064
B 0.5			0.081	0.138	
B 1					0.155
B 2	0.04		0.019	0.054	0.151
B 3			0.012		0.100

Source: Additional data are from Kilian and Nissen (1989).

S III (see the line at 4284) is present in early B-type stars, with a maximum around B 1 to B 2. A positive luminosity effect exists.

Emission lines of S III

The ultraviolet lines 1012–1021 of M.2 are seen in emission in the solar spectrum (Feldman and Doschek 1991).

Forbidden S III lines

The lines 9069 and 9532 are seen in *B[e] stars* in faint emission (Andrillat and

Swings 1976, Ciatti *et al.* 1974) and in *compact infrared sources* associated with H II regions (McGregor *et al.* 1984).

Lines of [S III] were also seen in at least one *nova* (Joy and Swings 1945).

Emission lines of S IV

The ultraviolet lines at 1062(1) and 1073(1) are visible in emission in the solar spectrum (Feldman and Doschek 1991).

Behavior in non-normal stars

S II lines are very strong in the spectrum of the *extreme He star* upsilon Sgr (Morgan 1935).

S behaves erratically in *Bp stars* of the Hg–Mn subgroup (Takada-Hidai 1991).

S is underabundant in two *Ap stars* of the Cr–Eu–Sr subgroup (Sadakane and Okyudo 1989).

In *lambda Boo stars* S is normal with respect to C, N and O, but overabundant with respect to Fe (Venn and Lambert 1990).

S seems to behave in a manner parallel to that of Mg (Spite 1992) in *metal-weak stars*.

S seems to behave in a manner parallel to that of Fe in stars of the *Magellanic Clouds* (Barbuy *et al.* 1981, Spite and Spite 1990), whereas other studies resulted in a slight overabundance with respect to Fe (Wheeler *et al.* 1989).

Bond and Luck (1988) detected one F-type *halo supergiant* in which S I is very strong, implying an overabundance by one order of magnitude. This object also has considerable overabundances of C, N and O by factors of ten or more with respect to iron. More objects of this type have been discussed by Waelkens *et al.* (1991, 1992). Some typical values for S I are $W(8694)=0.110$ and $W(4694)=0.052$. S is also overabundant in one F-type *globular cluster supergiant* (Gonzalez and Wallerstein 1992). The S overabundance is considered by Bond (1991) as a common feature of *post-asymptotic branch stars*, a name given to a group of objects that are evolving toward planetary nebulae. Similar ideas were put forward by Parthasarathy (1989).

In the nebular stage of *novae* [S VIII] lines sometimes appear (like that at 9911) in emission (Warner 1989).

Strong S II lines are present in the spectra of *supernovae* of type Ia (Branch 1990).

Isotopes

S has four stable isotopes, namely S 32, 33, 34 and 36, which occur in the solar system with frequencies 95%, 1%, 4% and 0.02% respectively. There also exist six short-lived isotopes.

Origin

All four stable isotopes can be produced by explosive nucleosynthesis, but S^{32} and S^{34} can also be produced by oxygen burning and S^{36} either by Ne burning or by the s process.

TANTALUM Ta $Z=73$

This element was discovered by A. G. Ekeberg at Uppsala, Sweden. The name comes from Tantalus, who was a god, father of Niobe.

Ionization energies

TaI 7.9 eV, TaII 15.5 eV.

Behavior in stars

TaI is present in some *Ba stars* (Lambert 1985, Smith 1984).

Isotopes

Ta has two stable isotopes, Ta^{180} and Ta^{181}, as well as 16 short-lived isotopes and isomers. In the solar system Ta^{181} occurs with a frequency of 99.99%.

Origin

Ta^{180} can be produced by the p process, the r process or the s process, Ta^{181} only by the r process and the s process.

TECHNETIUM Tc $Z=43$

This element was discovered by C. Perrier and E. Segre in Palermo, Italy, in 1937. Its name alludes to the Greek *technikos* (artificial).

Ionization energies

Tc I 7.3 eV, Tc II 15.3 eV, Tc III 29.5 eV.

No Tc lines have been observed in the sun.

Behavior in stars

Tc I has its resonance lines at 4238, 4262 and 4297. It was identified by Merrill (1952) in *S-type stars*. The interest of this element lies in the fact that it is unstable. Its longest lived isotope has a half life of 2.6×10^6 years. The element can be produced by a neutron flux operating on iron-peak elements (Fe, Co, Ni). Such a neutron flux can be either rapid (the r process) or slow (the s process) and the isotopic form in which Tc is formed depends on the time scale in which the neutron flux occurred. In any case, the presence of Tc denotes that its formation occurred a short time ago, on the cosmic time scale. The presence of Tc thus constitutes an indication of recent element formation. Since element formation is assumed to occur in the stars' interior, the material must have been forcibly circulated to the surface in a short time. Up to now Tc has been observed in stars of types *M, MS, S, SC, CS* and *C*. All of the stars having Tc show large-amplitude light variations. It should be stressed that not all stars of the types quoted exhibit Tc (Little-Marenin 1989). To be more specific, according to Smith and Lambert (1988) 40% of a sample of MS- and S-type stars do not show Tc although they exhibit enhanced lines of other elements formed by neutron capture.

There seems to exist a relation between Tc in S- and MS-type stars and the He I 10830 line (in absorption, emission or both): when one is present, the other is absent (Brown *et al.* 1990).

Typical values for $W(5924)$ of Tc I in SC stars are 0.100–0.400 (Smith and Wallerstein 1983). Typically, in an S 6 star, $W(4297)=0.490$. This is almost double the value for a normal giant of the same temperature (i.e. M 6) (Wallerstein and Dominy 1988). For a recent analysis stressing the uncertainties involved, see Kipper (1992a,b).

Tc has been searched for but not found in *Ba stars* (Warner 1965).

Isotopes

Tc has three long-lived isotopes, Tc^{97} (2.6×10^6 years), Tc^{98} (1.5×10^6 years) and Tc^{99} (2.1×10^5 years half life), as well as 20 short-lived isotopes and isomers.

TELLURIUM Te Z=52

This element was discovered by von Reichenstein in 1783 in Romania. Its name comes from *tellus* (earth).

Ionization energies

Te I 9.0 eV, Te II 18.7 eV, Te III 28.0 eV.

No Te lines have been observed in the sun.

Behavior in stars

Hartoog *et al.* (1973), Cowley *et al.* (1974) and Poli *et al.* (1987) found Te II in three *Ap stars*.

Isotopes

Te has eight stable isotopes, namely Te 120, 122, 123, 124, 125, 126, 128 and 130, as well as 21 short-lived isotopes and isomers. The respective frequencies of the stable isotopes in the solar system are 0.1%, 3%, 1%, 5%, 7%, 19%, 31% and 34%.

Origin

Te^{120} is a pure p process product, Te 122, 123 and 124 are s process products. Te^{128} and Te^{130} can only be produced by the r process and Te^{125} and Te^{126} can be produced by either the r process or the s process.

TERBIUM Tb $Z=65$

This element was discovered by C. Mosander in Stockholm in 1843. Its name comes from Ytterby in Sweden.

Ionization energies

TbI 5.8 eV, TbII 11.5 eV, TbIII 21.9 eV.

Behavior in stars

Tb lines have not been definitely identified in the sun. Reynolds *et al.* (1988) observed Tb in one F 0Ib star ($W(3874)=0.004$) and Davis (1947) observed Tb in one M 2III object.

TbII (3874, 3899, 4144) is seen in some *Ap stars* of the Cr–Eu–Sr subgroup (Adelman 1973b, Cowley and Henry 1979, Adelman *et al.* 1979) and in at least one late *Am star*, where $W(4144)=0.054$ (van 't Veer-Menneret *et al.* 1988).

Isotopes

Tb has one stable isotope, Tb^{159}, and 23 short-lived isotopes and isomers.

Origin

Tb can only be produced by the r process.

THALLIUM Tl Z=81

This element was discovered by W. Crookes in London in 1861. The name comes from the Greek *thallos* (green twig).

Ionization energies
TlI 6.1 eV, TlII 20.4 eV.

Behavior in stars
The presence of Tl in the sun is dubious. It seems to be present in a sunspot spectrum (Lambert *et al.* 1969).

Isotopes
Tl has two stable isotopes, Tl^{203} and Tl^{205}, which occur in the solar system with respective frequencies 30% and 70%. There exist also 26 short-lived isotopes and isomers.

Origin
Tl^{203} and Tl^{205} are both produced by either the r process or the s process.

THORIUM Th Z=90

This element was discovered in 1828 by J. Berzelius in Stockholm. The name comes from Thor, the Scandinavian god of war.

Ionization energies

Th I 6.1 eV, Th II 11.5 eV, Th III 20.5 eV.

Absorption lines of Th II

The equivalent width of Th II 4019(3) in the sun is 0.009 Å. Th II (4019) was observed in several G-type dwarfs and giants ($W \simeq 0.010$) by Butcher (1988).

Behavior in non-normal stars

Th II was identified by Jaschek and Brandi (1972) in one *Ap star* of the Cr–Eu–Sr subgroup and by Cowley *et al.* (1976) in other objects. Cowley *et al.* (1977) published a detailed study of Th and U and its isotopes in *Ap stars*.

In the solar system Th is more abundant than U, but in stars one generally finds the opposite.

Isotopes

Th occurs in the form of 12 isotopes and isomers. The longest lived isotope is Th^{232} with a half life of 1.4×10^{10} years. All Th in the solar system occurs in this form. Of the other isotopes, Th^{230} has a half life of 8×10^{4} years and the remainder are shorter lived.

Th^{232} can be used as a radio-chronometer, if one compares its abundance with that of a stable r-process element like Eu (Francois 1992).

Origin

Th can only be produced by the actinide-producing r process.

THULIUM Tm   Z=69

This element was discovered by P. Cleve in Uppsala, Sweden in 1879. The name comes from Thule, an island in the North Atlantic. Thule was also the first name for Scandinavia.

Ionization energies

Tm I 6.2 eV, Tm II 12.1 eV, Tm III 23.7 eV, Tm IV 42.7 eV.

Behavior in stars

The presence of Tm I in the sun is doubtful, but Tm II is present with *W*(3462)=0.011.

Reynolds *et al.* (1988) detected Tm II in one F 0 Ib star, with *W*(3462)=0.003. Tm is also present in one M 2 III star according to Davis (1947).

Behavior in non-normal stars

Tm is the least abundant element of the rare earth group. It was found in the form of Tm I in some *Ap stars* of the Cr–Eu–Sr subgroup by Jaschek and Brandi (1972) and Adelman (1973b). Cowley and Crosswhite (1978) and Poli *et al.* (1987) found Tm II in two Ap stars of the Si subgroup and Aikman *et al.* (1979) detected Tm III in the spectrum of another Ap star. Bidelman (1953) found Tm I lines in one *S-type star.*

Isotopes

Tm has one stable isotope, Tm^{169}, and 17 short-lived isotopes and isomers.

Origin

Tm^{169} can be produced by both the r process and the s process.

TIN Sn $Z=50$

This element has been known since antiquity. The Latin word for tin is *stannum*.

Ionization energies

Sn I 7.3 eV, Sn II 14.6 eV, Sn III 30.5 eV.

Absorption lines of Sn I

The equivalent width of Sn I 3801(2) line in the sun is 0.006.

Absorption lines of Sn II

Table 1. *Equivalent widths of Sn II 6454(1)*

Group	III
K 0	0.017
K 3	0.024

Note: Note that there exists an atmospheric absorption band near this line.

Behavior in non-normal stars

Adelman *et al.* (1979) announced Sn I to be probably present in one *Ap star* of the Cr–Eu–Sr subgroup.

Isotopes

Sn has ten stable isotopes, Sn 112, 114, 115, 116, 117, 118, 119, 120, 122 and 124, as well as 18 short-lived isotopes and isomers. In the solar system the respective frequencies are 1%, 1%, 0.4%, 15%, 8%, 24%, 9%, 32%, 5% and 5%. Among the unstable isotopes, Sn^{126} has a half life of 10^5 years.

Origin

Sn can be produced by the p, s and r processes. Pure r process products are Sn^{122} and Sn^{124}. Sn 116, 117, 118, 119 and 120 are produced by both the r process and the s process. Sn^{112} is a pure p process product, Sn^{114} is produced by either the p process or the s process and Sn^{115} by the r, s or p processes.

TITANIUM Ti Z=22

This element was discovered independently by W. Gregor in Creed, England in 1791 and by M. Klaproth in Berlin in 1795. The name alludes to the Titans, giants in Greek mythology.

Ionization energies

Ti I 6.8 eV, Ti II 13.6 eV, Ti III 27.5 eV.

Ti is a well-represented element in stellar spectra. In the sun it figures in third place as far as the number of lines is concerned.

Absorption lines of Ti I

Table 1. *Equivalent widths of Ti I*

	3998(12)		5020(38)		
Group	V	I	V	III	Ib
F 0	0.13	0.223(Ia)			
F 2		0.251(Ib)			
F 4			0.06		
F 6			0.06		
F 8			0.06		
G 0					0.17
G 1			0.11		
G 2	0.14				
S	0.110				
G 5	0.20		0.09		0.224
K 0			0.10	0.128	
K 2			0.10		0.28,0.41
K 5					0.48

Table 2. *Equivalent widths of Ti I*

	5300(74)		5367(35)		5426(3)	
Group	V	III	V	III	III	I
S	0.021		0.002			
K 0		0.060				
K 2				0.077		
K 3		0.132				
M 0				0.177	0.18	
M 2						0.53(Ia)

Table 2. (*cont.*)

	5300(74)		5367(35)		5426(3)	
Group	V	III	V	III	III	I
M 2.5				0.224		
M 3					0.30	
M 4					0.28	
M 5					0.34	0.38(Ib)
M 7					(0.44)	

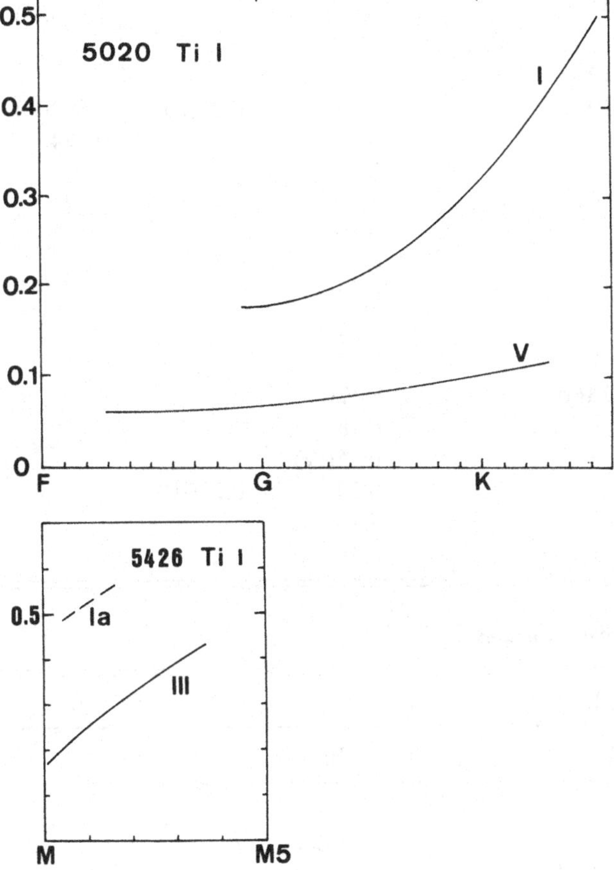

TiI (see for instance the line at 5020) appears in late A-type stars and grows continuously toward later types. A positive luminosity effect exists. The most intense infrared line of TiI is that at 9638(32), which in the sun has W=0.019.

Absorption lines of TiII

Table 3. *Equivalent widths of TiII*

	4300(41)		4563(50)		
Group	V	Ib	V	III	Ib
B 5	0.01				
B 7			0.005		
B 9	0.08				
B 9.5			0.042		
A 0	0.09	0.16			
A 1	0.11		0.084		
A 2	0.16		0.096		0.563(O)
A 3		0.18			
A 7	0.19		0.19		
F 0	0.34		0.22	0.27(II)	0.692(Ia)
F 2					0.524
F 4			0.23		
F 5		0.27	0.19		0.42
F 6			0.19		
F 8			0.15		0.54
G 0			0.14		
G 1			0.21		
G 2			0.12		
S	0.166		0.120		
G 5			0.18		
G 8			0.16(IV)		
K 0			0.19	0.21(III)	
K 2			0.16		
K 5			0.16		

Table 4. *Equivalent widths of TiII*

	5185(86)			5337(69)	
Group	V	III	Ib	V	III
G 0			0.25		
G 1			0.224,0.273		
S	0.058			0.071	
G 2			0.19		
G 5			0.19,0.194		
G 8	0.077(IV)		0.220		0.091

Table 4. (*cont.*)

	5185(86)			5337(69)	
Group	V	III	Ib	V	III
K 0		0.097			0.118
K 2		0.097	0.16,0.254		
K 3					0.135
K 5			0.18,0.263		
M 0		0.17			
M 2			0.346		

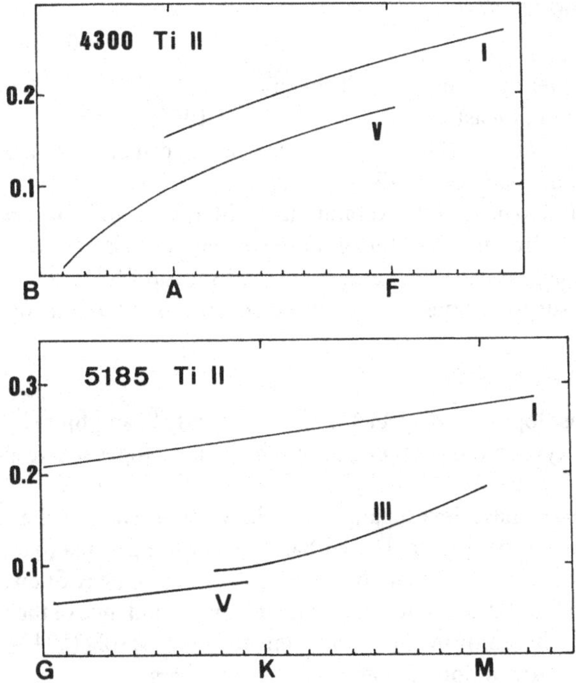

Ti II (see the lines at 4300 and 4563) appears in mid-B-type and increases toward a flat maximum for F-type, although the lines persist up to type M. There is a strong positive luminosity effect.

The resonance line is at 3349(1).

Emission lines of Ti II

Several multiples of Ti II are seen in emission in *T Tau stars* (Joy 1945). In *long-period variables*, the lines at 3759–61(13) appear in emission near maximum

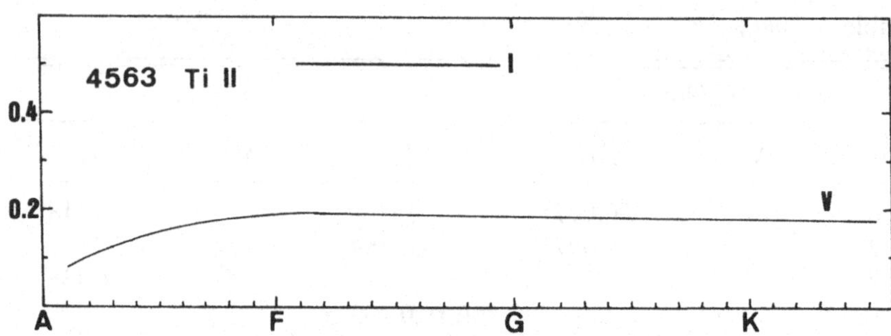

(Merrill 1947) and in *R CrB stars* (Merrill 1951b). Weak emissions occur frequently in stars that present strong emission lines of Fe II, for instance *B[e] stars* (Allen and Swings 1976).

Behavior in non-normal stars

Ti II lines are frequently strong in shell spectra.

Ti III is present in at least one *Ap star* (Bidelman 1966).

In the *lambda Boo stars*, Ti II lines are weaker than in normal stars by factors of about two (Venn and Lambert 1990).

Ti has been found to be slightly overabundant with respect to Fe in *metal-weak stars* (Magain 1989), by a factor of the order of two, and this also seems to be the case for *globular cluster stars* (Wheeler *et al.* 1989, Francois 1991).

Strong Ti II absorption lines appear in the spectra of *supernovae* of class II (Branch 1990).

Isotopes

Ti has five stable isotopes, namely Ti 46, 47, 48, 49 and 50, and four short-lived ones. In the solar system the frequencies of the stable isotopes are respectively 8%, 7%, 74%, 5% and 5%.

Clegg *et al.* (1979) have investigated the relative frequency of the isotopes through an analysis of the TiO molecular bands, using the gamma (0,0) lines in the region around 7070. They examined a sample of 11 late (>K 5) dwarfs and giants, including a few *MS* and one *S star*. The relative abundance of the isotopes seems to be solar. This confirms the earlier conclusion of Herbig (1948) of there being a solar-type distribution of isotopes in M-type stars.

Origin

Ti 46, 47, 48 and 49 are produced by explosive nucleosynthesis, whereas Ti^{50} arises from the nuclear statistical equilibrium process.

TUNGSTEN W Z=74

This element was isolated by J. and F. Elhuijar in Vergara in Spain, in 1783. The element has also been called wolfram. In Swedish *tung sten* means heavy stone. Wolfram comes from the German *Wolf* (wolf).

Ionization energies

WI 8.0 eV, WII 17.6 eV, WIII 26.1 eV.

In the sun, the equivalent width of WI 4009(6) is 0.013.

Behavior in non-normal stars

WI was found in one *Ap star* of the Cr–Eu–Sr subgroup (Jaschek and Brandi 1972) and in another one by Guthrie (1972). Sadakane (1991) found the 2030(5) line of WII in the ultraviolet spectrum of one *Am star*. Danziger (1965) and Warner (1965) found WI to be enhanced in *Ba stars*.

Isotopes

W has five stable isotopes, W 180, 182, 183, 184 and 186, and 17 short-lived isotopes and isomers. The stable isotopes occur in the solar system with the following frequencies: 0.1%, 26%, 14%, 31% and 29%.

Origin

W 182, 183 and 184 are all produced by the r process and the s process. W^{180} is produced only by the p process and W^{186} only by the r process.

URANIUM U Z=92

This element was discovered by M. H. Klaproth in Berlin in 1789. Its name alludes to the planet Uranus, discovered by Herschel in 1781, which in turn alludes to the Greek god Urania.

Ionization energies
UI 6.1 eV, UII 14.7 eV.

U is the heaviest stable element. It has not been observed in the sun.

Behavior in non-normal stars
UII was discovered independently by Guthrie (1969) and by Jaschek and Malaroda (1970) through the presence of 3859 in *Ap stars* of the Sr–Cr–Eu subgroup. The two groups each found this line in a different Ap star. It was also observed by Cowley *et al.* (1974) in other stars of this subgroup. Typically $W(3859)=0.030$.

A detailed study of U and its isotopes was made by Cowley *et al.* (1977).

Isotopes
The longest lived isotope of U is U^{238}, with a half life of 4.5×10^9 years. It is followed by U^{235} with a half life of 7.0×10^8 years and by 13 isotopes and isomers with shorter half lives. In the solar system U^{238} makes up 99.3% of all U. The isotopes of U can be used for radioactive dating.

Origin
The U isotopes are produced by the actinide-producing r process.

VANADIUM V Z=23

This element was discovered by A. del Rio in Mexico City, Mexico, in 1801 and rediscovered by G. Selfström in Sweden in 1831. The name comes from the Scandinavian goddess Vanadis.

Ionization energies

V I 6.7 eV, V II 14.6 eV, V III 29.3 eV, V IV 51.4 eV, V V 71.8 eV, V VI 141 eV, V VII 165 eV, V VIII 191 eV.

Many faint lines of V are present in stellar spectra. In the sun, V figures in fifth place among the elements as far as the number of lines is concerned.

Absorption lines of V I

Table 1. *Equivalent widths of V I*

	4379(22)		5727(35)		
Group	V	Ib	V	III	Ib
F 0	0.06	0.026(II)			
F 4			0.012		
F 5	0.10,0.04	0.024			
F 6			0.023		
F 8		0.070	0.013		
G 0	0.11				0.014
G 1			0.043		0.018
G 2	0.11		0.035		0.035
S	0.110		0.037		
G 5	0.15		0.047		0.054
G 8			0.037(IV)	0.107	0.098
K 0	0.24			0.123,0.170	
K 2				0.163	0.204
K 3				0.213	0.230
K 5	0.34				0.263
M 0				0.281	
M 2					0.269
M 2.5				0.288	

V I (for instance the lines at 4379 and 5727) appears in late A-type stars and increases toward later spectral types. A slight negative luminosity effect before type G may be present in the 4379 line.

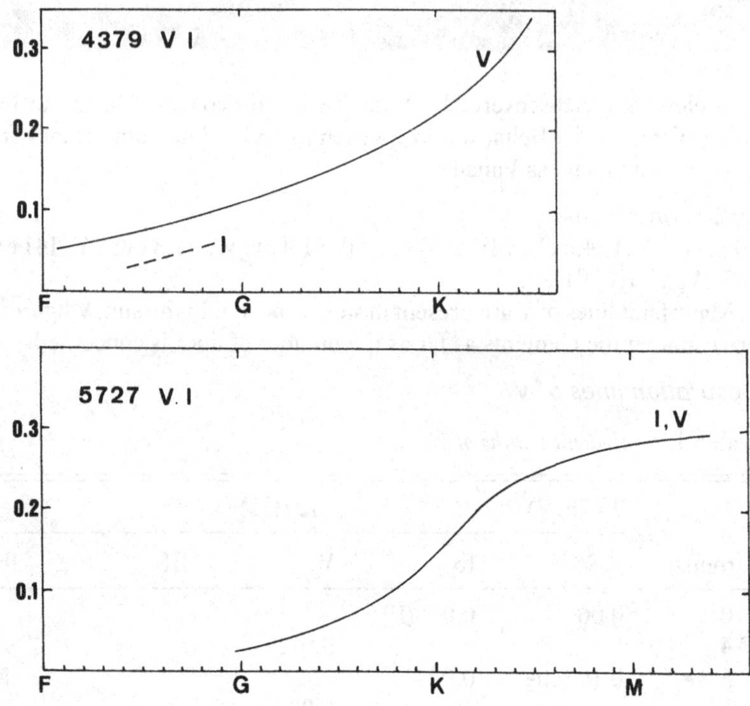

Emission lines of V I

V I lines from M.34 and 48 were observed in emission in one *Mira-type variable* by Grudzinska (1984).

Absorption lines of V II

Table 2. *Equivalent widths of V II*

	4005(32)		5929(98)		
Group	V	Ia	V	III	Ib
B 8		0.017			
B 9	0.024				
A 0	0.02	0.03,0.050			
A 1	0.061				
A 2	0.072,0.061	0.069(Ia)			
A 3		0.115(O)			
A 7	0.045				
G 0					0.039
G 1					0.055
G 2					0.047

Table 2. (*cont.*)

	4005(32)		5929(98)		
Group	V	Ia	V	III	Ib
S	0.085		0.004		
G 5					0.062
G 8					0.052
K 2				0.033	0.073
K 3					0.089
K 5					0.064,0.101

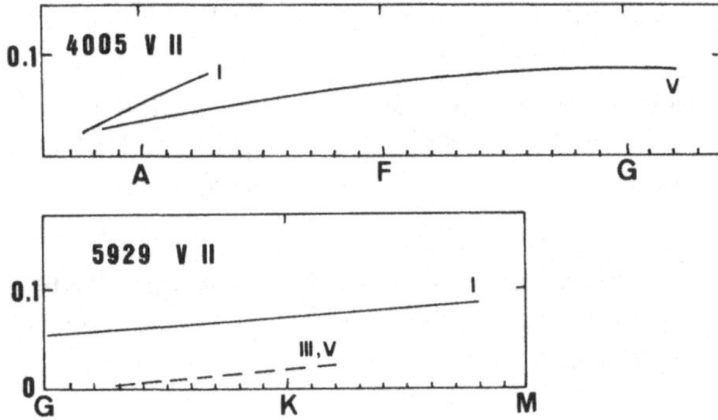

V II (for instance the lines at 4005 and 5929) appears in late B-type stars and increases toward later types. A positive luminosity effect exists.

Forbidden lines

V III forbidden lines are observed in one *recurrent nova* (Joy and Swings 1945).

Behavior in non-normal stars

V seems to behave in a manner parallel to that of Fe over the whole range of Fe abundances, and it can thus be regarded as a 'typical metal'. In *globular cluster stars*, V seems to be slightly overabundant with respect to iron by factors of the order of two (Wheeler *et al.* 1989).

Isotopes

V has two stable isotopes, V^{50} and V^{51}. In the solar system they occur with 99.75% and 0.25% abundances. There exist also seven short-lived isotopes. The isotopic effect in the VO bands due to the two stable isotopes would be very small and apparently no search has been undertaken.

Origin

V^{50} and V^{51} are produced by explosive nucleosynthesis, but V^{50} can also be produced by the nuclear statistical equilibrium process.

XENON Xe $Z=54$

This element was discovered by W. Ramsay and W. Travers in London in 1898. Its name comes from the Greek *xenos* (stranger).

Ionization energies

Xe I 12.1 eV, Xe II 21.2 eV, Xe III 32.1 eV.

Behavior in stars

No Xe lines have been observed in the sun.

Xe II was found by Bidelman (1962b, 1966) in two *Ap stars* of the Hg–Mn subgroup. Adelman (1987) measured $W(4415)=0.009$ for Xe II and Adelman (1992) found $W(4603)=0.016$. Jaschek and Brandi (1972) found Xe in one star of the Cr–Eu–Sr subgroup. See also Andersen *et al.* (1984).

Isotopes

Xe has six stable isotopes, namely Xe 129, 130, 131, 132, 134 and 136, which occur in the solar system with 27%, 4%, 21%, 27%, 10% and 9% abundances respectively. There exist also 25 short-lived isotopes and isomers.

Origin

Xe can be produced by the r, s and p processes. Xe^{130} is an s process product, Xe 129, 131, 134 and 136 are pure r products and Xe^{132} can be produced by either the r process or the s process.

YTTERBIUM Yb *Z*=70

This element was discovered by J. de Marignac in Geneva, Switzerland, in 1878. The name comes from the Swedish town Ytterby.

Ionization energies

Yb I 6.2 eV, Yb II 12.2 eV, Yb III 25.0 eV.

Absorption lines of Yb I

Table 1. *Equivalent widths of Yb I 3988 (M.2-resonance line)*

Group	V
S	0.018
K 5	0.077

Absorption lines of Yb II

In the sun the equivalent width of Yb II 3694 (M.1 resonance line) is 0.067.

Behavior in non-normal stars

Yb II was detached in at least one star of the Cr–Eu–Sr subgroup of *Ap stars* by Adelman *et al.* (1979).

Buscombe and Merrill (1952) found Yb I in one *S-type star.*

Isotopes

Yb occurs in seven stable isotopic forms, namely Yb 168, 170, 171, 172, 173, 174 and 176, and it also has nine short-lived isotopes and isomers. The stable isotopes occur in the solar system with frequencies 0.1%, 3%, 14%, 22%, 16%, 32% and 13% respectively.

Origin

Four isotopes – Yb 171, 172, 173 and 174 – can be produced by either the r process or the s process. Yb^{168} is a pure p, Yb^{170} a pure s and Yb^{176} a pure r process product.

YTTRIUM Y Z=39

This element was discovered by J. Gadolin in Abo, Finland in 1794. The name alludes to its origin, a mineral that came from Ytterby, a Swedish town.

Ionization energies
YI 6.4 eV, YII 12.2 eV, YIII 20.5 eV.

Absorption lines of YI

Table 1. *Equivalent widths of YI 6435(2)*

Group	V	III	Ib
G 0			0.020
G 2			0.011,0.015
S	(0.002)		
G 5			0.072
G 8			0.076
K 0		0.016	
K 2		0.032	0.148
K 3			0.182
K 5	0.031		0.214

YI (see for instance the line at 6435) appears in G-type stars and grows in intensity toward later types. There is a positive luminosity effect.

Absorption lines of YII

Table 2. *Equivalent widths of YII*

	4177(14)		5200(20)		
Group	V	Ib	V	III	Ib
A 2	0.14	0.29(Ia)			
A 7	0.20		0.04		
F 0	0.25	0.27(II),0.59(Ia)	0.07		
F 2		0.52			
F 4	0.24		0.044		
F 5	0.27	0.47	0.050		
F 6	0.25		0.036		
F 8			0.020		0.31
G 0					0.204
G 2			0.028		0.199

Table 2. (*cont.*)

	4177(14)		5200(20)		
Group	V	Ib	V	III	Ib
S	(0.075)		0.037		
G 5				0.066	
G 8			0.062(IV)		
K 0				0.078	
K 2				0.095	0.199
K 3					0.186
K 5					0.22

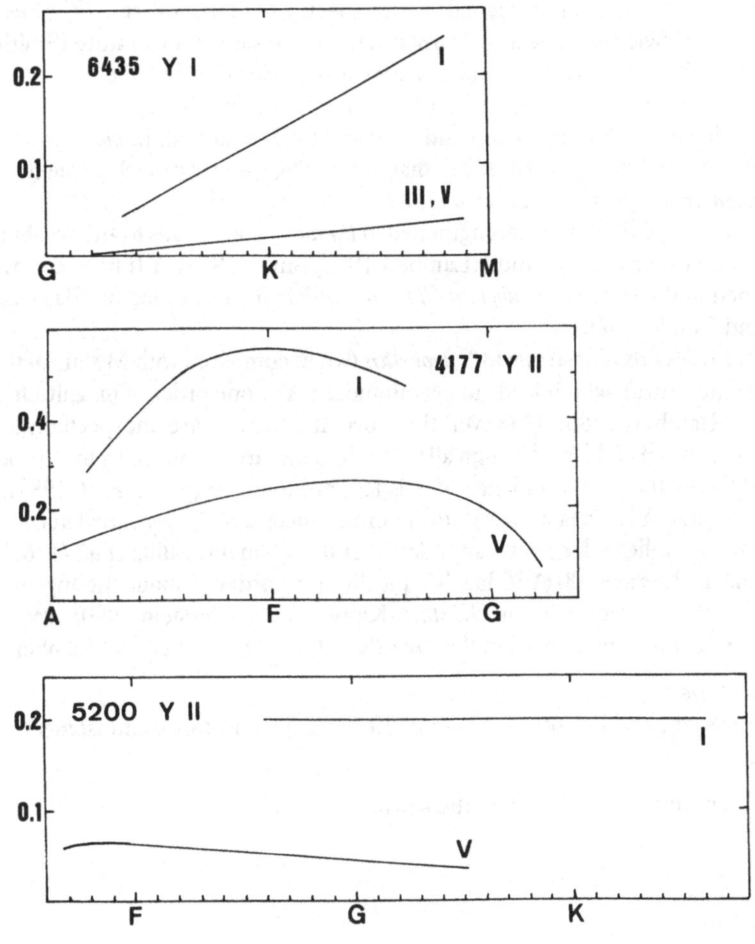

Y II (see for instance the lines at 4177 and 5200) appears in late B-type, grows toward a maximum for F-type and decreases toward G-type stars. According to Smith and Lambert (1985), the Y II 7450 line is still present in M 5 III stars. There is a strong positive luminosity effect.

Y II also has two strong lines, at 6614(26) and 6795(26) in the red spectral region. These lines are easily visible in M, *C* and *S* stars (Keenan 1957).

Behavior in non-normal stars

Y II lines are strong in the spectra of the *Bp stars* of the *Hg–Mn subgroup*. Typically, equivalent widths are about twice as large as for normal stars of the same spectral type (Kodaira and Takada 1978). Adelman (1974) found Y III in one star of this subgroup. The presence of this species has been confirmed by Redfors (1991) and studied in detail by Redfors and Cowley (1993).

Y II lines are somewhat stronger than normal in the spectra of *Ap stars of the Cr–Eu–Sr subgroup* (Adelman 1973b). Typically *W*(4398)=0.040 (Sadakane 1976).

Y II lines are strengthened in the spectra of *Am stars*, the *W* values being typically twice as large as in normal stars of the same temperature (Smith 1973, 1974). The *delta Del stars* show a similar behavior (Kurtz 1976).

Y II is weakened in *HB stars* (Adelman and Philip 1992a).

Magain (1989) and Zhao and Magain (1991) find that, in *metal-weak stars*, Y behaves in a manner parallel to that of Fe. The same is probably true for stars in *globular clusters* (Wheeler *et al.* 1989).

Y I and Y II lines are strengthened in *Ba stars*, which leads to an overabundance of one order of magnitude (Lambert 1985, Smith 1984). Y II lines are strengthened in the spectra of *subgiant CH stars*, which are also called hot Ba stars (Luck and Bond 1982).

Y lines are very strong in *S-type stars* (when compared with M stars of the same temperature), which leads to overabundances of one order of magnitude (Smith and Lambert 1986). However, the Y overabundances are unexpectedly less than those of Nd and La, although all three elements are produced by the s process. In *MS stars* the overabundance of Y is less pronounced (Smith *et al.* 1987).

Y I and Y II lines are very strong in the spectra of *C-type stars* later than C 3, which implies a large overabundance of this element. Fujita *et al.* (1963) found that the line at 4900 (Y II) has *W* typically of the order of about one ångström unit. Y is also overabundant in *SC stars* (Kipper and Wallerstein 1990).

Y seems to be normal in the *Magellanic Cloud stars* (Luck and Lambert 1992).

Isotopes

Y has one stable isotope, Y^{89}, and 20 short-lived isotopes and isomers.

Origin

Y can only be produced by the s process.

ZINC Zn Z=30

This element was known in antiquity and used in brass. The name comes from the German name *Zink.*

Ionization energies

Zn I 9.4 eV, Zn II 18.0 eV, Zn III 39.7 eV.

Absorption lines of Zn I

Table 1. *Equivalent widths of Zn I 4810(2)*

Group	V	III	Ib
A 2	0.03		
A 7	0.015		
F 0	0.075		0.036
F 4	0.050		
F 5	0.071		0.046
F 6	0.069		
F 8	0.060		0.114
G 0	0.060		
G 1	0.104		
G 2	0.071,0.076		
S	0.080		
G 8	0.068(IV)		
K 0		0.085	
K 2		0.082	
K 5	0.067		

Zn I (see for instance the line at 4810) appears in A-type stars and grows slowly toward late types. No luminosity effect is visible.

Absorption lines of Zn II

Table 2. *Equivalent widths of Zn II 2062(1)*

Group	V
B 5	0.095(IV)
B 7	0.115
A 0	0.150

Source: Data are from Sadakane *et al.* (1988).

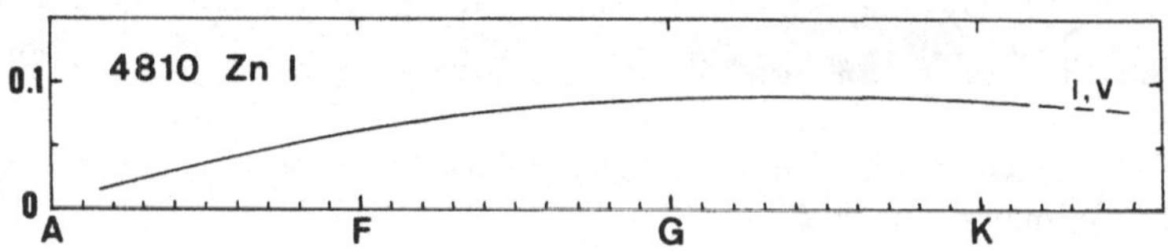

Behavior in non-normal stars

Zn II has its resonance lines at 2025(1) and 2062(1). Sadakane *et al.* (1988) studied the behavior of the 2062 line in *Bp stars* of the Hg–Mn subgroup. The line can be very strong (*W* up to 0.560), weak or absent. They find no clear relation to the behavior of the neighboring elements Cu and Ga. Cowley *et al.* (1974) detected Zn in *Ap stars* of the Cr–Eu–Sr subgroup in the photographic region.

Zn is strong in *Am stars*, the *W* values being stronger by a factor of about two than in normal stars of the same temperature (Smith 1973, 1974).

Sneden *et al.* (1991) found that Zn behaves like Fe and Ni in all *metal-weak stars* (disk and halo types). It can thus be considered a typical metal. Zn was detected in one iron-deficient (Fe/H=−5 dex) *population two A-type star*. This element does not share the iron abundance (*W*(4810, Zn I)=0.030) (van Winckel *et al.* 1992).

Isotopes

Zn has five stable isotopes, namely Zn 64, 66, 67, 68 and 70, which in the solar system occur with frequencies of respectively 48%, 28%, 4%, 19% and 1%. There also exist ten short-lived isotopes and isomers.

Origin

All five Zn isotopes can be produced by the statistical equilibrium process. Zn 67, 68 and 70 can also be produced by the s process and Zn^{64} by explosive nucleosynthesis.

ZIRCONIUM Zr $Z=40$

This element was first isolated by J. Berzelius in 1824 in Stockholm, Sweden. The name comes from the mineral zirconia. In Arabic *zargun* means golden colour.

Ionization energies

Zr I 6.8 eV, Zr II 13.1 eV, Zr III 23.0 eV.

Absorption lines of Zr I

Table 1. *Equivalent widths of Zr I 6127(2)*

Group	V	III	Ib
G 2	0.002		0.015
S	0.004		
G 5			0.033,0.101
G 8		0.038	0.082
K 0		0.032	
K 2		0.053	0.175
K 3		0.144	0.195
K 5	0.052		0.209
M 0		0.141	

Zr I (see for instance the line at 6127) appears in G-type stars and grows toward later types. There is a positive luminosity effect.

Emission lines of Zr I

Some Zr I lines appear in emission during the decreasing light of a *long-period variable.* Their great intensity is probably due to fluorescence (Merrill 1953).

Absorption lines of Zr II

Table 2. *Equivalent widths of Zr II*

	4149(41)		4211(15)	
Group	V	Ib	V	Ib
B 9.5	0.006			
A 0	0.015			
A 1			0.019	
A 2			0.025	
A 7			0.025	
F 0		0.104(Ia)	0.055	0.104(Ia)

Table 2. (*cont.*)

	4149(41)		4211(15)	
Group	V	Ib	V	Ib
F 2		0.083		0.083
F 4			0.042	
F 5	0.05	0.15	0.033	0.167
F 6			0.037	
F 8			0.044	0.29
G 0			0.075	
G 1			0.097	
G 2			0.061	
S	0.062		blend CH	
K 5	0.088		blend CH	

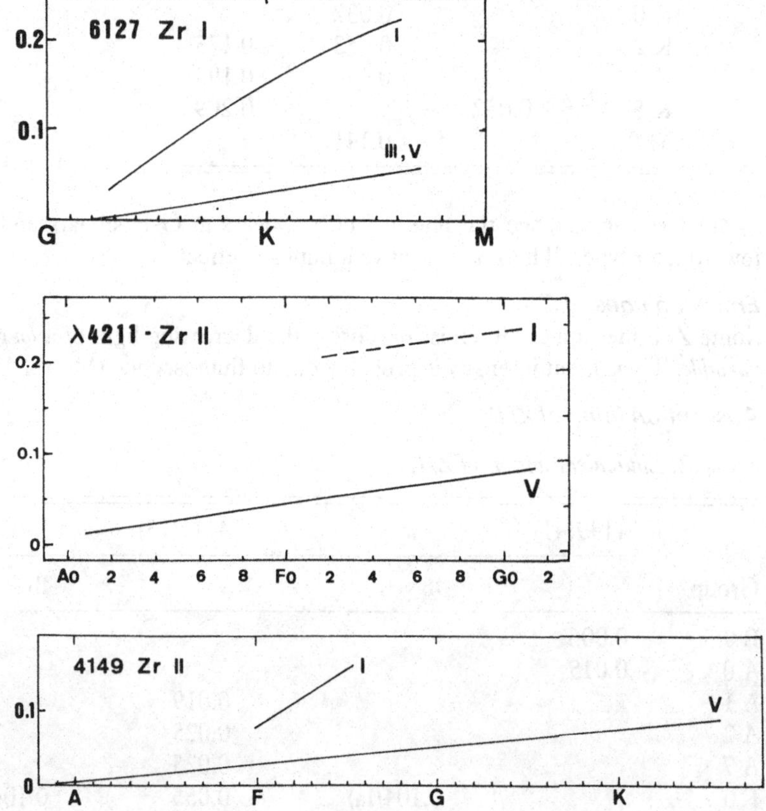

Zr II (see the lines at 4211 and 4149) appears in late A-type and grows monotonically toward cooler stars. There is a positive luminosity effect.

Zr has several strong lines in the red region – 5955, 6127, 6135 (M.2 and 3) – which are characteristic of late type stars (Keenan 1957).

Behavior in non-normal stars

Zr II lines are strong in the spectra of *Bp stars* of the Hg–Mn subgroup. A typical value is *W*(4149)=0.030 (Kodaira and Takada 1978). Redfors (1991) detected Zr III in the ultraviolet spectra of some Bp stars of the same subgroup and Redfors and Cowley (1993) added more stars.

Zr II lines are strong in the spectra of *Ap stars* of the Cr–Eu–Sr subgroup (Adelman 1973b). *W*(4719)=0.045 (Sadakane 1976).

In *metal-weak stars* Zr follows the behavior of the other metals. Magain (1989) found that, when Zr is compared with Fe, it tends to be overabundant. Typically *W*(4209)=0.034 for G 2 V (Zhao and Magain 1991).

In *globular cluster stars* Zr tends to behave parallel to Fe (Wheeler *et al.* 1989).

Zr lines are strengthened in the spectra of *Ba stars*, which leads to overabundances of one order of magnitude (Lambert 1985, Smith 1984). Typical *W* values for Ba stars are twice as large as for normal giants (Danziger 1965). Zr lines are also strengthened in the spectra of *subgiant CH stars*, which are also called hot Ba stars (Luck and Bond 1982).

Zr lines are strengthened in *S-type stars*, by factors of the order of 1.5 with respect to stars of the same temperature (Smith and Lambert 1986).

Zr I lines are very strong in *C-type stars* later than C 3, which probably indicates a large overabundance of Zr. *W*(5735) is typically about 1.0 Å (Fujita 1966). The overabundance of Zr was confirmed by Kilston (1975) and Utsumi (1984).

Zr I lines are also strong in the majority of the *SC stars* (Wallerstein 1989, Kipper and Wallerstein 1990).

Zr seems to be normal in the *Magellanic Cloud stars* (Luck and Lambert 1992).

Isotopes

Zr has five stable isotopes, namely Zr 90, 91, 92, 94 and 96, which occur in the solar system with the following frequencies respectively: 52%, 11%, 17%, 17% and 3%. There also exist 15 shorter lived isotopes and isomers. Among them Zr^{93} has a half life of 1.5×10^6 years.

Zook (1978, 1985) studied the band heads at 6930 of ZrO in two *S-type stars* and found about solar ratios for the stable isotopes. He found also a small amount of the unstable Zr^{93}, which is indicative of the operation of the s process. Lambert (1988) also found solar ratios, with perhaps an excess of Zr^{90}.

Origin

Zr 90, 91, 92 and 94 can only be made by the s process, whereas Zr^{96} can only be made by the r process.

Summary

The figure that follows provides a short summary of the behavior of the elements.

On the abscissae are given the spectral types and on the ordinates the different elements, specified by the formula (given on the left-hand side) and the atomic number (given on the right-hand side). The continuous line marks those spectral types in which the neutral species is visible, the broken line denotes the presence of the singly ionized species. No higher ionization stages are represented. Blanks indicate elements that are not observed. If a given species of a given element has only been detected in the sun, a black spot is entered in type G. A square in the next to last column indicates that this element has been found to be very strong in some type (or types) of non-normal stars.

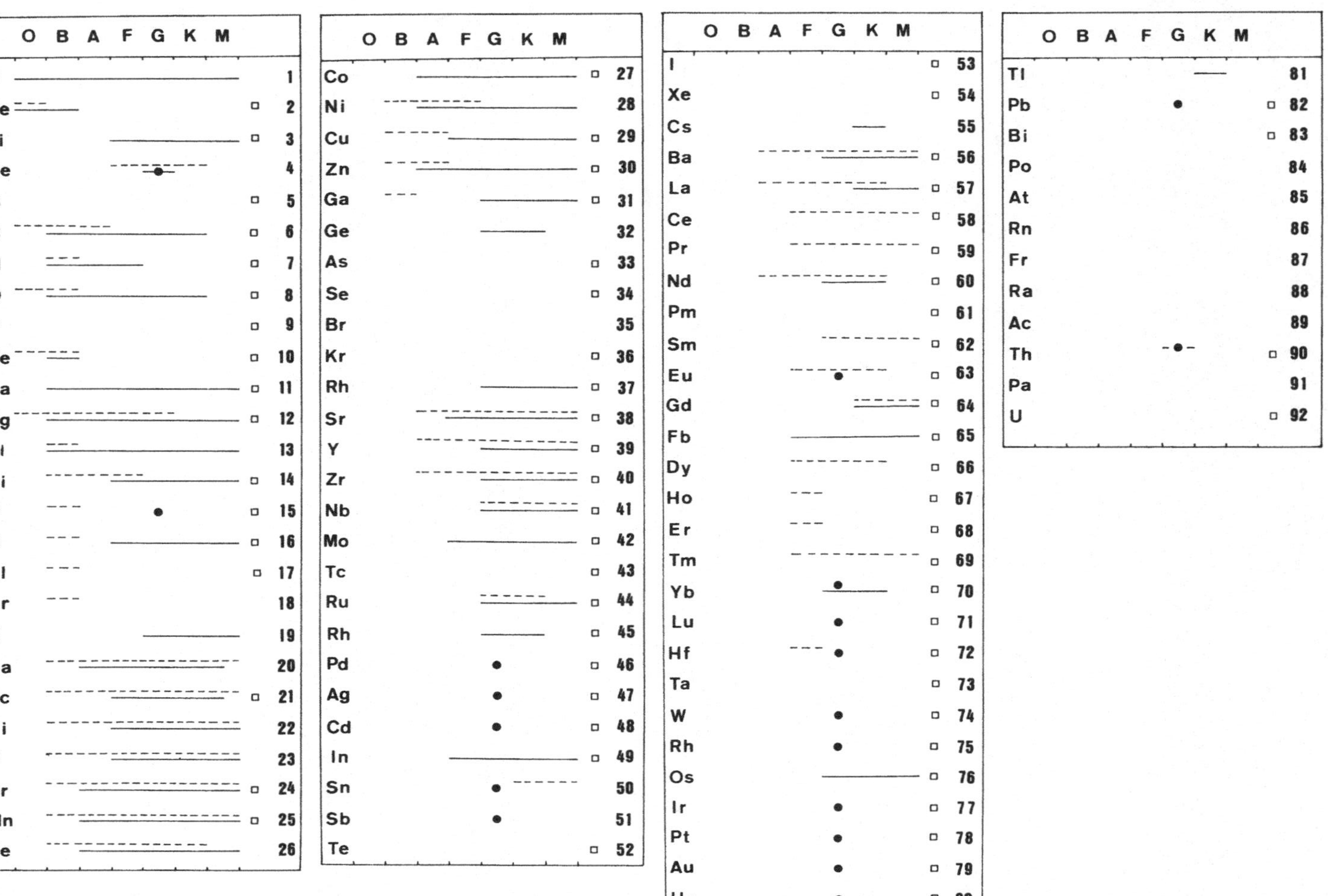

O B A F G K M
H 1
He 2
Li 3
Be 4
B 5
C 6
N 7
O 8
F 9
Ne 10
Na 11
Mg 12
Al 13
Si 14
P 15
S 16
Cl 17
Ar 18
K 19
Ca 20
Sc 21
Ti 22
V 23
Cr 24
Mn 25
Fe 26
O B A F G K M
Co 27
Ni 28
Cu 29
Zn 30
Ga 31
Ge 32
As 33
Se 34
Br 35
Kr 36
Rh 37
Sr 38
Y 39
Zr 40
Nb 41
Mo 42
Tc 43
Ru 44
Rh 45
Pd 46
Ag 47
Cd 48
In 49
Sn 50
Sb 51
Te 52
O B A F G K M
I 53
Xe 54
Cs 55
Ba 56
La 57
Ce 58
Pr 59
Nd 60
Pm 61
Sm 62
Eu 63
Gd 64
Fb 65
Dy 66
Ho 67
Er 68
Tm 69
Yb 70
Lu 71
Hf 72
Ta 73
W 74
Rh 75
Os 76
Ir 77
Pt 78
Au 79
Hg 80
O B A F G K M
Tl 81
Pb 82
Bi 83
Po 84
At 85
Rn 86
Fr 87
Ra 88
Ac 89
Th 90
Pa 91
U 92

Part Two

Content description

In the first chapter of part two we discuss the behavior of molecules in stars; in chapter two, we consider groups of elements and in chapter three, stellar chromospheres and coronas are discussed.

Chapter 1

In this chapter we provide a general description of the behavior of molecules in stars and in circumstellar envelopes. After a general introduction, the behavior of each molecule is described in detail, as we did for the elemental atoms. At the end of the chapter we provide a general summary of the behavior of molecules in stars, as we did for the individual atoms in stars.

Chapter 2

This chapter contains sections on the behavior of metals and on the behavior of rare earths.

Chapter 3

This chapter provides a summary of our knowledge on stellar chromospheres and coronas.

No detailed description of the content of chapters two and three is necessary.

1 MOLECULES OBSERVED IN THE STARS

To deal with molecules in a book on the behavior of the chemical elements seems a little bit awkward, since the information on molecules seems apparently unsuitable to add to our knowledge of the behavior of the elements. There are, however, several good reasons for dealing with molecules. The first is that, for some elements, we only know their existence through the molecules in which they participate. This is for instance the case for fluorine. In the second place, molecules permit easy study of the different isotopes of an element. As an example consider C, whose isotopes were first studied through the molecular bands of C^{12}/C^{12} and C^{12}/C^{13} (Sanford 1940). The study of isotopes is an extremely fruitful subject, which is by no means exhausted. In the third place, molecules are very sensitive indicators of spectral peculiarities. For instance the first indications of the non-uniform composition of stellar atmospheres came through the study of molecules in late type stars. It was found that the division of these stars into two families, one with carbonated molecules (then called R- and N-type stars) and the other group with metallic oxides (the M- and S-type stars) could not be explained by a uniform chemical composition. That such an abundance effect is more easily observed in molecules rather than in atoms or ions, is due to the fact that molecules are very sensitive to differences in chemical composition.

These three arguments largely justify the existence of a separate section on molecules. Another independent reason is that we have been unable to find an updated summary on molecules in stars. This is also true for the sun – the latest summary on molecules in the sun is that by Sinha (1991), who refers basically to other summaries published 20 years ago.

From what we have said, it is clear that we provide only a broad outline of the subject, concentrating on the essentials.

The interested reader who wishes to pursue these matters further is advised to consult the literature.

Books of note on molecular spectroscopy are

The Spectra and Structures of Simple Free Radicals by G. Herzberg, Cornell University Press 1971.

Molecules and Radiation edited by J. Steinfeld, Harper and Row, New York, 1974.

Reviews on molecular spectroscopy include

Spinrad and Wing (1969)

Tsjui (1986a)

Gustafsson (1989)

and various papers published in

IAU Symposia 122, *Circumstellar Matter*, and 150, *Astrochemistry of Cosmic Phenomena*.

1.1 Generalities

Molecular bands are seen in cooler, i.e. late type stars. This is because many molecules have small binding energies, so that they can be stable only at low temperatures. In the classic region of the spectrum, the first molecule seen is CH, whose bands appear in mid-F-type stars. Usually molecules become stronger in the spectra of cooler stars and habitually they are stronger in the spectra of dwarfs than in those of giants.

It can be argued that some molecules, like CO, have on the contrary a rather high binding energy, which makes them stable at higher temperatures. Nevertheless, in hotter stars the number of molecules, as compared with that of atoms, is rather small and one finds again that the molecule becomes easily visible only in cooler stars.

Most of the molecules present in stellar atmospheres are diatomic molecules.

The formation of a given molecule depends mostly on three factors, namely the abundance of the constituent atomic species, the temperature (more generally the physical conditions of the star's atmosphere) and the physical properties of the atoms. The abundance of the atomic species enters logically, since if the constituent atoms are only present in minute amounts, then only small quantities of molecules can be produced. With regard to the second factor – physical conditions – we have already said that high temperatures prevent formation of molecules. Other factors, like the electron pressure, may also influence the formation of molecules, but are generally less important. The third factor – physical properties of the molecule – enters through the stability of the interatomic bonds. We shall take the dissociation energy of the molecule as a fundamental parameter characterizing the stability of a molecule.

Another excellent place to observe molecules is the circumstellar envelopes. If a star loses matter, this matter replenishes the circumstellar regions before being lost into space. Far away from the star the temperature is rather low, and we thus have all the conditions favorable for formation of large molecules, that is, those grouping more than two atoms (up to a dozen or so).

Molecules radiate by three different kinds of transitions, which produce three kinds of spectra, namely rotation, rotation–vibration and electronic spectra.

Since the energies involved are different for each type of transition, one observes these transitions in different wavelength regions, namely rotational spectra in the radio, microwave and infrared, rotational–vibrational spectra mostly in the infrared and electronic spectra in the ultraviolet, visible or near ultraviolet. Since a molecule may be detected in more than one spectral region, one can obtain additional information on the physical conditions in the different layers of the star and/or its surroundings. This makes the situation now much better than it was some decades ago, when the only source of information on molecules was the visual spectral region, which corresponds mostly to electronic transitions.

It should also be mentioned that most of the molecular lines produced in stellar photospheres are *absorption lines*. The only exceptions – molecular emission lines – are produced by fluorescence and are observable in long-period variables. The situation is very different in the radio region, where most of the lines appear in *emission*. It should be added that infrared transitions of the same molecule are often seen in absorption and emission.

Once the presence of a given molecule in stars is established, studying its behavior in stars of different types is complicated by the fact that fixation of the continuous spectrum is problematic in all late type stars. As a consequence it is difficult to measure equivalent widths. In some cases authors have measured (both in the ultraviolet and in the classical region) the depth of the intensity jump produced by the molecule (see part three, chapter 4 on equivalent widths). However, in the majority of cases no quantitative measurements exist at all and description of the behavior of molecules is made in purely descriptive terms like 'strong' or 'weak' bands, 'traces of' and so on. For some of the more abundant molecules, the equivalent widths can be substituted by narrow band photometric measurements of regions in which the molecule is present. Usually the difficulty resides in the fact that one needs in addition a comparison band free of molecules. A description of the technique and some of the results obtained is provided by Wing (1991).

Another important fact for molecular studies is that the appearance of the bands depends very much on the plate factors used. That which is a band at 120 or 200 Å mm^{-1} dissolves into an array of lines at 5 or 10 Å mm^{-1}. If not stated otherwise, the descriptions of the behavior provided here correspond to intermediate plate factors between 40 and 80 Å mm^{-1}, except for the sun, for which very high plate factors can be used. This of course introduces an observational bias. If a molecule is seen in the sun and in stellar spectra only in late K types, then this apparent discontinuity in behavior is simply attributable to the different plate factors (and resolving power) used. Furthermore, when one deals with the sun, a distinction has to be made between the photosphere – whose spectral type corresponds to G2V – and the sunspots, whose spectral type corresponds to (about) K5V.

Molecular bands and/or lines are very numerous, if an adequate plate factor is used. For instance in the solar spectrum one finds 2300 unblended molecular lines in the 3000–8700 Å region, corresponding to six molecules. Molecular are thus responsible for about 20% of all spectral lines. If this happens in a relatively warm star, it is easy to imagine what happens in a cooler star in which the number of molecules (and of molecular lines) is considerably larger. In other wavelength regions the situation is similar. In the region 2–16 μm two thirds of all lines of the solar disk spectrum correspond to four molecules, and this represents some 10000 lines.

The existence of a large number of lines has of course also a very positive aspect. The fact that numerous lines do exist, with a particular band structure and a regular spacing, is a great help in identifying molecular spectra. In this sense one is much better off than with atomic spectra, where the number of lines is usually much smaller.

For molecules observed radioastronomically in stellar envelopes the situation is quite different, first because the number of observed lines is much smaller and second because very often molecules are observed in just one object (for instance the C-type star IRC + 10216). This of course does not imply strongly that the molecule is present in other stars of the same type.

The general behavior of the more important molecules in the 4000–8000 Å region is well described in Keenan (1957) and illustrated in the spectral Atlas of Keenan and McNeil (1976).

Other atlases that can be used to follow the overall behavior of molecules are

An Atlas of Digital Spectra of Cool Stars (Types G, K, M, S and C) by Turnshek *et al.* (1985). Tracings of 75 MK standards in the region 4200–7900 or 4500–6200 at 8–12 Å resolution.

An Atlas of the 4 μm region (2400–2778 cm^{-1}) by Ridgway *et al.* (1984). This contains the spectra of an M supergiant, a K giant, a C star, an S star and a late M-type Mira variable.

An Atlas of S Type Stars and a List of Positions of More than 200 Molecular Bands in the Region 4600–8100Å by Wyckoff and Clegg (1978).

1.2 The behavior of particular molecules

In what follows we will deal with the behavior of different molecules. For each molecule we quote its dissociation energy, the types of stars in which it is seen – both normal and non-normal stars – and the bibliographic references.

Following the procedure of Tsuji (1986a) the molecules are ordered by the position of the element in the periodic table. The periodic table is given in table 1 of part four. The numbers at the top of each group of elements are the numbers given on the top line of the periodic table. These numbers are also given in the general summary, which follows at the end of this chapter.

Dissociation energies are given next, in electron-volts. Dissociation energies are not always as accurate as the number of decimal places seems to indicate. Whereas the best values are accurate to better than 1%, the situation is less satisfactory for many molecules, which have errors of the order of 2–3%.

It should be added that, for molecules with more than two atoms, the dissociation energy quoted is that of the whole molecule and not of a partial dissociation.

After the value of the dissociation energy is indicated the source, according to the following code:

a – Sauval and Tatum (1984)
b – Sauval, private communication
c – Kerr and Trotman-Dickenson (1972).

We express our thanks to Dr Sauval for providing many additional values.

The lack of dissociation energies for some of the more complex molecules is one example of the lack of laboratory data on molecules, which has been stressed by many authors – see for instance the proceedings of the Montpellier Colloquium. The lack of laboratory work is one of the important difficulties encountered when identifying molecules. For an illustration of the problems that appear, see for instance the discussion by Lambert (1983) of the three molecules FeH, ZrS and CN.

At the end of the chapter, table 1 provides a general summary of the presence of molecules in different types of stars.

Wavelengths are given in ångström units (10^{-8} cm) or micrometers (10^{-4} cm).

Infrequently reciprocal centimeters or gigahertz are also used. The wavelength can be obtained through the formula

$$\text{wavelength (cm)}=30/\text{frequency (GHz)}$$

The different units have been kept to facilitate comparison with the bibliography quoted. The units are a function of the observational technique used.

Group H

The H_2 molecule (4.48 eV – a) is fairly abundant in cool stars. The molecule is present in the solar photosphere and in sunspots (Jordan *et al.* 1977) through its ultraviolet electronic transitions. Hall and Ridgway (1978) established its presence in *S-type stars* and Johnson *et al.* (1983) found it in *C stars* through quadrupole vibration–rotation transitions at 4713 and 4498 cm^{-1}. These features are present in *Mira variables* and in *HH objects* (Curiel 1992).

Group 1A (Li, Na, K, Rb, Cs, Fr)

NaCl (4.24 eV – c) has been detected in the envelope of the *C-type star* IRC + 10216 through five transitions between 91.1 and 169.2 GHz (Cernicharo and Guelin 1987). The discovery was confirmed through the detection of one predicted $NaCl^{37}$ line.
KCl (4.42 eV – c) was detected by Cernicharo and Guelin (1987) in the circumstellar envelope of the *C-type star* IRC + 10216 through five lines in the 99.9–161.3 GHz domain.

Group 2A (Be, Mg, Ca, Sr, Ba, Ra)

Among the molecules of this group, hydrides of Mg and Ca are conspicuous in late type stars.
MgH (1.34 eV – a) bands are present in the spectrum of the solar photosphere (Laborde 1961, Grevesse and Sauval 1973) and of sunspots (Sotirovski 1971). Bands at 4780, 5211 and 5621 appear at about K 7 and strengthen toward M-type stars, becoming insensitive to temperature after M 1 (Pettersen and Hawley 1989).

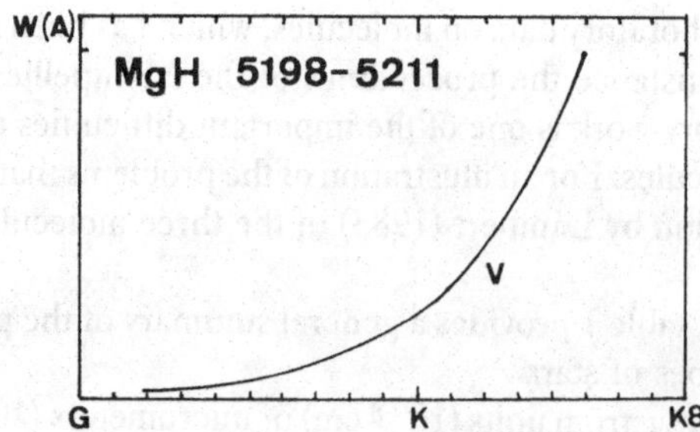

Table 1. *Residual intensity of MgH 4780, calculated as the ratio* I*(4780)/*I*(4750)*

Group	V
K 3	0.91
K 5	0.83
M 0	0.77
M 1.5	0.72
M 3.5	0.64
M 5	0.55

Source: Data (smoothed) are from Petterson and Hawley (1989).

MgH is observed only in dwarfs. Its behavior parallels that of CaH.

MgO (4.48 eV – c) bands are probably present in the spectra of sunspots (Babcock 1945).

MgNC was identified as being responsible for six previously unidentified lines observed radioastronomically in the spectrum of the circumstellar envelope of the *C-type star* IRC + 10216 (Kawaguchi *et al.* 1993).

CaH (1.70 eV – a) bands are present in the spectra of sunspots (Engvold 1973). Bands at 6385, 6880, 6908 and 6946 are present in M-type dwarfs. They are enhanced in dwarfs and are not seen in giants (Ohman 1936, Jones 1973, Kirkpatrick *et al.* 1991). The bands have a flat maximum around M 4 (Keenan and McNeil 1976) and are insensitive to temperature after M 2–3 (Pettersen and Hawley 1989).

Table 2. *Residual intensities of CaH 6385 for dwarfs, calculated as the ratio of intensities at 6385 (the band head) and 6400 (the continuum)*

Group	Residual intensity
K 6	0.92
M 0	0.86
M 1	0.83
M 3	0.71
M 4	0.71
M 5	0.71

Source: Data from Petterson and Hawley (1989).

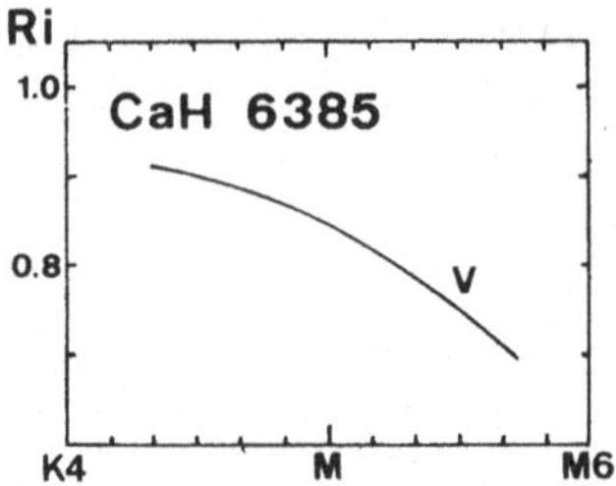

The strengths of the CaH bands (and of those of TiO) have been measured photometrically by Mould and McElroy (1978), who use the results to discriminate between subdwarfs of different populations (halo and old disk). There exists a small number of M-type dwarfs in which the CaH bands are very strong (Stephenson 1986).

CaH bands are seen in CS stars (Bidelman and Irvine 1974, Catchpole 1975). **CaOH** was discovered by Pesch (1972) through a band at 5500–5560. This molecule is visible in dwarfs later than K 7 and becomes prominent from M 4 onwards. It is absent in giants. Other useful bands are those at 6158 and 6230. The latter saturates at M 4 (Pettersen and Hawley 1989).

Table 3. *Residual intensities of the 6230 band in dwarfs, calculated as the ratio of intensities at 6230 (the band head) and the average of those at 6150 and 6350 (the continuum)*

Group	Residual intensity
K 6	0.84
M 0	0.76
M 1	0.68
M 3	0.56
M 4	0.45
M 5	0.33

Source: Data are from Pettersen and Hawley (1989).

CaCl (4.09 eV – a) was discovered by Sanford (1940). Bands at 5900, 6066, 6185 and 6212 are observed in *C-type stars* and are very strong in *CS Mira variables* (Greene and Wing 1975, Wyckoff and Wehinger 1976, Wallerstein 1989). Usually the strongest CaCl bands are observed in stars with strong Na lines. The molecule has also been observed in the ultraviolet (Bennett and Johnson 1985).

This molecule is absent in pure S and M stars (Clegg and Wyckoff 1979).

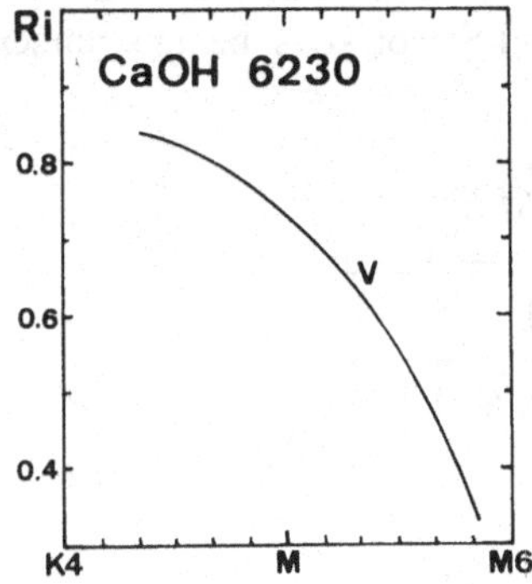

Group 3B (Sc, Y, rare earths and actinides)

The presence of monoxides of the type ScO, YO and LaO is well established. Furthermore, theory predicts also the presence of dioxides (of the elements of this group and of group 4B), but unfortunately they are difficult to observe.
ScO (6.96 eV – a) is present as two bands at 4858 and 6036 in the spectra of sunspots (Babcock 1945). In M-type stars one observes in addition bands at 5097, 6064 and 6079.

Table 1. *Equivalent widths of ScO 6064*

Group	*W*
M 2	0.6
M 3	0.86
M 4	2.3
M 6	2.7
M 8	2.8

Source: Data are from Fujita *et al* (1963).

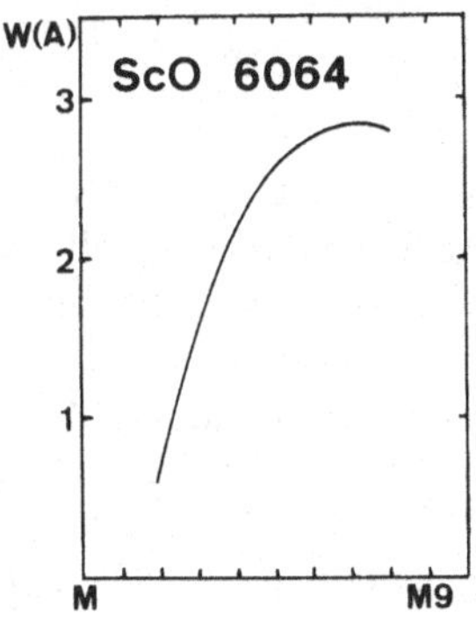

LaO (8.23 eV – a) bands at 7403 and 7910 were detected only in *S-type stars.*
YO (7.29 eV – a) bands are seen in M- and S-type stars, being stronger in the latter.

Table 2. *Equivalent widths of YO 6132*

Group	III	I
M 2		0.68(Ia)
M 3	0.85	
M 4	1.4	
M 5	1.8	1.5(II)
M 8	2.8	

Source: Data from Fujita *et al.* (1963).

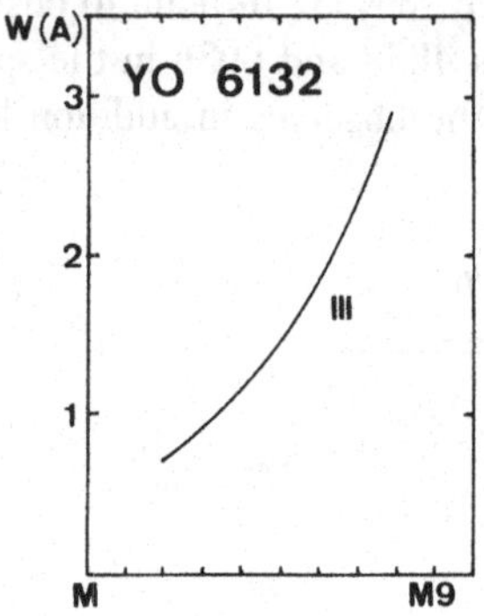

CeO (8.18 eV – a) has been established by Wyckoff and Wehinger (1977) in one *S-type star.* Clegg and Lambert (1978) through the 7235 and 7275 bands identified the molecule in one *MS*- and another S-type star. Further CeO bands were found at 8219 and 8235 (Lambert and Clegg 1980).

Group 4B (Ti, Zr, Hf)

The presence of ZrO and TiO is well established in late type stars.
ZrO (7.85 eV – a) has bands at 5552, 5718, 6229, 6345, 6474, 6597, 9299 and 9315. The presence of the molecule in the solar spectrum is dubious. Babcock (1945) identified two bands (at 6345 and 6474) in sunspot spectra. ZrO bands are weakly present in M-type stars and become characteristic features of *S-type stars.* They are also present in *SC stars* (Wallerstein 1989). For additional bands see Murty (1980) and Davis and Hammer (1981).
ZrS (5.95 eV – c) has been identified by Murty (1980) and Hinkle *et al.* (1989a,

1989b) in *S-type stars*. The bands have heads degraded to longer wavelengths at 7877, 7957, 8030, 8379, 8459 and 8530 cm^{-1}. These bands are absent in M-type stars.

TiO (6.87 eV – a) bands are present in the spectra of sunspots (Sotirovski 1971, Engvold 1973). TiO bands grow in intensity from K 5–7 to M 4 and become relatively insensitive to temperature in later M-type stars (Keenan and McNeil 1976). The most important TiO bands are listed in table 1.

Table 1. *The most important TiO bands (denoted by *) in the blue–red region*

4422	5448*	6158	10150
4584*	5497	7054	
4626	5759	7589	
4760*	5810	7672	
4954	5847*	8420*	
5167	5862	8860	

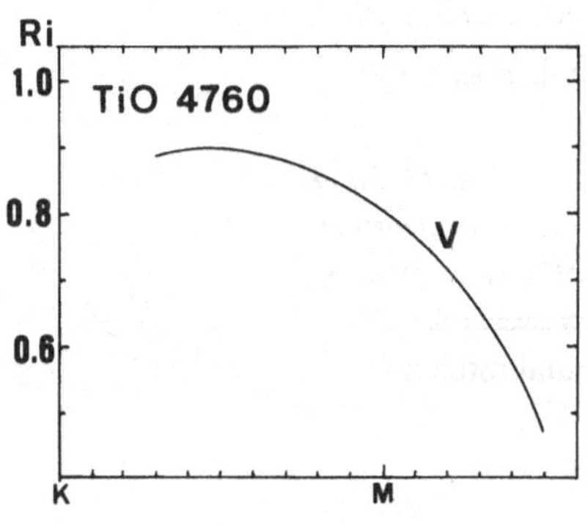

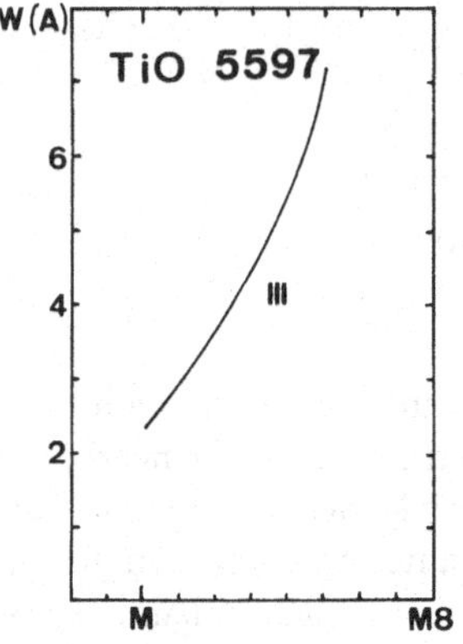

If examined in detail, each band of table 1 has a slightly different behavior. It is not useful to give details, since the aspect depends very much on the plate factor and resolution used, as can be seen in Boyartchuk (1969), Kipper (1976) and Solf (1978).

Pettersen and Hawley (1989) have observed residual intensities of different TiO bands in a sample of K and M dwarfs and find a clear correlation of these with temperature for the bands at 4761, 4954, 5448, 7054, 8420 and 8860, between K 7 and M 6.

Table 2. *Equivalent widths of the TiO band 5597*

Group	III	I
M 0	2.7	
M 2		2.7(Ia)
M 3	4.9	
M 4	5.1	
M 5	6.9	6.0(II)

Note: No luminosity effect is observable.

Table 3. *Residual intensity of the 4760 band in dwarfs, calculated as the ratio of intensities at 4760 (the band) and 4750 (the continuum)*

Group	Residual intensity
K 3	0.89
K 6	0.84
M 0	0.80
M 1	0.78
M 3	0.67
M 4	0.54
M 5	0.48

Source: Data are from Pettersen and Hawley (1989).

The TiO bands can also be measured photoelectrically with narrow band filters; one centered upon one TiO band and the other on the nearby continuum. This is the technique used by Wing (see for instance McConnell *et al.* (1992)). The difficulty with the procedure is to find a nearby continuum band, which is not affected (too much) either by TiO or by other molecular bands, so as to have a 'clean' continuum. The figure reproduces the behavior of the TiO index with spectral type.

In S-type stars the TiO bands show a somewhat irregular behavior (Keenan 1957).

TiO bands are observed in many *symbiotic stars*. Since symbiotic systems contain a late type giant (plus a hot plasma), the presence of the bands is normal. The most extensive work on TiO in symbiotic stars has been carried out by Kenyon and Fernandez-Castro (1987).

Other molecules of this group are ZrS, TiH, TiCl and TiS.

TiH (1.64 eV – c) was identified in M-type stars and in sunspots, through bands at 5305, 5309 and 5354. Most of the bands are faint, with $W<0.07$ (Yerle 1979).

TiCl (5.11 eV – c) was analyzed by Phillips and Davis (1989), who suggest that the molecule could be present in M-type stars.

TiS (4.75 eV – c) has been identified by Jonsson *et al.* (1992) in *S-type Mira variables* on the basis of some band heads (12451–11718–8625–8283) mistakenly attributed to ZrS.

Group 5B (V, Nb, Ta)

Of this group only VO (6.41 eV – a) bands have been observed in very late type stars at 5469, 5635, 5737, 7372, 7865, 7896, 7939, 7961, 7993 and 10459.

Table 1. *Equivalent widths of VO 5635*

Group	III
M 6	4.6
M 8	7.5

Source: Data are from Fujita *et al.* (1963).

The bands appear at about M 6 and intensify toward M 9. For a detailed description of the behavior with spectral type, see Solf (1978) and Kirkpatrick *et al.* (1993). The presence of this molecule in sunspots is dubious. The behavior of VO in *symbiotic stars* has been studied by Kenyon and Fernandez-Castro (1987).

Group 6B (Cr, Mo, W)

Of this group only CrH (2.85 eV – c) was identified tentatively in sunspot spectra (Engvold *et al.* 1980) and probably in one *CS Mira variable* by Wyckoff and Wehinger (1976). The most important band is located at 8610.

Group 7B (Mn, Tc, Re)

Molecules with atoms of this group have never been detected, probably due to the small dissociation energy of the different compounds.

Group 8 (iron group: Fe, Co, Ni, Ru, Rh, Pd, Os, Ir, Pt)

CoH (1.70 eV – a) and **NiH** (3.07 eV–a) were found in sunspot spectra by Woehl (1971). The identification of CoH is doubtful.
FeH (1.6 eV – a) was detected in the solar photosphere by Carroll *et al.* (1976). Woehl *et al.* (1983) found it also in the sunspot spectrum through bands at 8690 and 9890. It is present in late K and M dwarfs and giants (Nordh *et al.* 1977), with a maximum at about M 5 (Clegg and Lambert 1978, Kirkpatrick *et al.* 1993). The molecule has a negative luminosity effect. Jorgensen (1991) found that it is stronger in *S-type stars* than in M-type stars.
FeO (4.23 eV – c). Woehl (1971) announced the possible identification in the spectra of sunspots.

Group 1B (Cu, Ag, Au)

CuH (2.73 eV – a) has been found in sunspot spectra (the band at 4280) (Hauge 1971) and in a *C-type star* (Wojslaw and Peery 1976).

Group 2B (Zn, Cd, Hg)

It is possible that ZnH (0.85 eV – a) is present in one *C star* (Wojslaw and Peery 1976).

Group 3A (B, Al, Ga, In, Tl)

AlH (3.06 eV – a) is present in sunspot spectra (Sotirovski 1972). It is also present in M giants (Davis 1940), in one *S-type long-period variable* and in one *irregular variable C star* (Branch and Peery 1970). AlH bands are observed in emission in some *Mira variables of the S or MS type* (Herbig 1956). However, not all lines of the band are present, which helps us understand why the identification was difficult. The explanation of the lines present was given by Herbig in terms of the formation of the molecule.
AlO (5.27 eV – a) bands at 4842 and 4866 were observed in M-type giants.

Emissions of AlO were observed by Joy (1926) and identified by Baxandall (1928) in one *Mira variable* at one abnormal low maximum. Kipper and Kipper (1979) find, however, that such emissions are an extremely rare event.
AlCl (5.11 eV – c) was detected by Cernicharo and Guelin (1987) in the circumstellar envelope of the *C-type star* IRC + 10216 through four lines between 87.5 and 160.3 GHz. The identification is confirmed by the observation of a predicted line of $AlCl^{37}$.
AlF was detected by Cernicharo and Guelin (1987) in the circumstellar envelope of the *C-type star* IRC + 10216 through three lines in the 98.8–164.9 GHz domain. The identification should be confirmed.

Group 4A (C, Si, Ge, Sn, Pb)

Two elements of this group – C and Si – are extensively involved in the formation of molecules. Let us recall that, in stars, one finds either that carbon is less abundant than oxygen or that it is more abundant. Since a large percentage of the atoms goes into the formation of CO, it is clear that one can expect carbonated molecules to be abundant in stars where carbon predominates, and oxygen-containing molecules in stars where oxygen predominates. This same consideration is true for the circumstellar envelopes, which have mostly been explored radioastronomically. It needs to be pointed out that, in many such studies, complex molecules have only been identified in one object, namely IRC + 10216, which is taken as 'the' prototype of an object with a C-rich envelope. Despite our rule of not quoting individual objects, we have made an exception here, because we do not know whether IRC + 10216 is really a typical object.

Diatomic molecules involving C

CH (3.46 eV – a) bands are a common feature in stars of the solar type and in cooler stars. In the sun, CH is seen in the spectrum of the disk and of sunspots. The following table provides the equivalent widths of two bands.

Table 1. *Equivalent widths of CH*

	4210		4218
Group	V	III	V
F 5	0.012		0.007
F 6	0.022		0.021
F 7	0.038		0.051
F 8	0.021		0.015
F 9	0.060		0.041
G 0	0.082		0.065
G 2	0.100		
S	0.100		
G 5	0.135		0.100
K 0	0.137		
K 5	0.116		
K 0		0.083	
K 2		0.090	

The band intensity changes slowly from F 5 to K, with a maximum between G 8 and K 0. The bands dissolve at about K 4–5, depending upon the plate factor. The CH bands have a negative luminosity effect, being weak in supergiants and stronger in giants and dwarfs. Since between giants and dwarfs the variation is

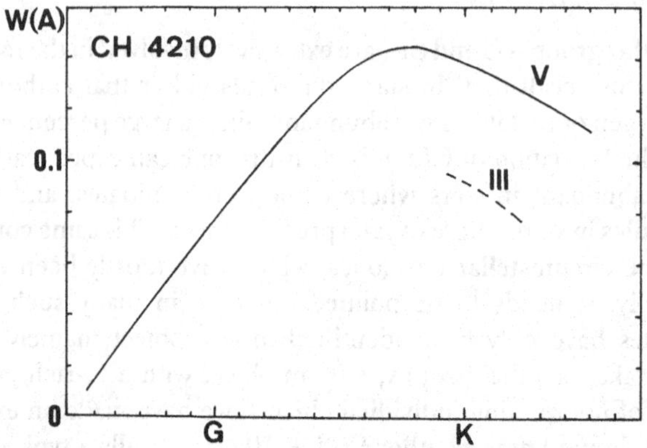

not very large, CH band anomalies are easily visible on spectra having plate factors between 40 and 100 Å mm^{-1}.

CH bands are also present in the ultraviolet at 3144 and 3156 (Eaton *et al.* 1985).

CH bands were observed by Polosukhina and Khoklova (1978) in the spectrum of one *Ap star* of the Cr–Eu–Sr subgroup.

In *C stars*, the CH bands are very strong and have a maximum at about C 4. Since the CH bands are easily visible in G- and K-type stars, it is easy to pick out groups with *CH anomalies*, i.e. with strong or weak CH features. For details see Jaschek and Jaschek (1987a). We retain here only the main characteristics of these groups.

CH stars are G (G 5–K 5) giants with very strong CH bands, weak metals and enhanced lines of Ba II and Sr II. They are also defined as halo carbon stars.

Subgiant CH stars are F- and G-type subgiants and dwarfs with enhanced bands of CH, weakened metals and strong lines of Ba II and Sr II. Some authors also call them hot Ba stars.

Weak G-band stars are G-type giants (G 5 to K 5) with very weak or absent CH bands (see for instance Cottrell and Norris (1978)).

CH bands were also detected in the 3.3 µm region by Hall and Ridgway (1978) in the circumstellar envelope of the C star IRC + 10216.

CH^+ (4.08 eV – b) is present in the solar photospheric spectrum (Grevesse and Sauval 1971, Sauval 1972). All lines are very weak.

C_2 (6.21 eV – a) is present in the sun. It presents several bands, all degraded towards shorter wavelengths, with heads at 4382, 4737, 5165, 6191 and 6535. The Swan bands at 5165, 5635, 6122 and 6191 are very strong in *C-stars*. The bands attain maximum strength in C 5–6 stars and decline rather sharply thereafter. In the infrared the electronic C_2 bands at 1.4–1.767 and 2.5 µm are among the strongest features of *C-type stars*. C_2 bands are also seen in *Ba stars* but not in the hotter Ba stars, which are also called *subgiant CH stars* (Luck and Bond 1992).

The C_2 band at 4737 is seen in emission in some *semiregular* and *long-period variables* (Lloyd Evans 1989). These stars also have very strong SiC_2 bands in absorption.

CN (7.76 eV – a) is present in stars later than G 0. The main bands in the classical region are 3883, 4150 and 4216 (the violet system). These bands intensify toward later stars and have a maximum for dwarfs at early K-type stars. After about K 3 they dissolve into isolated lines. CN bands have a strong luminosity effect (Lindblad 1922). The maximum intensity is reached for supergiants at G 9.

In *C stars*, CN bands are very strong. They have a maximum at about C 5–6 and decline slowly thereafter. The infrared CN bands are only visible in C stars (Hyland 1974). The CN bands are also strong in *CS* stars and in *Ba stars*.

Table 2. *Equivalent widths of CN*

	3883	4150	4216	1.19 μm
Group	V	V		III
G 0		0.125		
G 2	0.096			
G 5		0.143		
K 0	0.117		2.0	
K 3			3.0	
K 5	0.095		3.9	0.013
K 7			4.0	
M 0			4.0	0.007
M 1				0.012
M 2				0.011
M 3				0.021
M 4				0.013
M 5				0.019

Source: Data in the last column are from Smith and Lambert (1985), *W* is in reciprocal centimeters.

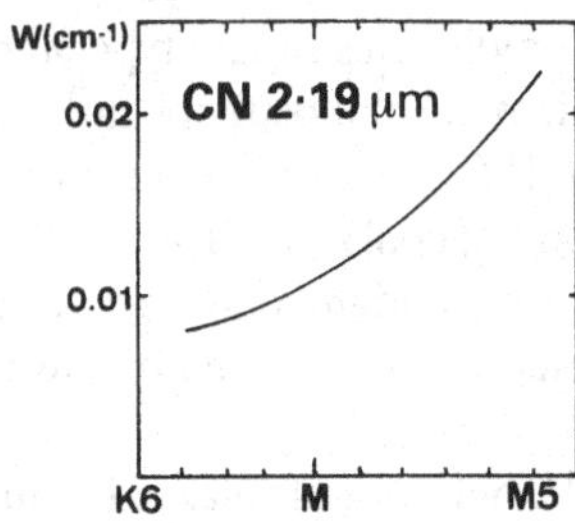

At longer wavelengths the bands at 6910, 6926, 6943, 7850, 7873, 7894, 8100, 9150 and 10900 are also present, all degraded towards longer wavelengths. The infrared electronic transitions in the region 0.75–2.5 μm have been confidently identified in the sun; more than 1000 lines have been observed. Quite generally, in cool stellar spectra, the infrared spectral region is dominated by the red system of the CN bands.

CN was found in *emission* in *R CrB stars* by Herbig (1949).

CN anomalies

Stars with both weak and strong CN bands are known; for details see Jaschek and Jaschek (1987a). At the present time, *CN-weak stars* are denoted simply as metal-weak stars. For instance analysis of a sample of 30 giants has shown that, when CN bands are weakened, all metals are weakened; the abundance anomalies of both C and/or N are marginal. The stars are thus metal-weak giants with perhaps slightly weakened CN bands (Luck 1991a).

CN bands are also weak in the *weak G-band stars* (Cottrell and Norris 1978).

CN-strong stars by analogy are metal-strong stars. They are also called *super-metal-rich stars*. If all elements in a star are strengthened, then C and N are also intensified and as a result the CN bands are enhanced. CN bands have a strong luminosity effect, so that even slight enhancement of the CN bands in giants becomes easily visible. Because of this the star was called 'CN-strong' rather than 'metal-strong'. Detailed analysis has shown that the metals are intensified by factors of the order of 2–3. The CN strength can also be measured photometrically, for instance with narrow filters centered upon a prominent band. For more details, see Golay (1974). Langer *et al.* (1992) have made such measurements on globular cluster giants, for instance.

CN bands are very strong in *C stars*, moderately strong in *S-type stars* (Keenan 1957) and of intermediate strength in *SC stars* (Wallerstein 1989) and in *Ba stars*. In C-type stars the CN bands have a maximum at about C 5–6 and decline slowly thereafter.

CN can also be observed at radio wavelengths, where a large number of transitions exist, for instance at 226.6 and 340.0 GHz. These transitions have been observed in the circumstellar envelope of a *C star* (Jefferts *et al.* 1970, Avery *et al.* 1992, Nyman 1992) but also in an *M-type Mira star* with an oxygen-rich envelope, where its presence was rather unexpected (Olofsson *et al.* 1991).

CO (11.09 eV – a) is the most stable of all diatomic molecules and as can be expected it is also one of the most abundant molecules in late type stars.

The molecule was detected in the sun, both in the photosphere and in sunspots (Goldberg and Mueller 1953, Goldberg *et al.* 1965). This molecule produces the most prominent features in the infrared spectra of cool stars (Tsuji 1986a, 1991). The bands that have been observed are the fundamental first and second overtone bands at 1.6–2.3 and 4.6 μm, which are transitions of the rotation–vibration type.

Smith and Lambert (1985) have studied the behavior of weaker features of the second overtone band and find that it increases toward later types, doubling its intensity between K 5 and M 8 (see also Mould 1978 and Kleinmann and Hall (1986)).

Table 1. *Equivalent widths of CO 6167 cm^{-1} (in reciprocal centimeters)*

Group	III
K 5	0.058
M 0	0.066
M 1	0.092
M 2	0.073
M 3	0.085
M 4	0.083
M 5	0.104

Source: Data are from Smith and Lambert (1985).

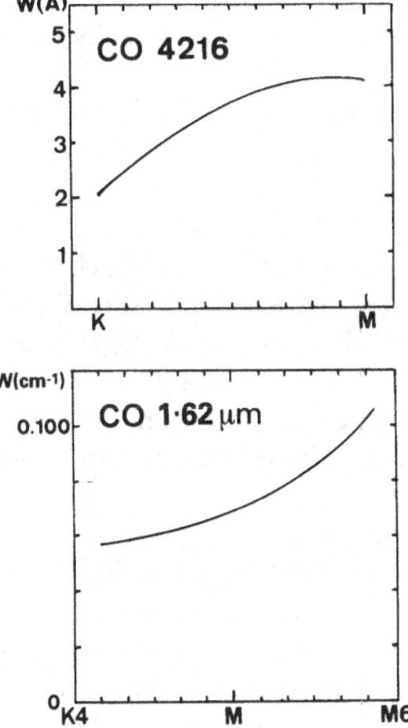

The bands show a positive luminosity effect (Hyland 1974, Wing 1979). The strength of the CO infrared bands has also been measured photoelectrically by Baldwin *et al.* (1973) by means of intermediate-band photometry. Their conclusions confirm the increase in strength toward later types and the positive luminosity effect mentioned above.

The bands of CO (as well as those of all carbonated molecules) are very strong in *C stars*, but the maximum strength is reached in *S-type stars* (Hyland 1974).

Ayres *et al.* (1981b) showed also that CO ultraviolet bands are probably responsible for a number of faint emissions (at 1340, 1380, 1500 and 1600 Å), which appear in the ultraviolet solar spectrum and in that of a K 2 III star. The origin of these bands is due to fluorescence excited by chromospheric emissions of O I, C I and H I. These emissions were also studied by Jordan (1988). The *supernova* 1987A has also shown CO in emission (Arnett *et al.* 1989), as have *novae* of the C–N–O type (Rawlings 1992).

CO bands at 2.294 μm appear in some *high-luminosity objects* (McGregor *et al.* 1988).

CO vibrational transitions were observed in the circumstellar envelope of the *OH maser* by Fix and Cobb (1987) in the 4.6 μm spectral region.

CO radio lines in emission were detected by Wilson *et al.* (1971) at 1.282 GHz. Knapp and Morris (1985) performed a study at 1.153 GHz and found the lines in *Mira variables* of types M, *C* or *S* as well as in some supergiants, and *OH/IR objects*. Since the emissions are strongly related to the infrared radiation of an object, a search of radio CO emitters was made in the luminous infrared objects observed with IRAS; Zuckerman *et al.* (1986) observing at 115.3 and 230.5 GHz detected in this way a large number of new sources. Similar results were obtained by Nguyen-Q-Rieu *et al.* (1987).

The isotopic $C^{13}O$ lines were also observed in *C-type stars* (Nyman 1992, Kahane *et al.* 1992).

Polyatomic molecules involving C

HCN (13.15 eV – b) was found by Ridgway *et al.* (1978) and Goebel *et al.* (1978, 1981) in *C-type Mira variables*. Ridgway *et al.* (1978) studied especially the feature at 3.1 μm, composed of HCN bands of the vibration–rotation type, and Goebel *et al.* (1981) concentrated on the bands at 1.85–3.56 and 4.8 μm. Fujita (1990) found that *hydrogen-deficient stars* show particularly strong lines of this molecule. HCN is present in emission in many C stars. Jura (1991) found that the emission has an erratic behavior for a given spectral type.

This molecule increases in intensity toward the coolest stars.

HCN was also observed radioastronomically in the circumstellar envelopes of *C-type stars* (Nguyen-Q-Rieu *et al.* 1987). Afterwards it was also observed in *M-type Mira variables*, which was quite unexpected since it is a molecule containing C (Lindqvist *et al.* 1988, Nercessian *et al.* 1989, Olofsson *et al.* 1991). In fact many of the M stars showing HCN also have OH, H_2O or SiO maser lines, which classes them unambiguously in the oxygen-rich group. There is a line at 88.6 GHz (see for instance Morris *et al.* (1987)). Lucas *et al.* (1988) have made a systematic search for the transition at 89.09 GHz, which is a vibrationally excited HCN maser line in the envelopes of C-type stars. They found that about

10% of sources showed this type of emission. Temporal variations occur on time scales of the order of one month. The HCN maser is the first maser to be found in carbon-rich envelopes.

HNC was also observed through a transition at 90.6 MHz by Morris *et al.* (1987) and Nercessian *et al.* (1989). It is found in the same type of circumstellar envelopes as HCN, but its occurrence is less frequent.

HC_2N was detected radioastronomically at 109.77 and 131.8 GHz in the circumstellar envelope of the *C-type star* IRC + 10216 (Guelin and Cernicharo 1991).

HC_3N was detected in the circumstellar envelopes of *C-type objects* (Henkel *et al.* 1985, Lucas *et al.* 1986, Nyman 1992).

The molecules with the formula $H(C_2n)CN$ with $n=1$, 2, 3, 4, 5 and 6 are called cyanopolyynes.

HC_5N was detected radioastronomically through the line at 23.96 GHz in the circumstellar envelope of the *C-type object* IRC + 10216 (Winnewisser and Walmsley 1978).

HC_7N was observed radioastronomically in the circumstellar envelope of the *C-type object* IRC + 10216 (Winnewisser and Walmsley 1978) and in one *oxygen-rich object* by Nguyen-Q-Rieu *et al.* (1984). There is a line at 23.69 GHz.

HC_9N was observed radioastronomically in the circumstellar envelope of the C-type object IRC + 10216 at 14.52 and 10.46 GHz (Broten *et al.* 1978).

$HC_{11}N$ was detected radioastronomically in the circumstellar envelope of the *C-type object* IRC + 10216 at 23.70 GHz (Bell *et al.* 1982).

$HC_{13}N$ has been observed in the circumstellar envelope of two *C-type stars* (Avery *et al.* 1992, Nyman 1992) and of one *oxygen-rich object* (Morris *et al.* 1987).

HCO^+ was discovered radioastronomically at 89.188 GHz in the spectrum of the circumstellar envelope of one *oxygen-rich object* (Morris *et al.* 1987).

H_2CO was detected radioastronomically in the circumstellar envelopes of several *oxygen-rich* stars by Lindqvist *et al.* (1992).

CH_4 was discovered by Hall and Ridgway (1978) in the spectrum of the circumstellar envelope of the *C-type star* IRC + 10126. Clegg *et al.* (1982) made a study of the vibration–rotation lines near 3 μm and confirmed the identification. It was also found in other C-type stars by Ridgway and Hall (1980). Guelin *et al.* (1987a) found it radioastronomically.

C_2H_2 (16.8 eV – b) was found by Ridgway *et al.* (1978) on the basis of the 3.1 μm feature, and by Bregman *et al.* (1978) and Goebel *et al.* (1978) on the basis of the 3.9 μm feature in *C-type Mira variables*. The identification was confirmed by Goebel *et al.* (1981) on the basis of bands at 1.04, 1.53, 1.85, 2.8, 3.56 and 4.8 μm. Some of these bands coincide with HCN bands. The bands increase in strength toward later spectral types.

C_3H_2 has been observed radioastronomically at 85.34 GHz in the circumstellar envelope of a *C-type star* (Cernicharo *et al.* 1991a, Nyman 1992).

C_4H_2 has been detected radioastronomically through transitions between 80 and

135 GHz in the spectrum of the circumstellar envelope of the *C-type object* IRC + 10216 (Cernicharo *et al.* 1991b).

Molecules obeying the formula HC_nH with n=2, 3, 4 . . . belong to the *closed shell families*.

C_2H (12.0 eV – b) was found in at least one late *C-type star* (Goebel *et al.* 1983) through an electronic absorption band at 2.9 μm.

It has also been detected in the circumstellar envelope of C-type stars (Lucas *et al.* 1986, Nyman 1992) through lines at 87.3 GHz (see also Avery *et al.* (1992)).

C_3H has been observed radioastronomically at 98.0 GHz in the circumstellar envelope of *C-type stars* (Nyman 1992).

C_4H was detected through the lines at 95.15 GHz in the circumstellar envelopes of *C-type stars* (Lucas *et al.* 1986, Nyman 1992). It was further studied by Guelin *et al.* (1987a) and Avery *et al.* (1992).

C_5H was detected by Cernicharo *et al.* (1986a, 1986b) in the circumstellar envelope of the *C-type star* IRC + 10216 through rotational transitions at 74.5, 84.1, 88.9 and 98.5 GHz.

C_6H was detected by Suzuki *et al.* (1986), Saito *et al.* (1987) and Geulin *et al.* (1987b), who observed it in the spectrum of the circumstellar envelope of the *C-type star* IRC + 10216 through transitions at 23.6, 40.2 and 43.0 GHz.

The molecules C_nH, with n=2, 3, 4, 5 and 6 are called *acetylenic radicals*.

C_3N was detected in the circumstellar envelope of a *C-type object* through lines at 89.06 GHz (Guelin and Thaddeus 1977, Lucas *et al.* 1986, Henkel *et al.* 1985, Nyman 1992).

C_2H_4 was detected by Betz (1981) and confirmed by Goldhaber *et al.* (1987) on the basis of several lines around 10.53 μm in the circumstellar envelope of the *C-type star* IRC + 10216.

C_2H_3N was detected radioastronomically in the circumstellar envelope of the *C-type star* IRC + 10216 (Guelin and Cernicharo 1991).

C_3 (14.00 eV – b) was found by Goebel *et al.* (1978) in one *C-type star*. The molecule is believed to contribute to the violet absorption in *cool C stars*. The molecule was also detected by Hinkle *et al.* (1988) in the circumstellar envelope of the *C-type star* IRC + 10216, through transitions in the 4.9 μm region.

C_5 has been detected through transitions at 4.6 μm by Bernath *et al.* (1989) in the spectrum of the circumstellar envelope of the *C-type object* IRC + 10 216.

Molecules involving Si

In O-rich atmospheres, Si goes mostly into SiO, whereas in C-rich atmospheres, Si goes mostly into SiC.

SiH (3.06 eV – a) is present in the spectrum of the solar photosphere (Sauval 1969) with some very faint bands (W<0.02). It is also present in the spectrum of sunspots, with bands at 4142, 4128 and 4184 (Babcock 1945). Davis (1940) states that the molecule has its maximum at M 1 III. It is seen in emission in *Mira variables* of types *M* and *S* (Merrill 1955).

SiH$^+$ (3.17 eV – b) is present in the spectrum of the solar photosphere (Sauval 1972).
SiH$_4$ has been detected in the circumstellar envelope of the *C-type star* IRC + 10 216 through three transitions at 11.1 μm (Schrey *et al.* 1987) and by Goldhaber and Betz (1987) through a series of transitions in the region 26.56–27.76 GHz.
SiC (4.50 eV – c) has been detected radioastronomically through rotational lines at 157.5 and 236.3 GHz in the spectrum of the circumstellar envelope of the *C-type object* IRC + 10216 (Cernicharo *et al.* 1989).
SiC grains (4.64 eV) have a strong feature at 11.15 μm, which appears in emission in *C-type stars* and is due to circumstellar dust grains. These bands are also seen in *nova shells* (Rawlings 1992).
SiC$_2$ (13.0 eV – b) bands were detected by Wyckoff and Wehinger (1976) at 4866 and 4977 in one *SC star*. The bands can be very strong in some *semiregular or Mira variables* of low temperature (Keenan, private communication).

The molecule has also been detected radioastronomically by Thaddeus *et al.* (1984), Lucas *et al.* (1986) and Nyman (1992) in the circumstellar envelopes of *C-type stars* through lines at 95.58 and 115.3 GHz (see also Avery *et al.* (1992)).

Nguyen-Q-Rieu *et al.* (1987) detected this molecule in several more *C-type stars* selected from IRAS photometry.
SiC^{13}C has been detected in the spectrum of the circumstellar envelope of the *C-type star* IRC + 10216 by Cernicharo *et al.* (1991c). The authors used the molecule for a study of the C^{12}/C^{13} ratio.
SiC$_4$ has been detected radioastronomically at 42.9 GHz in the spectrum of the circumstellar envelope of the *C-type star* IRC + 10216 by Ohishi *et al.* (1989).
SiN (4.54 eV – c) has been detected radioastronomically through five transitions in the millimeter region in the circumstellar envelope of the *C-type object* IRC + 10216 (Turner 1992).
SiO (8.26 eV – a) is present in the solar photosphere (Grevesse and Sauval 1991) and in sunspots (Sinha 1991). Rinsland and Wing (1982) analyzed the behavior of the bands at 4 μm (4.003–4.043 μm) in a number of late type stars.

Table 2. *Equivalent widths* W *(cm^{-1}) of the 4.0 μm feature of SiO*

Group	III,I
K 2	2
K 4	10
M 0	18
M 2	26
M 4	34
M 6	41

Source: Data from Rinsland and Wing (1982).

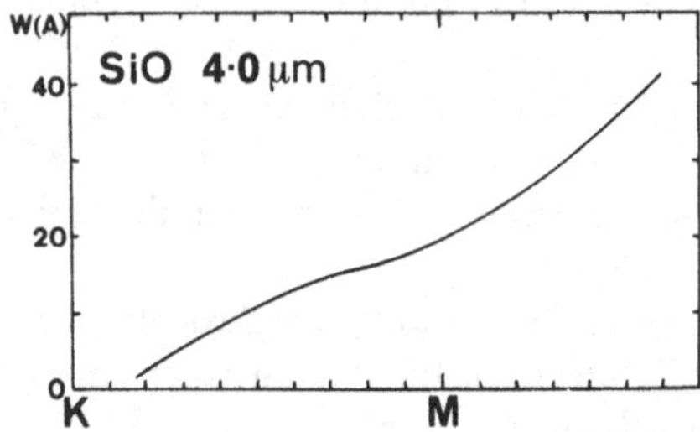

No luminosity effect is present, but measurements of supergiants show a large dispersion, which could be due to the existence of an SiO emission. Rinsland and Wing also propose that the SiO band at 4.028 μm due to $Si^{29}O^{16}$ could be used for determination of the isotopic ratio Si^{28}/Si^{29}.

No SiO was found in *C-* and *SC-type stars*.

SiO appears in emission in the spectrum of the *supernova* 1987A (Arnett *et al.* 1989) and in *novae* (Rawlings 1992).

Besides the SiO (thermal) emission lines one observes also SiO masers at 43 and 86 GHz. SiO is the only molecule that produces maser radiation in its excited rather than in its ground vibrational state.

The SiO masers were observed in *M-type Mira variables*, *semiregulars* and *irregulars* as well as in M-type supergiants and giants. In Mira variables the maser line intensity is well correlated with the near infrared intensity of the star and, as a consequence, with the visual light curve (Hjalmarson and Olofsson 1979, Bujarrabal *et al.* 1987), but time lags do exist between the visual light curve and the others. Bujarrabal *et al.* (1987) found that, in *C-type stars*, the emission intensity is about 100 times less than that in M-stars. The emission in *S-type stars* is of intermediate strength. Some SiO masers are associated with *OH/IR stars* (Gomez *et al.* 1991). Some 200 masers are known at present (Benson *et al.* 1990).

Both $Si^{29}O$ and $Si^{30}O$ have been observed radioastronomically in the circumstellar envelopes of *C-type objects* (Johansson *et al.* 1984, Nyman 1992) and in *evolved stars* (Alcolea and Bujarrabal 1992).

SiO grains emit at 9.8 μm. Emissions at this wavelength were found in the spectra of M and *S giants* (i.e. in *O-rich stars*), where they are produced in circumstellar dust grains. Usually a small fraction (1%) of the SiO is in the gaseous phase. Curiously, SiO emissions also appear in a number of *C stars* (Little-Marenin 1986, Barnbaum *et al.* 1991, Chen Peisheng *et al.* 1992, Lorenz Martins and Codina 1992). Willems and de Jong (1986) proposed that all stars showing this emission are *J stars*. Later studies by Lambert *et al.* (1990) and Lloyd Evans (1991) have confirmed this hypothesis.

In *C-rich stars*, Si is no longer locked up in SiO and hence other Si compounds may be formed. So far two compounds, SiH and SiC_2, have been detected.

SiS (6.41 eV – c) was detected radioastronomically by Ziurys *et al.* (1984) through a transition at 87.55 GHz in the circumstellar envelope of the *C-type star* IRC +

10216. Henkel *et al.* (1985) and Olofsson (1988) observed this molecule in the circumstellar envelopes of several *C-type stars*, but also in the O-rich envelope of an *M-type Mira star* (Lindqvist *et al.* 1988).

In the *C-type object* IRC + 10216 the lines of SiS are of variable strength, its intensity being correlated with the infrared variability of the star (Olofsson 1988).

Ziurys *et al.* (1984) examined the lines corresponding to $Si^{28}S$ and $Si^{30}S$ and found a Si^{28}/Si^{30} ratio compatible with the solar system ratio.

Molecules involving Ge and Sn

GeH (3.32 eV – c) and **SnH** (2.73 eV – c) were detected by Wojslaw and Peery (1976) in one *C-type star.*

Group 5A (N, P, As, Sb, Bi)

NH (3.47 eV – a) is present in the sun, both in the photosphere (Grevesse and Sauval 1973) and in sunspots (Woehl 1971), with bands at 3240, 3253, 3360 and 3370 (Babcock 1945). Ridgway *et al.* (1984) detected infrared vibration–rotation bands of NH in *O-rich giants* and *supergiants*. A band at 2850 cm^{-1} is also present in M and *early C stars* (Gustafsson 1989).

$\mathbf{NH_3}$ has been detected by Betz *et al.* (1979) in the circumstellar envelope of the *C-type object* IRC + 10216 through the vibration–rotation band at 10 μm. This molecule was observed at 23.69 GHz in this same object by Kwok *et al.* (1981). Betz and McLaren (1980) and Morris *et al.* (1987) observed the molecule through transitions at 23.69 and 23.72 GHz in *oxygen-rich objects*.

CP (5.31 eV – c) has been detected radioastronomically at 238.8, 238.3 and 143.4 GHz in the circumstellar envelope of the *C-type star* IRC + 10216 (Guelin *et al.* 1990).

PN (6.39 eV – c) was searched for radioastronomically at 234.93 and 93.98 GHz in circumstellar envelopes and detected in those of some *oxygen-rich objects* (Turner *et al.* 1990).

Group 6A (O, S, Se, Te, Po)

If the number of atoms of O is larger than the number of atoms of C, then oxygen-containing molecules are formed. If, furthermore, the number of atoms of O is larger than the number of atoms of C plus that of Si, then one finds a large amount of CO and SiO molecules. Of the remaining O atoms, a large part goes into the formation of OH and H_2O.

Molecules involving O

We start by mentioning two molecules, which do not exist in stars, but are listed here because of the interferences that they produce in ground-based observations.

$\mathbf{O_2}$ (5.12 eV – a) occurs in the terrestrial atmosphere, with bands at 6277, 6867, 7594, 7688 and 8645.

O_3 (6.16 eV – a) is present in the terrestrial atmosphere, where it blocks the ultraviolet wavelength region 2975–3400.

OH (4.39 eV – a) is present in the sun (both in the ultraviolet and in the infrared spectrum), and in the solar disk, and also in sunspots and K and M giants. The band intensities increase toward later types (Mould 1978, Smith and Lambert 1985). The most important ultraviolet bands are those at 3063, 3078, 3428 and 3432 (Babcock 1945). In other regions one can add the band at 1.20 μm (Kirkpatrick *et al.* 1993), which is prominent in late M-type dwarfs, and those in the region 1.8–4.0 μm (Smith and Lambert 1985).

Table 1. *Equivalent widths* W *(cm^{-1}) of OH 5842 cm^{-1}*

Group	III
K 5	0.091
M 0	0.100
M 1	0.111
M 2	0.126
M 3	0.119
M 4	0.133
M 5	0.147
M 6	0.166

Source: Data are from Smith and Lambert (1985).

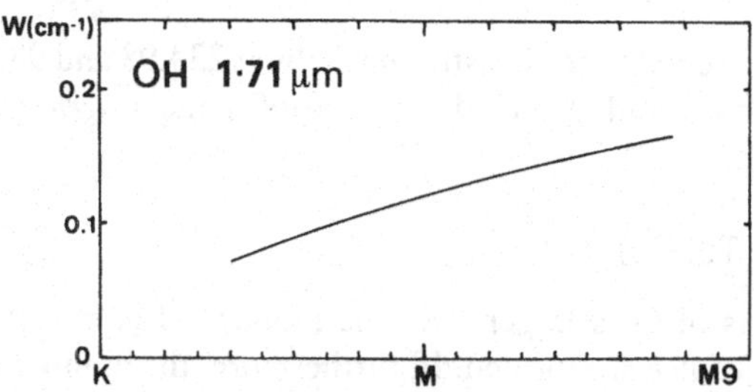

OH masers were detected by Weinreb *et al.* (1963) through lines at 1.667, 1.612 and 1.665 GHz. The OH lines originate from the ground vibrational state and are pumped by infrared radiation, as is shown by the variations in phase of the infrared and OH emission. These maser lines are present in *M- and C-type Mira variables*, in *semiregular variables* and in *M-type supergiants*. Systematic surveys have permitted detection of more than 700 OH masers, collected in the catalog of Benson *et al.* (1990).

Usually these masers are divided into two types. Type II, called radio-luminous *OH/IR stars*, have a very bright line at 1.612 GHz. Type I maser stars have emission 100 times weaker at 1.612 GHz, but stronger emission in the main lines. The latter group comprises the classical *Mira variables*. Many of the type II objects have pulsation periods of very large amplitude (600–2000 days) (Jones 1987).

Most of the masers are variable sources whose intensity is correlated with the intensity of the infrared spectrum of the source. In *Mira variables* there exists thus a general correlation with the light curve (see for instance Le Squeren and Sivagnanam (1985)).

Furthermore, type II masers usually have excess radiation at 10 μm and type I masers have H_2O masers associated with them.

Some of the objects are associated with infrared objects that have no visible counterpart. These are called the '*OH/IR objects*' (*or sources*). Usually the infrared source exhibits silicate features in its spectrum at 10 and 20 μm. Because of the infrared association, the masers can be selected from infrared surveys, like the one carried out by the infrared satellite IRAS (te Lintel Hekkert *et al.* 1991).

H_2O (9.51 eV – b) was found in the spectra of sunspots (Woehl 1971). Hall (1974) has detected H_2O in the sun, through bands near 0.9, 1.9 and 2.7 μm.

H_2O is found in very cool M giants (Ridgway *et al.* 1984) and in M-type dwarfs. The band at 1.35 μm becomes the most prominent feature of the spectrum at type M 9. Another important band is that at 1.14 μm (Kirkpatrick *et al.* 1993).

The strength of the H_2O bands has also been measured photometrically with an intermediate band system in the infrared by Baldwin *et al.* (1973). These authors find an increase toward the latest stars and a negative luminosity effect. The bands are visible from about M 0V and M 3 I onwards and become really strong only after M 7 (Spinrad and Wing 1969).

H_2O bands are very strong at maximum light of *long-period variable stars* of type M. In *S-type stars* the bands are generally absent (Hyland 1974).

H_2O masers were detected at 22.2 GHz. Such masers were observed in M-type *Mira variables* and in *semiregular variables* (Schwartz and Barrett 1970). They are of variable intensity, correlated with stellar pulsations (Little-Marenin *et al.* 1991). H_2O masers are also observed in a large fraction of the *OH/IR objects* (Engels 1987, Gomez *et al.* 1991).

About 400 H_2O masers are known at present (Benson *et al.* 1990). Since they were thought to be associated with *O-rich objects*, it was quite unexpected that they are also present in *C stars* with O-rich envelopes (Benson and Little-Marenin 1987).

Menten *et al.* (1990) have also observed H_2O masers at 321 and 325 GHz. They find that the 321 and 325 lines are observed in one third of the stars that show the 22 GHz maser line. Apparently the 321 and 325 lines are produced nearer to the star than is the 22 GHz line.

Molecules involving S

HS (3.55 eV – c) is probably present in the spectrum of the solar photosphere (Sauval *et al.* 1980).

H_2S was detected radioastronomically by Ukita and Morris (1983) on the basis of the line at 1.66 GHz in the circumstellar envelope of an oxygen-rich star and Omont *et al.* (1993) found it in several more *oxygen-rich* circumstellar envelopes. It is also observed faintly in the *C-type object* IRC + 10216 (Cernicharo *et al.* 1987).

CS (7.35 eV – a) has been detected at 3.95 μm by Bregman *et al.* (1978) and by Ridgway *et al.* (1984) in the spectra of *C-type stars*.

CS was also observed radioastronomically through rotational transitions in the circumstellar envelopes of several *oxygen-rich stars* (Morris *et al.* 1987) and in *M-type Mira variables* (Lindqvist *et al.* 1988, Olofsson *et al.* 1991). The molecule is also present in the circumstellar envelope of *C-type stars* (Wilson *et al.* 1971, Nyman 1992, Henkel *et al.* 1985, Avery *et al.* 1992).*

C_2S was observed radioastronomically through several transitions (for instance those at 81.50, 93.87 and 131.55 GHz in the circumstellar envelope of the *C-type object* IRC + 10216 (Cernicharo *et al.* 1987)).

C_3S was observed radioastronomically at 109.82 GHz (as well as in other lines) in the circumstellar envelope of the *C-type object* IRC + 10216 by Cernicharo *et al.* (1987).

SO (5.36 eV – a) has been detected radioastronomically by Guilloteau *et al.* (1986) in the circumstellar envelope of an M-type star. Sahai and Howe (1987) have shown that the 219.9 GHz line is fairly frequent in circumstellar *oxygen-rich envelopes* (see also Omont *et al.* (1993)).

SO_2 (11.12 eV – c) has been detected by Lucas *et al.* (1986) through the line at 104.03 GHz in the circumstellar envelope of an M-type star and by Guilloteau *et al.* (1986) in other M-type stars and in OH masers. It is possible that the abundance of the molecule varies from one object to another (see Omont *et al.* (1993)).

OCS has been detected radioastronomically through the 97.30 GHz line in one *oxygen-rich* circumstellar envelope by Morris *et al.* (1987).

YS (5.47 eV – d) has been identified by Lambert and Clegg (1980) in one *S-type star* through bands at 7503 and 7509 Å.

Group 7A (F, Cl, Br, I, At)

HF (5.87 eV – a) has been found in the spectrum of sunspots, through the rotation–vibration lines of HF around 2.3 μm (Hall and Noyes 1969). Later it was also observed in stars.

Table 1. *Equivalent widths* W *(2,278 μm) of HF*

K 5 III	7×10^{-3} cm^{-2}
M 0 III	14×10^{-3} cm^{-2}
M 4 III	38×10^{-3} cm^{-2}

* Note added in proof. Bell *et al.* (1993) *Ap. J.* **417**, L37 have also detected C_3S and C_5S in IRC 10216.

In *S*, *SC* and *C stars*, HF is overabundant by up to 0.6 dex (Jorissen *et al.* 1992). **HCl** (4.43 eV – a) has been detected in the spectrum of sunspots (Hall and Noyes 1972) and later in the spectrum of one *S-type star* (Ridgway *et al.* 1984). These authors were surprised that the HCl feature is strong, in view of the fact that Cl is a minor constituent in the sun. Two bands of HCl^{35} and HCl^{37} can be used for isotope analysis. A preliminary evaluation suggests that the abundance ratios are consistent with the solar system ratio.

1.3 Summary

We have summarized the general behavior of the molecules in stars of different spectral types in table 1, which follows.

The table provides the spectral types in which the molecule is seen. We have listed separately G-type stars and the solar photosphere (Ph) and K-type stars and the sunspots (Sp) because of the problems of detection mentioned in the introduction to this chapter.

The first column provides the formula of the molecules. The following columns give the spectral types in which the molecule is present. Presences are signaled with a ×, insecure detections are given within parentheses. The last column provides the designation of the group (of elements) to which the molecule pertains.

Table 2. *Spectral types in which molecules are seen*

Molecule	G	Ph	K	Sp	M	C	S	CS	Group
H_2		×		×	×	×	×		
NaCl						×			1A
KCl						×			1A
MgH		×	×	×	×				2A
MgO				(×)					2A
MgNC						×			2A
CaH				×	×				2A
CaOH			×		×				2A
CaCl					×	×		×	2A
ScO				×	×				3B
LaO							×		3B
YO					×		×		3B
CeO							×		3B
ZrO		(×)		×	×		×	×	4B
ZrS							×		4B
TiO			×	×	×		×		4B
TiH				×	×				4B
TiCl					(×)				4B
TiS							×		4B

Table 2. (*cont.*)

Molecule	G	Ph	K	Sp	M	C	S	CS	Group
VO				(×)	×				5B
CrH				(×)				×	6B
CoH				(×)					8
NiH				×					8
FeH		×	×	×	×		×		8
FeO				(×)					8
CuH				×		×			1B
ZnH						(×)			2B
AlH				×	×	×	×		3A
AlO					×				3A
AlCl						×			3A
AlF						(×)			3A
CH	×	×	×	×		×			4A
CH^+		×							4A
C_2		×		×		×			4A
CN	×	×	×	×	×	×	×	×	4A
CO		×	×	×	×	×	×		4A
HCN					×	×			4A
HNC					×	×			4A
HC_2N						×			4A
HC_3N						×			4A
HC_5N						×			4A
HC_7N					×	×			4A
HC_9N						×			4A
$HC_{11}N$						×			4A
$HC_{13}N$					×	×			4A
HCO^+					×				4A
H_2CO					×				4A
CH_4						×			4A
C_2H_2						×			4A
C_3H_2						×			4A
C_4H_2						×			4A
C_2H						×			4A
C_3H						×			4A
C_4H						×			4A
C_5H						×			4A
C_6H						×			4A
C_3N						×			4A
C_2H_4						×			4A
C_2H_3N						×			4A

Table 2. (*cont.*)

Molecule	G	Ph	K	Sp	M	C	S	CS	Group
C_3						×			4A
C_5						×			4A
SiH		×		×	×				4A
SiH^+		×							4A
SiH_4						×			4A
SiC						×			4A
SiC_2					×			×	4A
SiC_4						×			4A
SiN						×			4A
SiO		×	×	×	×				4A
SiS					×	×			4A
GeH						×			4A
SnH						×			4A
NH		×		×	×	×			5A
NH_3					×	×			5A
CP						×			5A
PN					×				5A
OH		×	×	×	×	×			6A
H_2O				×	×	×			6A
HS		×							6A
H_2S					×	×			6A
CS					×	×			6A
C_2S						×			6A
C_3S						×			6A
SO					×				6A
SO_2					×				6A
OCS					×				6A
YS							×		6A
HF			×	×	×	×	×	×	7A
HCl				×			×		7A

2 THE BEHAVIOR OF GROUPS OF ELEMENTS IN THE STARS

This chapter contains two sections, which deal with metals and rare earth elements respectively. The behavior of both groups of elements presents certain general characteristics, which would be completely lost if they were described in the sections on individual elements.

2.1 The behavior of metals

The term 'metals' is used by astronomers in an ill-defined way. It may refer to 'all elements other than He' or to 'elements with $Z>8$' or to 'elements with $Z>21$'. One should always try to ascertain what a given author means when he uses the word 'metals'. If the term 'metals' is ill-defined then the same happens with 'metal abundances'. The abundances of metals are usually defined with respect to hydrogen, in the sense $N(\mathrm{m})/N(\mathrm{H})$, normalized with respect to the same ratio in the sun. More often one uses the decimal logarithm of this quantity in order to have small numbers without exponents. So for instance [Fe/H]$=-0.6$ means that the abundance of iron with respect to hydrogen is four times less than in the sun. Usually this is abbreviated to -0.6 dex.

Another possible definition, due to Kudritzki (1987), is to consider the ratio $N(\mathrm{m})/N(\mathrm{H})+4N(\mathrm{He})$. This has the advantage that the denominator remains constant if the star is in its hydrogen burning phase.

The two preceding definitions are clear and unambiguous. However, very often the discussions based upon them become ambiguous because authors speak of 'metals' when in reality they are referring to the abundance of one individual element. One discovers for instance that the author defined the iron abundance, but speaks of metal abundance, as if the two terms were completely equivalent. All this stems from careless use of the term 'metals'.

In the 1950s it was assumed that metallicity was a simple function of age. Young stars have normal line intensities and old stars have weak metallic lines. For some years this was implicit in all work on stellar populations. More recently it has become clear that the different elements grouped under the name 'metals' do not behave in a strictly similar way. This implies that if, for instance, Fe weakens by a factor of ten, then we must not necessarily expect that (for example) Cu weakens by the same factor. It must be said that detection of a differential behavior is difficult in practice. The difficulty comes from the fact that, in normal stars, most of the metal lines are not very strong and this implies that they are even weaker in the metal-weak stars. This makes it very difficult to find differences between sets of weak lines. Let it be remembered that, with the lines of medium or weak intensity that are used for stellar atmosphere analysis, an underabundance of metals by a factor of ten implies a weakening of the equivalent widths by a factor of about three. Correspondingly a weakening of equivalent widths by a factor of ten implies an underabundance by a factor of 100. This means in practice that analysis of metal-weak stars is based upon the measurement of very small equivalent widths. Obviously such measurements should be done only on the best material, having both a high signal/noise ratio and high resolving power. For a discussion of the systematic effects present, see Cayrel de Strobel (1983).

Leaving aside these practical difficulties, let us have a look at the terminology used. Although no strict definition is accepted by all astronomers, over the years a general terminology for metal-weak stars has emerged. Usually one speaks of

'*young*' or '*young disk population*' with iron abundances differing by not more than a factor of two from that of the sun,
'*intermediate*' or '*old disk population*' with iron underabundances of 2–4 times less than that of the sun,
'*old*' or '*halo population*' with iron underabundances of 6–100 times less than that of the sun, and
'*extreme halo*' or '*population III*' with iron underabundances of larger than a factor of 100 (and up to 30000) of that of the sun.

In many cases authors often substitute the precise term 'Fe abundance' by the ill-defined term 'metal abundance'.

So far the most extreme metal-deficient objects found have an Fe/H ratio of 10^{-4} of that of the sun (Molaro and Bonifacio 1990).

Among the young stars, in which iron is supposedly normal, there exist also some objects in which iron is overabundant. These objects are usually called '*super-metal-rich*', following a designation introduced by Spinrad and Taylor (1969). By the way, the existence of these stars generated a long controversy, reviewed by Taylor (1982). At the present time it seems that a (relatively small) number of objects does exist that have an overabundance of a factor of two (and up to four times) with respect to the sun (Cayrel de Strobel 1991). Rich (1988) has shown that a large number of such objects exists among the *K-type giants* in the bulge of our galaxy.

Since the terminology of 'metal-weakness' is the more frequently used, we have preferred not to substitute 'Fe-weakness'. This would have been clearer – or at least less equivocal – but would have introduced more confusion regarding what authors originally meant.

Since the metal abundance – whatever the word 'metal' means – of a star can be obtained only through a lengthy series of measurements (and reductions) and since on the other hand one would like to get a large number of 'metallicities' of stars for studies of galactic evolution, it is clear that several shortcuts have been sought over the years.

Such shortcuts use either photometric or spectrophotometric techniques. In photometric measurements one defines a 'metallicity index', which can be imagined as being a measure of the absorption produced by the spectral lines of all elements considered together. Since this quantity can be measured with a photometric precision of ±1%, it is clear that one has a very precise measure of the integrated effect. The next step is then to calibrate this index with something determined by reference to the stellar atmosphere – for instance the abundance of iron. This procedure is equivalent to assuming that there exists a strict proportionality between the strength of iron and that of the other elements that produce line absorptions. This is called a 'lockstep variation'. As we have said already, this is not the case, and in consequence the result cannot be confidently relied upon.

Typically the error of an abundance determination is of the order of 0.3 dex, whereas the error in determination of the total effect of the lines of all elements is of the order of ±1%, as remarked above. One is thus in the situation that the photometric metallicities are apparently more precise than the spectroscopic determinations. This is, however, only true if the variation is lockstep – if not, one measures very accurately an ill-defined quantity.

A similar procedure can be envisaged with spectrophotometry. In such a case one takes a strong line and measures its intensity in a sample of stars. Then one correlates the strength of this feature (of one particular element) with something determined spectroscopically – for instance the metallicity, or the iron abundance. Such a procedure is adopted for instance by Beers *et al.* (1990), who use the Ca II line at 3933 and find, over a large range of [Fe/H] (between −1 and −4.5 dex), a close correlation between the two quantities. This again is equivalent to assuming a constant ratio of Ca/Fe abundances, i.e. a lockstep variation.

Studies of metal abundances in old stars have shown that the picture of a lockstep variation must be abandoned in favor of a more complex picture (see for instance Spite and Spite (1986) and Gratton and Sneden (1988)), which can be summarized as follows.

The light even elements O, Mg, Si, Ca and Ti are overabundant with respect to Fe.

The light odd elements Na, Al and probably also the odd iron-peak elements Sc and Mn are underabundant with respect to Fe.

The rare earths and also Y and Ba are underabundant with respect to Fe.

This situation is illustrated in figure 1. Observe that the differences between the abundances of many elements are rather small. Taking into account the error bars, many differences are at the limit of what is attainable at present.

There is no doubt that this picture approximates reality better than the previous one, which we called lockstep variation. Reality is surely still more complex, for several reasons. First of all because this picture includes only a handful of elements – less than 15 – and ignores all others. Secondly, only a small number of stars was analyzed and general conclusions are based upon this small number. Finally, one assumes that stars follow a general composition pattern, with no individual degree of freedom.

To see whether an individual degree of freedom exists, one may examine stars in clusters to find out whether the metallicity is identical in all of the stars, or, if not, how large the scatter is. Of the many papers on the subject we shall quote only a few referring to two points, namely a discussion of open cluster stars and a discussion on globular cluster stars.

Boesgaard and Friel (1990) have determined the [Fe/H] ratio for a number of F-type stars in *open* and/or *moving clusters*. They find that the scatter in [Fe/H] of the stars in each cluster is very small – of the order of ±10%, but that there are much larger differences between the different clusters. The same problem is discussed by Cayrel de Strobel (1990), who also provides a list of individual determinations of [Fe/H] abundances of stars in clusters. She concludes, on the

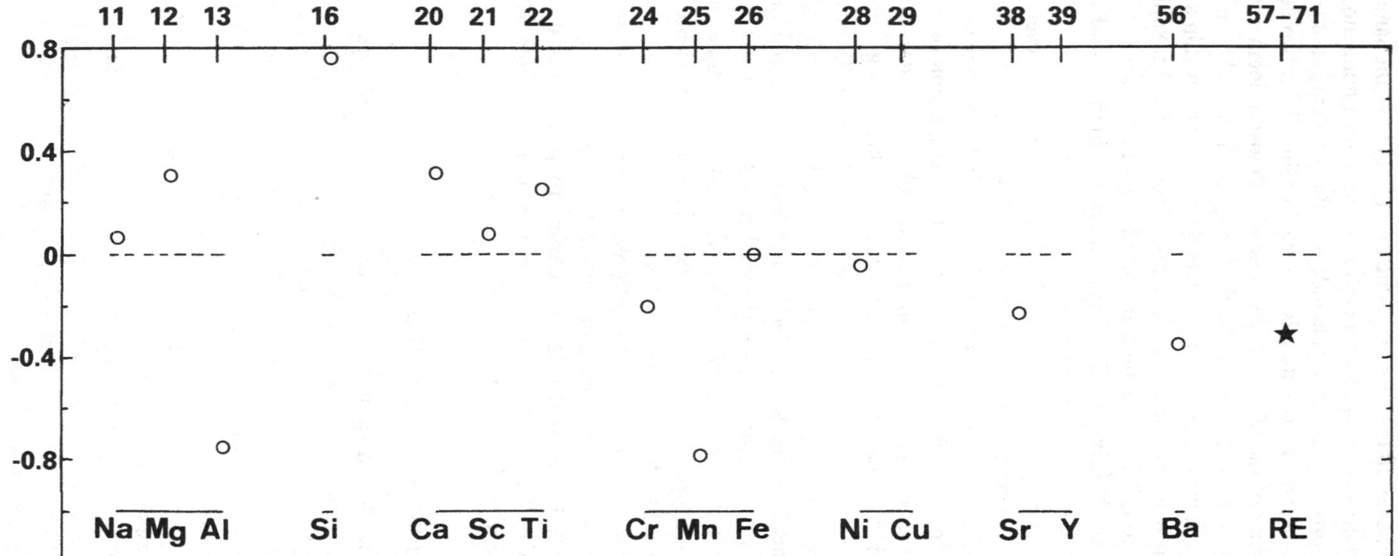

Fig. 1. Abundances of elements relative to Fe in weak metal stars. Abscissae: elements (bottom) and atomic number (top). Ordinates: abundances relative to Fe, in dex. Data are from Gratton and Sneden (1988), except for the rare earths.

contrary, that practically all clusters (with one exception) have the same average [Fe/H], but that there exist real differences in composition among the stars of a given cluster. It is probably safest to conclude that the problem is not yet solved.

For *globular cluster stars* the situation is in a state of flux. After two decades in which the uniformity of metallicity in clusters was almost a dogma, it became clear that this is an oversimplification. Smith (1987) has discussed the problem at length and he concludes that, in some clusters, [Fe/H] varies within a factor of two. The situation is worse for other elements. Another fact that supports the idea of large variations is that one finds in globular clusters a variety of CH and CN anomalous stars. This implies a variety of anomalies in the abundances of C and N. For other elements there exist similar indications.

For *field stars*, Holweger *et al.* (1986) illustrate the case of several normal A-type stars, which exhibit a significant diversity of line strengths. This subject was considered in more detail by Lemke (1989, 1990), who analyzed the Fe abundance in 15 'normal' A-type stars. He found a dispersion of ± 0.16 dex (implying differences of up to 0.5 dex in the sample). If field stars come from dissolved associations or clusters, then we can expect that among field stars we should find all the differences that exist between the different stellar groups. The dispersion observed agrees with what Boesgaard and Friel found in clusters.

Metal abundances in the *Small Magellanic Clouds* seem to converge to an underabundance by a factor of four (Russell and Bessell 1989, Luck and Lambert 1992). For the *Large Magellanic Cloud* the underabundance seems to be smaller – on average probably by a factor of two. The iron peak elements Ni, V, Mn and Ti behave like Fe – but Co seems to be underabundant (Luck and Lambert 1992). Since the evidence for differential behavior is not very strong, one can only conclude that the problem is not yet solved.

As a final remark, let us say that the term 'iron group elements' includes the elements V, Cr, Mn, Fe, Co and Ni, but sometimes also the lighter elements Sc and Ti and the heavier elements Cu and Zn.

2.2 The behavior of rare earths

Let us start with the definition of the group of rare earths. The rare earths are a group of elements from $Z=57$ to $Z=71$, namely La 57, Ce 58, Pr 59, Nd 60, Pm 61, Sm 62, Eu 63, Gd 64, Tb 65, Dy 66, Ho 67, Er 68, Tm 69, Yb 70 and Lu 71. Chemists call this group the 'lanthanum series'. Chemists call the elements 58–71 (Ce to Lu inclusive) 'lanthanides' and use also the term 'rare earth metals' for the group of rare earths plus Sc and Y.

Rare earths are found in two different places on the HR diagram: among A-type stars and among late type giants.

In the region of the A-type stars one finds them in *Ap stars* of the Cr–Eu–Sr subgroup and in sharp-lined late *Am stars*, in the form of strong lines of the ionized species. Their presence is unexpected, since these lines are normally present as weak lines in late type stars.

In the cooler part of the HR diagram one finds the rare earths enhanced in *Ba stars* and in *S-type stars*. What is unexpected in these stars is not the presence of rare earths, but their intensity.

For the behavior of rare earths, there exist three general rules. First, elements with an even atomic number are more abundant than those with odd numbers. Second, one does not find all even (or all odd) elements in a given star. Third, lighter elements are more abundant than heavier ones.

It must be remarked that there exist exceptions to all three rules.

Detection of rare earth elements is hindered by the fact that many elements have no outstanding strong lines, but a wealth of lines of medium or weak intensity, which are often blended with lines of abundant elements like Fe, Cr and Ti. As a consequence, identification of a given rare earth may pose considerable difficulties and many proposed identifications have been challenged afterwards. As a rule, identifications should be carried out using common sense precautions, which are discussed in part three, chapter 3. To the general rules one should add two more, namely

(1) identify all lines of all elements, besides the element in which one is interested, to make sure that other contributors can be eliminated, and
(2) pay due attention to hyperfine structure (Mathys and Cowley 1992).

Rare earths in *Bp* and *Ap stars* have been studied by Cowley, using statistical methods (Cowley 1976, Cowley and Arnold 1978, Cowley and Henry 1979). Their conclusions may be summarized as follows.

(a) There exists a group of Ap stars in which the even–odd rule is well obeyed, if Eu is left aside, because this element presents specific problems of hyperfine splitting and Zeeman broadening (Hartoog *et al.* 1974). In this group La is usually weak, Ce is not the strongest element and one often observes medium- and high-mass lanthanides such as Dy, Ho and Er.
(b) There exists another group in which La and Ce are strong, as well as Eu and Gd. However, the other even elements, Nd and Sm, which one would also expect to be strong, are weak or absent. This is known as the 'Nd–Sm anomaly'.
(c) Another group is constituted by stars in which the iron peak elements are overabundant, whereas the rare earths are less well represented. The strongest element is often Ce.

In stars with strong rare earth lines, one usually – but not always – finds the singly ionized species and also the doubly ionized species. (See for instance Ryabchikova *et al.* (1990).)

It should be added that some stars do not fit into the preceding scheme (see for instance Cowley and Crosswhite (1978)). There exist stars in which just one rare earth is greatly enhanced – for instance Eu or Gd (see the corresponding sections) and there exist other Ap stars that show no rare earths at all (Cowley 1991).

All the rare earth anomalies become weaker or disappear both in the hottest Bp and in the coolest Ap stars.

Rare earths are slightly enhanced in the *delta Del stars*, in which the even–odd effect is preserved (Berthet 1990). The even–odd effect is also preserved in A and F giants (Berthet 1991).

Apparently no rare earth anomalies are observed in F-type stars; they reappear in G-type stars.

Rare earths are enhanced in *Ba stars* by about one order of magnitude, but not all rare earths are present (Lambert 1985, Smith 1984). Pm, Tb, Ho, Tm and Yb have not been detected, and Er and Dy were detected only in some stars.

Rare earths are also enhanced by factors of about three in the *subgiant CH stars* (Luck and Bond 1992), which are closely related to the Ba stars. Rare earths are enhanced by one order of magnitude or more in *CH stars* (Luck and Bond 1992).

Rare earths are present in *SC stars* (Wallerstein 1989), but only the light rare earths can be definitely identified (La, and to a lesser degree, Ce and Pr). Eu seems to be present, but Gd and Lu are absent.

Rare earths are highly enhanced in *metal-poor carbon stars* (Kipper 1992b).

Rare earths are very strong in some unusual objects like *FG Sge*, which has been changing its spectrum from O 5 to G 5(!) (see Wallerstein 1990).

From the scattered data given in the literature it seems that, in stars of the two *Magellanic Clouds*, rare earths are more abundant than what could be expected from the general weakening of the elements (Luck and Lambert 1992). This situation is similar to that of *galactic halo stars*.

Gilroy *et al.* (1988) have analyzed the rare earths in very *metal-poor stars* of population II. They found that the lines of ionized species are present but rather weak (equivalent widths of tenths of a milli-ångström unit). In general, rare earths (La to Dy) are overabundant with respect to iron by factors of 2–3, but there exist genuine star-to-star differences. The even–odd pattern is well obeyed in the *extreme metal-deficient stars*. Gonzalez and Wallerstein (1992) analyzed one *globular cluster supergiant* and found that the rare earths (La, Ce to Nd) are overabundant with respect to Fe.

3 CHROMOSPHERES AND CORONAS

Traditionally the absorption line spectrum of a star has been explained as originating in a (cooler) layer, which surrounds the (hotter) stellar photosphere. The observation of many (unexpected) emission lines from various kinds of objects contradicts such a simple picture. Through the improved observational facilities that have become available in recent decades we now have a much clearer picture of the outer layers of the stars. Since these layers were first observed in the sun, the best introduction is to start with a description of what happens in the sun. This book is certainly not the place for a detailed discussion of the chromosphere and the corona. We shall provide only some information concerning matters that are of interest for the study of stellar spectra. The reader can find more detailed treatment of the solar chromosphere (and also the corona) in many books, like for instance *Astrophysics of the Sun* by Zirin (1988) and *Spectroscopy of Astrophysical Plasmas* edited by Dalgarno and Layzer (1987).

3.1 The solar chromosphere

The solar chromosphere is the layer enclosing the sun's photosphere. Temperatures diminish in the outer layers of the sun and reach a minimum at what is conventionally called the photosphere. Beyond this layer, temperatures rise again with distance above the sun – first slowly and later very rapidly, until a standstill is reached at about 300 000 kelvin. The first part of the atmosphere, where the temperature rise is slow, is called the 'chromosphere'. The part where temperature rises rapidly is called the 'transition region'. *Grosso modo* the chromosphere extends up to 10 000 kelvin and the transition region up to 300 000 kelvin. Temperatures then rise again, but much more slowly, and this outer region is called the 'corona'. The temperatures in the corona go beyond one million and may reach several million kelvin.

The spectrum of the chromosphere can best be observed during total solar eclipses, when the moon covers the bright photosphere, i.e. just before complete darkness (or at the end of the total phase). Preceding total eclipse, a spectrum composed of bright lines flashes up for a few seconds; this is the so-called 'flash spectrum'. Since the moon is rapidly obscuring the sun, one obtains a high spatial resolution, and one can study the behavior of the spectrum at different heights.

Another way of observing the chromospheric spectrum is to observe the spectrum of the solar limb. This can be done without any eclipse but only the lower part of the chromosphere can be observed in this way. In order to observe the upper chromosphere and the transition region one must observe from a spacecraft (see for instance Firth *et al.* (1974) and Doschek *et al.* (1976)).

One can also study parts of the chromospheric spectrum, namely short-wavelength stretches corresponding to single lines. This produces a monochromatic image of the sun – for instance the picture of the sun in the Ca II K line, or in H alpha.

The spectrum of the chromosphere is very rich and more than 10 000 emission lines have been observed. A fact that hinders considerably its study is that the spectrum varies with height above the photosphere because of the variation of the physical conditions. Table 1 provides a summary of the places where the different solar emission lines are formed. The data are taken from Dupree (1986). See also figure 1.

If one groups the elements of figure 1 by their excitation level, one finds at the lowest level lines of Mg II (2800), C I (1993), O I (1305), H (Lyman alpha) (1215)

Table 1. *Average temperature and height of the line-forming layers*

Line	Wavelength	Temperature	Height (km)
Si	1575	4500	500
Si	1524	5500	700
Ca II	3933	6500	1200
Mg II	2800	6500	1200
H	6562	6500	1400
H	1215	23000	2200

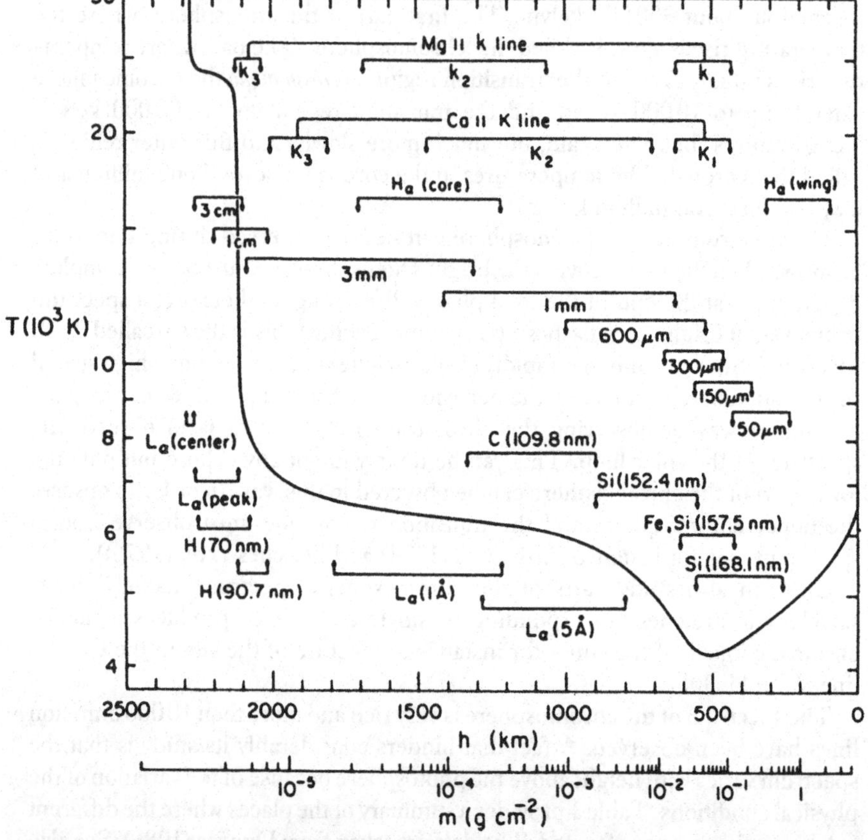

Fig. 1. The temperature distribution in the average quiet solar atmosphere. This semi-empirical model was constructed from line and continuuum observations at many frequencies. The approximate depth of formation of these features is noted. (Figure from Vernazza *et al.* (1981). Courtesy *Ap. J. Suppl.*)

and Si II (1815). Then follow C II (1335) and Si III (1892). The third group is formed by C III (1175, 1909), Si VI (1394), C IV (1550), He II (1640), N V (1238, 1242) and O V (1371). The first group corresponds to the lower chromosphere, the second to the higher chromosphere and the third to the transition region.

If we now turn our attention to stars, we find as a fundamental difference that we cannot yet study details of the surface of the stars. This also means that limb observations for studying the chromosphere are completely out of the question. What one sees is thus a mixture of radiations coming from all parts of the stellar photosphere and chromosphere, as well as from the corona.

As a result the many chromospheric lines observable in the sun are reduced – in solar-type stars – to only the strongest lines, as listed in table 2.*

To the lines listed in table 2 one must add the lines visible at longer wavelengths, as listed in table 3 (from Linsky (1980)).

Table 2. *The strongest chromospheric lines in stellar spectra*

Element	Wavelength
Chromosphere	
Mg II	2800
H I	1215
O I	1305
Si II	1304,1309,1527,1533,1815
Top of the chromosphere	
C II	1334,1335
Si III	1206,1892
Transition region	
He II	1640
C III	1175,1909
Si IV	1394,1403
C IV	1548,1551
N V	1238,1242
O V	1371
O VI	1032,1037

Note: In addition one also observes a large number of fainter lines of Fe II (van der Hucht *et al.* 1979, Judge and Jordan 1991) (see also the discussion on Fe II).

* Note added in proof. Dupree *et al.* (1993) *Ap. J.* **418**, L 41 found numerous emission lines of Fe XV–XXIV in the chromosphere of a G-type giant in the wavelength range 70–740 Å.

Table 3. *Stellar chromospheric lines at longer wavelengths*

Element	Wavelength	Notes
Ca II	3934 and 3968	
H I	6562	Particularly in M-type dwarfs
Ca II	8498,8542 and 8662	
Na I	5890 and 5896	See Pettersen (1989)
He I	5876,10830	Absorption lines

As stated before, in stars we cannot separate the lines originating in different parts of the outer atmosphere since we see only the integrated spectrum. There exists, however, one exception to this rule – a group of stars in which one can follow the behavior of lines with height. This is a group of eclipsing binaries, the so-called *zeta Aurigae stars*.

Such systems are composed of a small hot object (a B-type star) and a late type supergiant star with an extended atmosphere. During an eclipse, the hot object passes behind the extended atmosphere of the supergiant. Since its radius is small compared with that of the supergiant, its light acts as a 'point source', which illuminates the chromosphere of the supergiant. At this moment the ultraviolet spectrum shows a number of (chromospheric) emissions superposed upon the spectrum of the B-type star. Since in the zeta Aur systems the eclipse is very long, it is possible to follow the variation of the spectrum in the chromosphere and even in the transition region toward the corona. For a general overview see Reimers (1987) and Carpenter (1992) and for detailed studies see for instance Schroeder (1985), Schroeder *et al.* (1990) and Schroeder and Huensch (1992). However, the group of zeta Aur stars is very small – less than ten systems – so that in the vast majority of the stars we see only the combined spectrum, as explained above.

All stellar chromospheric lines are variable, both in strength and in profile, on time scales of months and years (Pettersen 1989). The variations frequently have a cyclic pattern and in some variable stars – for instance in *long-period* and in *semi-regular variables* – the changes are closely related to the general light curve (Querci 1986). The same is valid for *R CrB stars* (Clayton *et al.* 1992).

In what follows we shall comment on the behavior of some of the chromospheric lines. Before doing so we provide in table 4 a general overview of the intensities of the different chromospheric lines in different types of stars in the ultraviolet spectral region. The figures should be taken just as indicative values.

The most important chromospheric lines are the *CaII* and *MgII emissions* and most diagnostics of chromospheres have been based upon these two pairs of lines. Normally these emissions become visible after spectral type F 0. In earlier type stars the emissions are rather weak and can only be seen if the rotation of the star does not smear them out. Bidelman (1954) provides a very useful list of objects with *CaII emissions*. These are infrequent in F-type, become more

Table 4. *List of ultraviolet chromospheric lines in stars*

Element	Wavelength	G 2V	K 2V	K 8V	M 2V	G 5III	K 2III	F 8Ia	K 5Ib
HI	1215	95	180	210	150				
OI	1305	1	4	8		7	2	2	5
CII	1335	2	4	10		2			
SiIV	1394	2	2						
CIV	1550	3	9	8	3?	4			
HeII	1640	2	6	6	<3	2			
SiII	1815	9	11	15		10	1		1
SiIII	1892	1		2	2				
CI	1993	0	1						
MgII	2800	280	600	>360	49	380	95	62	110

Note: The figures are the monochromatic fraction of the stellar bolometric magnitude at that particular wavelength, times 10^{-7}.
Source: Data from Ayres *et al.* (1981a, 1983) and Byrne and Doyle (1990).

frequent in G-type and are very common in K- and M-type stars. This does not imply that emissions are always present in all late type stars, since in some of them they are very weak or absent. This absence may not be real, but may simply be due to the fact that, at the time of observation, the chromospheric activity was low.

Dempsey *et al.* (1993) have investigated the infrared Ca II triplet (8498, 8542 and 8662) in a sample of 45 late type stars. They find that the triplet is a good indicator of chromospheric activity.

MgII emissions at 2796(k) and 2803(h) are observed essentially from the same spectral types as the Ca II emissions. Stencel *et al.* (1980) and Doyle *et al.* (1990) have shown that Mg II emissions persist down to dM 6 e stars.

The Mg II emission is generally the strongest ultraviolet emission, so that all other lines can be compared with its strength. A plot of the Mg II emission strength against that of other emission lines shows that there exists a tight correlation, if both lines originate in the chromosphere. Correlations become looser when one compares the strength of Mg II with that of lines formed in higher regions.

H alpha emissions are more frequent from M-type dwarfs than for other types. Also, in the hydrogen emissions one finds a large variety of equivalent widths, which may simply reflect the degree of activity of the star. When H alpha emission is very strong it is usually accompanied by emissions in other Balmer lines and the star is a '*flare star*'. Pettersen and Hawley (1989) and Mathioudakis and Doyle (1991) observed a number of *K and M flare dwarfs* and found, in a sample of 19 stars, that emission is usually visible up to H delta, but may go up to H 9. Hydrogen emission is always accompanied by Ca II emission.

Smith and Dupree (1988) have observed a sample of *metal-deficient field giants* and found that about 20% of them exhibit H alpha in emission. The emission strength varies both from star to star and for the same star over time.

More extended surveys by Robinson *et al.* (1990) and Strassmeier *et al.* (1990) permit one to conclude that H alpha and Ca II emissions are usually related, but not in a one-to-one fashion. The most important point is that virtually all K and M dwarfs do show signs of chromospheric activity.

The Lyman alpha emissions have also recently been cataloged by Landsman and Simon (1993). These authors found that the Lyman alpha emission is well related to coronal X-ray emission. Lyman alpha can thus be considered a chromospheric indicator, at the same level as the emissions of Mg II and Ca II.

Other important chromospheric lines are *HeI absorptions* at 5876 and 10 830. In the sun both lines are seen in plages, with 10 830 about ten times stronger than 5876. This suggests that, if chromospheres were present in late type stars, then these lines should also be observed. Their presence would be completely unexpected, since He lines are absent in spectra of stars later than B type. The 10 830 line was effectively observed by Zirin (1968) in late type stars. Several authors (Vaughan and Zirin 1968, Wolff *et al.* 1985, O'Brien and Lambert 1986, Lambert 1987) later made systematic surveys of He lines in giants and supergiants. The 10 830 line first appears in mid- and late-F-type supergiants

($W \simeq 0.1$ Å), increases in strength at G-type, is present in all K stars and disappears toward M 1. The $W(10\,830)$ value in G-type stars may reach up to 1 Å, but is usually of the order of a few tenths of an ångström unit.

If the stars are plotted on the HR diagram, there exists a border line running roughly from G 0 II to K 0 III. To the right of this line the profiles are highly variable over time scales of weeks or months. Occasionally the W values may be as high as 1.5 Å or the line may appear in emission. Left of the border line the profiles are usually constant. Analogous border lines appear for Ca II and Mg II emission cores (O'Brien and Lambert 1986).

The He I 5876 absorption line is much weaker, with $W(5876)$ typically of the order of 0.04 Å. The line appears in F-type stars and disappears at about K 2 (Wolff *et al.* 1985, García López *et al.* 1993). The intensity ratio between the two He I lines seems to be normal (i.e. ten) for G- and K-type dwarfs, whereas it is much higher for supergiants.

Both He lines are correlated with other chromosphere indicators like Ca II emissions, C IV emissions and X-rays.

For completeness let us now mention data collections on chromospherically active stars.

There is a catalog of stars exhibiting Ca II emissions by Bidelman (1954).

There is a catalog of Ca II observations carried out at the Mount Wilson Observatory, which contains 65 000 measurements for 1300 stars, by Duncan *et al.* (1991).

There is a catalog of chromospherically active binaries (including RS CVn and BY Dra binaries) by Strassmeier *et al.* (1993) (second edition).

There is a catalog of stellar Lyman alpha fluxes by Landsman and Simon (1993) listing about 300 stars.

3.2 The solar corona

As mentioned before, the solar corona is the outmost layer of the solar atmosphere. It is usually divided into two parts – the internal and the external (or outer) corona.

The brightness of the corona is such a tiny fraction of the brightness of the photosphere that, under normal observing conditions from the ground, observation of the corona is made impossible by stray light from the solar photosphere. The inner solar corona can only be observed with coronographs installed at high mountain stations, where the scattered light is very much reduced. The outer solar corona, on the other hand, can be observed only during a solar eclipse or (in the absence of an eclipse) from spacecraft.

The emission lines that are visible in the classic region of the coronal spectrum defied identification for a long time and astronomers even attributed them to a hypothetical element, which was called 'coronium'. The solution of the problem came when Grotrian (1939) pointed out that some lines like 6374 and 7892 could be attributed to forbidden transitions of highly ionized atoms, namely Fe X and

Fe XI. Afterwards, Edlen (1942) provided more identifications and solved the identification problem.

In the following decades, observations from spacecraft permitted the list of coronal lines to be extended, especially in the ultraviolet.

We provide in table 5 the most intense lines of the solar corona. The data are compiled from Sandlin *et al.* (1977) for the region 975–3000 and from Jefferies *et al.* (1971) for the lines beyond 3000.

A large number of coronal lines, both permitted and forbidden, is also observed in the region below Lyman alpha (<1216). For the region 50–300 see Malinovsky and Heroux (1973); for the region 150–870 Å, Firth *et al.* (1974); and for the region 1175–1710 Å, Sandlin *et al.* (1986).

If we turn now to stars, we can expect to observe – at best – the strongest coronal lines, but in fact no line of table 5 is detected in normal stars (for exceptions see below). Evidence for existence of stellar coronas comes from X-ray emission, a subject that is outside the scope of this book. Other (indirect) evidence comes from observation of characteristic lines of the transition region between the chromosphere and the corona, which suggests that a corona exists farther away.

'Classical' coronal lines have so far only been observed for a few stars. Bowen and Swings (1947) observed forbidden emission lines of Ca XIII, Ni XII and Fe X. Reimers *et al.* (1990) have observed Si IV and C IV lines in strong absorption in the inner corona of a G-type giant. Other coronal lines in emission have been observed in *novae.*

We may now summarize our knowledge of stellar chromospheres and coronas. It is now generally accepted that one can divide the late type stars into three classes, namely coronal stars, non-coronal stars and hybrid stars (Judge 1988). These groups are shown in figure 2.

Coronal stars are dwarfs, subgiants and giants from F 0 to K 2. The ultraviolet (IUE) spectra correspond, if scaling factors are applied, to those of the sun's chromosphere and transition region, with temperatures up to 2×10^5 K (inferred from the observation of N V 1240). Although no lines originating in the corona have been observed so far, the existence of the corona is a (reasonable) extrapolation. The strongest argument for its existence is the X-ray emission.

The hotter limit of coronal stars can be placed at A 7 from indicators like Lyman alpha, Ca II, C II and C IV emissions (Freire Ferrero *et al.* 1992).

The cooler limit of the coronal stars can be fixed by different criteria such as X-ray emission, He I absorption lines, C IV emission or Lyman alpha emission. Since these criteria do not agree completely, the limits are slightly different when different criteria are used (Haisch 1987). On average one can put the limit at K 2.

Non-coronal stars are dwarfs cooler than K 2 and stars more luminous than K 0 III. The ultraviolet spectra show strong chromospheric lines (especially of the neutral species) but no lines corresponding to temperatures above 2×10^4 K (see Linsky *et al.* (1982)). Asymmetric line profiles indicate strong mass outflow. No X-ray emission has been detected.

Table 5. *The strongest forbidden coronal lines in the sun*

Wavelength	Species	Intensity	Ionization potential	Wavelength	Species	Intensity	Ionization potential
1216	Fe XIII	6	331	3388	Fe XIII	37	331
1242	Fe XII	10	290	4086	Ca XIII	22	657
1349	Fe XII	6.7	290	5303	Fe XIV	190	361
2085	Ni XV	6.7	430	5694	Ca XV	28	817
2125	Ni XIII	12	352	6374	Fe X	40	235
2169	Fe XII	29	290	6701	Ni XV	27	430
2405	Fe XII	150	290	7891	Fe XI	50	262
2565	Fe XII	43	290	10746	Fe XIII	100	331
2578	Fe XIII	160	331	10798	Fe XIII	50	331

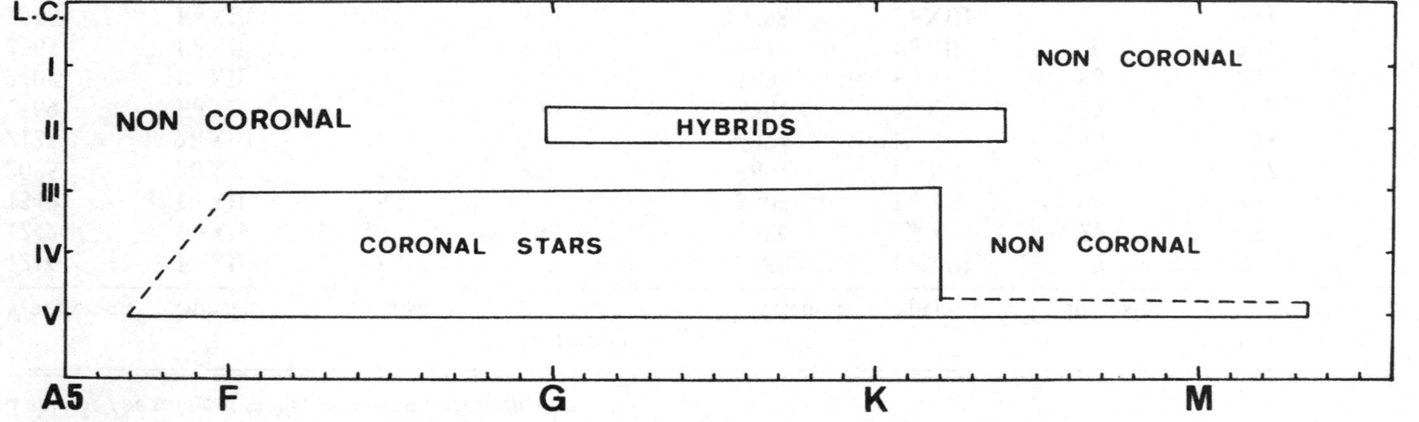

Fig. 3. Regions occupied by stars of different types of chromospheres and coronas. Abscissa: spectral types. Ordinate: luminosity classes.

Hybrid stars are mostly stars of luminosity class II that exhibit both hot plasma lines (1×10^5 K) and mass outflow.

3.3 Stellar groups with active chromospheres and/or coronas

In what follows we list the stellar groups that exhibit signs of an extended atmosphere.

For *T Tau stars* (see for instance Brown *et al.* (1984), Basri (1987) and Bertout (1989)) all lines characteristic of the so-called 'coronal stars' are present. However, the NV lines are weaker than expected. Other (chromospheric) emissions that may be present include lines of Na I, Mg I, He I, Fe I and Fe II in the optical region. The ultraviolet spectrum also shows lines of H_2, apparently excited by fluorescence with Lyman alpha.

Carter and O'Mara (1991) introduced the idea of an emission sequence for pre-main-sequence flare stars, going from active T Tau stars to weak-lined T Tau stars and finally to stars that are in a quiescent stage.

In late type peculiar star groups, according to Querci (1986), *Ba stars* have normal chromospheres, early *C-type stars* have no indicators of chromospheres, later *C-type stars* do have indicators of chromospheres and *S-type stars* show a wide range of ultraviolet emission features. To this general summary we may add some details. The ultraviolet spectrum of C stars differs from that of M stars in the sense that they show neither H alpha nor Ca II emissions (Judge 1989). For C stars see also Eaton *et al.* (1985). S-type stars, on the other hand, seem to have a more active chromosphere than M stars (Judge 1989).

R CrB stars, during deep extinction minima of the light curve, show a spectrum that corresponds to the lower chromosphere (Merrill 1951b, Evans *et al.* 1985).

Symbiotic stars exhibit in the ultraviolet region an emission line spectrum, which resembles that of a chromosphere and a transition region.

Dwarf novae in outburst present many lines, which correspond to the lower chromosphere and the transition region, with the addition of the N IV 1729 and Al III 1850 lines (Verbunt *et al.* 1984).

In the eclipsing binary *VV Cep*, Hagen Bauer *et al.* (1991) detected some lines corresponding to a lower chromosphere, but no high excitation lines corresponding to higher temperatures. See also Hack *et al.* (1989).

In a sample of *FK Com stars*, Bianchi and Grewing (1987) found high excitation chromospheric and transition region lines.

In the *recurrent nova* RS Oph a large number of coronal lines have been observed by Joy and Swings (1945). These authors remark that there exist some differences with the solar corona, in the sense that forbidden lines of Fe IV, Fe V, Fe VI and Fe VII are also present. Swings and Struve (1941a,b) detected in the M 5 star RX Pup a large number of high excitation lines corresponding to forbidden Fe VI and Fe VII (6085), Ne V and Ca VII. This object is associated with a planetary nebula.

Part Three

Content description

In this part of the book we shall provide short summaries (but no detailed treatment) on different topics, which are necessary for a better understanding of the book. We have grouped these matters in five chapters, namely

1 Terminology of spectral lines
2 The selection of stars
3 Line identification
4 Equivalent widths
5 Abundances

We have added also chapter 6, which contains some general thoughts on the matters covered in this book, which constitutes the epilogue.

1 TERMINOLOGY OF SPECTRAL LINES

Most of the terminology used by astronomers for spectral lines follows the definitions of the physicists. Let us recall briefly the meaning of some terms that are used in this book. For more details, the reader can consult any book on spectroscopy, like *Atomic Spectra and Atomic Structure* by Herzberg (1945) or *Structure and Spectra of Atoms* by Richards and Scott (1976).

An atom can have all its electrons, in which case it is said to be in the neutral state. If it has lost one electron it is said to be singly ionized and if it has lost n electrons it is said to be n times ionized. Degrees of ionization are indicated by Roman numerals – Na I is neutral sodium, Na II singly ionized sodium and so forth. If one wants to refer to any ionization stage of an atom then one speaks of species.

A spectral line is the result of the transition of an electron between two energy levels. Among the different energy levels available for an electron, some transitions are permitted by the selection rules. Among these one calls *fundamental, resonance or ultimate lines* those connecting a level to the lowest energy level. Usually these lines are the most intense ones.

Forbidden lines are formed by transitions between levels that are not allowed by the selection rules. Forbidden lines are indicated by the element enclosed in a square bracket – for example [Fe II]. If the bracket appears only on the right-hand side – for example O IV] – this corresponds to a (forbidden) intersystem line.

Forbidden lines appear in a multitude of astronomical objects. Usually they originate either in low-density, optically thin media or in very hot plasmas. Examples of objects of low density are the solar corona, nebulae, planetary nebulae, nova envelopes or extended stellar atmospheres. Examples of hot plasmas are the coronal flares. Some of the most common forbidden lines have been listed in table 1 (see the discussion below on nebular lines), although it should be observed that not all nebular lines are forbidden lines.

Besides the classical list of forbidden lines given by Moore (1945), a useful list of forbidden lines present in hot astronomical objects is that by Eidelsberg *et al.* (1981). The latter authors also provide a general introduction to the subject.

Fluorescent lines are described as follows. When a line is much stronger than expected, this may be due to overpopulation of its upper level. This can arise because a well-populated level of another atomic transition lies close to the upper level. One speaks then of fluorescence. Because of this explanation it is clear that fluorescence affects mostly less abundant elements and originates because a highly populated level of some abundant species exists. Hydrogen induces, for instance, numerous fluorescent lines of less abundant elements, like Fe.

For examples of fluorescent lines in stellar spectra see for instance

Willson (1975) for Fe I lines in *T Tau stars*,
Wallerstein *et al.* (1991) for lines in *symbiotic stars*,
Feibelman *et al.* (1991) for the ultraviolet (1360, 1776, 1869, 1881, 1884 and 1975) Fe II emissions induced by fluorescence of O VI, C IV and H I and Harper (1990), for a study of Fe I emissions (2823 and 2843) induced by Mg II.

A very useful table of predicted fluorescent lines is given by Gahm (1974). Gahm *et al.* (1977) have complemented this list with one of fluorescent molecular lines.

If the overpopulation is very strong, then one speaks of a *maser* effect. Maser effects are observed in molecular lines of OH, H_2O and SiO (see part two, chapter 1).

Nebular lines (table 1) are all those lines that are found in a variety of low-density astrophysical plasmas, heated by a hot radiation source. Examples are furnished by the *planetary nebulae, novae* in the later stages of an outburst (the 'nebular stage'), *supernova remnants* and *gaseous nebulae*. Nebular lines are also found in *symbiotic stars* (Wallerstein *et al.* 1991). In almost all these objects the line strengths are subject to temporal variations.

Table 1. *Nebular lines*

Element	Lines
H	6562,4861,4340,4101
He I	5876,6678
He II	4686
[N I]	5199
[N II]	5754,6548,6583
[O I]	5577,6300,6364 (auroral lines)
[O II]	3726,3729,7325
[O III]	5007,4959,4363
[Ne III]	3869,3968
[Ne IV]	4714,4716,4724,4726
[Ne V]	3346,3426
[S II]	4068,4076,6716,6730
[S III]	6312
[Ar III]	7006,7136

Source: The table lists only the most frequently observed (emission) lines, compiled from Acker *et al.* (1991).

Under such physical conditions one usually finds a number of characteristic lines, whose intensity is a function of the temperature of the exciting source and of the density and temperature of the plasma. Exciting sources of higher temperature produce in general higher excitation stages of the atoms in the surrounding plasma.

The lines of table 1 can be used as diagnostic tools for the density and temperature of the medium where they are formed. The relative intensity of the different lines of a given element is mostly explained by its transition probability. Table 2 provides the transition probabilities of some commonly observed nebular lines.

Circumstellar (absorption) *lines* are those lines that originate at some distance from the star's surface, but sufficiently close to it for them not to be interstellar. Usually the line profiles are either of the P Cygni type (a line with a deep absorption core displaced toward the violet and bordered on the red side by an emission component) or have a blue-shifted absorption line with one or more components. Usually the excitation potential of circumstellar lines is of the order of 3 eV or less.

Typical circumstellar lines are Ca II lines (H and K), Na I (D line), Sr II 4078 and Ca I 4226. If the stellar atmosphere itself produces a line of this element, the circumstellar profile is superposed upon that of the stellar line.

Table 2. *Forbidden line transition probabilities (s^{-1})*

Element	Wavelength	Transition probability
[O I]	6300	0.0051
	6364	0.00164
[O II]	3728	4.8×10^{-5}
	3726	1.7×10^{-4}
[O III]	4363	1.6
	4959	0.0071
	5007	0.021
[Ne III]	3868	0.17
	3967	0.052
[Ne IV]	4714	0.11
	4716	0.40
	4724	0.44
	4726	0.389
[Ne V]	3426	0.382
	3346	0.138
[S II]	4068	0.34
	4076	0.13
	6716	0.0005
	6730	0.0043

Source: Values are from Wiese *et al.* (1966) except for those for [S II], which were taken from Allen (1973).

2 THE SELECTION OF STARS

The selection of objects included in this book is of considerable importance and we shall make some comments on this point.

To trace the curves of equivalent widths for objects of different spectral types and luminosity classes we have relied only upon MK standard stars. This was done for three reasons. First of all, because MK standards have well-established classifications. Second, they have also been closely scrutinized for spectrum anomalies (both for general line weakening and/or for abnormal line strengths of individual elements). Third, they have been observed so often that spectrum variability can be excluded. If, in the papers from which the equivalent widths were taken, the authors have observed other stars besides the MK standards, then these were not taken into account. When nothing else was available, in a few explicitly mentioned cases, we have accepted stars that are not MK standards.

The use of spectral types and luminosity classes, rather than temperatures and log *g* values, is justified by the fact that the former are observable characteristics of stellar spectra and are thus stable, whereas temperatures and gravities are fitted parameters and are thus subject to changes in calibration.

The restriction to MK standards has been applied consistently with regard to the behavior of atomic lines. They could, however, not be applied in the chapter on molecules, because the number of stars in which molecules have been measured is too small to allow such stringency in the selection of stars. So we have taken whatever data were available, provided that the classification of the stars was in the MK system.

For the non-normal stars we have used the designations of groups that are given in the book *The Classification of Stars* (Jaschek and Jaschek 1987a). In order not to oblige the reader to look up cross references, we provide short definitions of the different groups in what follows. We have also added definitions of variable star groups, which are not given in our previous book.

In all cases we have preferred observational definitions, i.e. definitions based upon observable characteristics. This is important since many groups are defined in the literature by a mixture of observation and theory – if the theory changes the definition becomes useless or equivocal. Take for instance the group of so-called 'old stars'. Since age is not an observable parameter, when one refers to old stars one is really using theory based upon some kind of observation. Since the time scale of evolution has varied over the years (and is probably not yet definitive) even a numerical definition like an 'old star is a star older than 5×10^9 years' is subject to change. It is obvious that, with such a definition, we are obliged to re-assign objects into different categories each time calibration changes. Observational definitions on the other hand, like 'S stars are stars that show bands of ZrO', are stable and therefore more useful.

The designations of the groups are given in alphabetical order.

Ae or *A emission stars*. A-type stars with emission in one or several Balmer lines.

A shell stars. A-type stars in which two different types of line profiles co-exist.

Am or *metallic line stars*. A- or F-type objects to which no unique spectral type can be assigned. Usually the classifier provides a classification according to the hydrogen, metallic and calcium lines.

Asymptotic branch stars. Globular cluster stars, which are found in that part of the HR diagram that connects the top of the giant tip with the horizontal branch.

Ba or *barium stars*. Late type giants (G 2 to K 4) with a very strong Ba II 4554 line. Main sequence stars with strong Ba II lines have also been discovered recently.

Be or *B emission line stars*. Non-supergiant B-type stars, which have shown emission in at least one of the Balmer lines at some time.

B[e] stars Be stars exhibiting forbidden lines in emission.

Blue stragglers. Globular or open cluster stars, which are located on the main sequence but at such a place that they are disconnected from the cluster main sequence.

Bp, Ap stars or *peculiar B-* or *A-type stars*. B- or A-type stars in which the lines of one or several elements are abnormally enhanced. Traditionally the most important subgroups are Si λ 4200, Hg–Mn and Cr–Eu–Sr stars. The latest objects of the latter group correspond to early F-type.

Bright blue variables. Early type high-luminosity stars with peculiar spectra and large-amplitude light variations over a long time scale (e.g. eta Car).

C stars. Late type giants with strong bands of carbonated molecules (C_2, CN, CH) and no metallic oxide bands. Formerly they were called R or N types, the R types being the hotter and the N types the cooler C stars.

Cataclysmic variables. A collective name for stars in which the brightness increases suddenly because of an explosive event. The class comprises supernovae, novae, recurrent novae, dwarf novae and flare stars.

Cepheids. Strictly periodic variables with periods 1–50 days, of spectral types F, G and K.

CH stars. G-type giants (G 5 to K 5) in which the molecular bands of CH are very strong.

CN-weak stars. High-velocity stars with both weak metallic lines and weak CN bands.

CN-strong stars or super-metal-rich stars. Late type giants with strong CN bands. Metallic lines are also stronger than in normal giants.

CNO stars. Late O- or early B-type stars (O 8 to B 4) in whose spectrum the lines of some of the elements C, N and O are weaker or stronger than in the standard stars.

Compact infrared sources. Strong compact infrared sources embedded in nebulosity.

Composite spectrum stars. Objects with a spectrum due to superposition of the spectra of two different stars.

CS or *SC stars*. Stars exhibiting combined characteristics of C- and S-type stars – i.e. the presence of both C_2 and ZrO bands.

D or *degenerate* or *white dwarf stars*. Objects lying five or more magnitudes below the main sequence. Usually the following subgroups are distinguished:

DC continuous spectrum, no lines clearly visible,
DB only strong helium lines present,
DA only strong hydrogen lines present,
DO both He and H lines present,
DZ no He nor H lines, but metallic lines present and
DQ carbon lines (atomic or molecular) present.

Delta Del stars. A group of late A- and early F-type stars with very weak Ca II lines. Some stars of this group are delta Scu variables (dwarf Cepheids or AI Vel type objects).

dKe, dMe stars. K or M dwarfs with hydrogen emission lines.

Dwarf novae or *U Gem stars*. Cataclysmic variables in which the brightness increases suddenly at intervals ranging from several days to years.

Field horizontal branch stars. High-velocity metal-weak stars of either B or A spectral type.

Flare stars or *UV Cet stars*. Stars undergoing erratic jumps in brightness (up to a few magnitudes) on time scales of the order of minutes. During the quiescent phase the spectrum is that of an M dwarf with emissions in the Ca II and Balmer lines.

FU Ori stars or *fuors*. A subgroup of T Tau stars with considerable changes in brightness. The post-eruption spectrum is that of a late supergiant.

Halo stars. Stars that have high spatial velocity and low metallicity. For more details see Part Two, Section 2.1. This is not an observational definition.

HB or *horizontal branch stars*. Globular cluster stars defined by their position in the color–magnitude diagram. They are located on both sides of the RR Lyr gap and one speaks therefore of blue or red HB stars.

Helium-strong stars. B-type stars in which the helium lines are stronger than in normal stars. One distinguishes usually the *extreme helium stars* (also called *hydrogen-deficient stars*), in which no trace of hydrogen is seen, and the *intermediate helium-rich stars*, in which the hydrogen lines are still visible, but weaker than in normal stars. Related to these objects are the *hydrogen-deficient C stars*.

Helium-weak stars. (*Bp helium-weak stars*). B-type stars in which the helium lines are weaker than in normal stars.

Helium variable stars. Bp stars in which the strength of the helium lines varies periodically. At the extreme phases the objects appear as helium-rich, whereas at other phases He can be very weak or absent.

Herbig Ae, Be stars. Be or Ae stars associated with nebulosity.

H and K emission line stars. Late objects (F 4 to M), which exhibit emission features in their H and K lines of Ca II.

High-luminosity early type objects. A collective designation for some early type stars with very peculiar spectra, like S Dor and P Cyg.

High-velocity stars. Late type stars whose spatial velocity is greater than 100 km s^{-1}. Other authors prefer the definition, 'with radical velocity greater than 60 km s^{-1}'.

Hydrogen-deficient early type stars. Early type stars of type O, B or A in which the hydrogen lines are very weak or absent.

Hydrogen-deficient C-type stars. A subgroup of high-luminosity C stars with weak or absent hydrogen lines, mostly of types F and G. Variable stars having such characteristics are called *R CrB stars*.

J stars. A subgroup of C stars, with strong C^{12}/C^{13} bands.

Lambda Boo stars. A-type stars with weak metallic lines, low rotational velocity and low radical velocity.

Late type stars. Stars of spectral type later than the sun (G 2).

Li stars or *Li-rich stars*. A subgroup of C stars, with a very strong Li I 6078 line.

Luminous blue variables or *S Dor variables or Hubble–Sandage variables*. A variable-star designation for the *high-luminosity early type objects*.

Metal-weak stars. Weak-line stars.

Metal-rich stars or *metal-strong stars* or *CN-strong stars* or *super-metal-rich Mg-strong stars*. A small subgroup of A-type stars in which the lines of MgII are very strong.

MS stars. Stars sharing the M and S characteristics. They thus exhibit bands of both TiO and ZrO.

Mira variables or *long-period variables*. Cyclic variables with cycles 100–500 days, and of spectral types K, M, S and C.

N stars. In the old terminology, the cooler C-type stars.

Neutron stars. Objects composed of neutrons, except perhaps in the outermost layers. The objects have very high density and a very small radius. This is not an observational definition.

Novae. Stars that undergo an explosion during which their brightness increases by up to ten magnitudes. Usually the following phases are distinguished (in order of time): pre-maximum, principal, diffuse enhanced, Orion, nebular and post-nova (Warner 1989).

Oe. O-type stars with emissions in the Balmer lines.

Of. O-type stars exhibiting emissions at both 4630–34 of NIII and 4686 of HeII.

O(f). O-type stars in which NIII is present in emission and HeII is weakly present in absorption or emission.

O((f)). O-type stars in which NIII is present in emission and He is strong in absorption.

Old stars. Stars that, according to contemporary stellar evolution theory, have an age comparable to that of the galaxy to which they belong. This is not an observational definition.

Oxygen-rich giants. A collective designation for giants showing metal oxide molecules – thus M, MS and S stars.

P-strong stars. A small subgroup of B-type stars in which P lines are very strong.

P Cyg stars. High-luminosity early type stars, in which all lines have a P Cyg type profile (an emission component on the red side of the absorption line).

PG 1159 stars or *pre-degenerates*. See *pre-degenerates*.

Planetary nebulae. Objects composed of an easily observable expanding shell surrounding a central star.

Post-asymptotic branch stars. F-type supergiants with strong sulfur lines.

Pre-degenerates. Very hot stars with strong O VI and C IV lines, which are X-ray emitters. Probably these stars are the central stars of planetary nebulae that have dissipated their envelopes.

R stars. In the old terminology, the hotter C-type stars.

R CrB stars. See hydrogen-deficient C-type stars.

RR Lyr. Strictly periodic variables with periods less than one day, and of spectral types A to early F.

RS CVn stars. Close binaries, which show H and K emissions.

Runaway stars. Early type stars (O and early B) outside the galactic plane, which reached large distances (from the galactic plane) because of their high velocities.

RV Tau variables. Periodic variables with periods 60–100 days, and of spectral types G and K.

S-type stars. Late type giants (K 5 to M) showing distinct bands of ZrO.

sdB or *subdwarf B.* B-type stars with very broad and shallow Balmer lines; fewer lines of the Balmer series are visible than for normal dwarfs.

sdO or *subdwarf O.* O stars showing few very broad and shallow Balmer lines and a very strong He II 4686 line.

Subdwarfs. Late type objects whose observed colors and absolute magnitude place it below the main sequence.

Subgiant CH stars. Hot Ba stars (spectral type <G 5).

Supernovae. Stars that have undergone an explosion during which their brightness increases by up to 20 magnitudes.

Symbiotic stars. Objects exhibiting a spectrum corresponding to a low-temperature star (generally a giant) plus emission lines corresponding to a hot plasma.

T Tau stars. Late type irregular variables associated with bright or dark nebulosity. The spectrum exhibits emission in both Ca II and H lines.

Ultraviolet Ga stars. A small group of Bp stars, which in the ultraviolet spectrum exhibit a strong 1414 line of Ga II.

VV Cep stars. A subgroup of composite spectrum stars. One observes a spectrum of a K or M supergiant, showing emissions of hydrogen and [Fe II] plus the spectrum of the secondary, which is generally of type B.

Weak G-band stars. A G-type giant (G 5 to K 5) with a very weak or absent G band of CH and weak CN bands. These stars are C-deficient.

Weak line stars or *metal-weak stars.* Late type objects in which the lines of all metals are weakened when compared with normal stars of the same temperature.

WR or *Wolf Rayet stars.* Hot stars characterized by wide emission lines of highly ionized elements, standing out distinctly from the continuous spectrum. There exist three varieties:

WN exhibit emission lines of N III, N IV or N V,
WC exhibit emission lines of C II, C III or C IV and
WO exhibit emissions of O IV, O V or O VI.

[WR]. WR stars that are the central object of a planetary nebula.

YY Ori stars. A subgroup of T Tau stars with inverse P Cyg type profiles in the Ca II and H line emissions.

3 LINE IDENTIFICATION

Line identification is one of the basic operations in stellar spectroscopy, which aims at determining the element responsible for each observed spectral line.

The process of identification is based upon

(a) measurement of accurate wavelengths of the lines,
(b) measurements of the intensity of the lines (or line intensity estimates on a personal scale) and
(c) use of line identification tables.

With modern detectors and reduction techniques points (a) and (b) offer no great difficulty, if carried out on adequate observational material – i.e. small plate factors, high resolving power and high signal-to-noise ratio. We should mention an important point concerning the resolving power of the spectrograph, namely that one can only see as separate features two lines that are more separated than the resolving power. The latter is usually fixed by the size of the photographic grain or the pixel size and is normally much larger than the accuracy with which one can measure the position of the barycenter of the (unresolved) line.

On the other hand, point (c) presents a certain number of problems. Very often identifications start with use of the classical paper 'A multiplet table of astrophysical interest' (Moore 1945). Such a practice leaves out all laboratory work that has been done since then and that often represents a considerable improvement over Miss Moore's tables. The fact that most of this recent work is not used by astrophysicists is due to the absence of an updated version of wavelength tables, arranged in a similar way to that of Moore. The following compilations are available.

'Line spectra of the elements' by Reader and Corliss (1982). About 40000 lines of all elements in many ionization stages. Accurate wavelengths for calibration are also indicated.

'Tables of spectral line intensities' by Meggers *et al.* (1975). About 39000 lines between 2000 and 9000 Å.

'An ultraviolet multiplet table' by Moore (1950, 1952, 1962).

'Atomic and ionic spectrum lines below 2000 Ångströms' by Kelly (1987).

'Tables of spectral lines of atoms and ions' by Striganov and Odintsova (1982).

Bibliographic references of recent laboratory work can be found in the *Reports of IAU Comm.* 14 'Atomic and molecular data'. These reports are included in the 'Reports on Astronomy' series, published in the *Transactions of the IAU* (Kluwer Academic Publishers). The reports refer to line identifications, line classifications, intensities, energy levels, transition probabilities, atomic collision data, line broadening and also molecular structure and transition data.

A point to which attention should be paid is that published wavelengths do sometimes have systematic errors (Learner *et al.* 1991, Jorissen *et al.* 1992, Vitrichenko and Kopylov 1962). If many lines of a given element are available for study, then systematic errors are not too important. In contrast, when only a few lines are studied, the systematic wavelength error can easily lead to identification errors.

In general, line identifications in a star are made with the guidance of another previously established identification list of a similar object accepted as being correct. One should always remember that this can be done only if the resolving power of the spectroscopic material used in both studies is identical – otherwise blending problems appear. A list of *normal* stars of different spectral types and luminosity classes in which good identifications have been made is given in table 1 of the next chapter (chapter 4). That table provides references to the papers used for this book. For *peculiar* spectra, it is more difficult to find a good 'reference star'. Data centers (like the CDS at Strasbourg) can provide a list of objects of the groups in which one is interested, as well as the bibliographic references. From the latter one can select the identification list that is needed.

As a rule, identifications should be carried out using the common sense precautions, namely

1 use only high-resolution spectrograms,
2 if possible, use sharp-lined stars,
3 measure several spectra of the same object,
4 work on an extended wavelength region and not just on a small one,
5 use the latest identification lists for each element,
6 try to identify all lines of the wavelength region, not just the lines of one element, to eliminate other possible contributors and
7 make sure that the strongest laboratory lines of a given element are present in the spectrum of the star.

With regard to the *first point*, it is clear that high-resolution means less blends and improves considerably the probability of carrying out a good identification. Since the sun is the object that can be studied with the highest resolution, the solar spectrum is the best for line identifications. The identifications of elements in the sun, and the measurements of their equivalent widths, are one of the basic anchoring points for all work using equivalent widths. In this book the solar values are marked 'S' in all tables – values quoted for the sun's spectral type (G 2 V) refer to stellar spectra.

The line identifications and equivalent widths quoted for the sun were taken from Moore *et al.* (1966) for the photographic region (2935–8770) and from Swensson *et al.* (1970) for the 7498–12016 region. For line identifications in the solar chromosphere and corona, see part two, chapter 3.

With respect to the *second point*, one should always use a typical object of the group, which has a low rotation. Stars with exceptionally low rotation should, however, be avoided. For instance slow rotators are very scarce among early B-type stars, which rotate typically with $V\sin i = 200$ km s^{-1}. Because of its slow rotation (10 km s^{-1}) the B 0V star tau Sco was used very often as a standard. However, it is not clear whether a star that has an abnormally low rotation really behaves as a typical B-type star with respect to abundances.

Point three constitutes an elementary precaution against spectrum variations (which may exist) and also a useful check for the correctness of the reduction procedure. Radial velocities should be determined separately for each spectrogram to find out whether the star has – or does not have – radical velocity variations. In the first case it is interesting to watch out for phenomena that usually occur in close binaries.

Point four is another safeguard. Identifications become more secure if the wavelength region studied is large, because this permits one to base the identification of an element on many lines and not just on the one (or two) lines that happen to lie in the (small) wavelength interval that one is studying.

Point five has already been discussed previously.

Points six and seven are crucial for good identification work. It is only after identifying all common elements (Fe, Cr, Mn, etc.) that one can expect to detect the exotic species, which are represented usually by a few weak lines. Difficulties usually arise with the line intensities, since laboratory sources are seldom photometrically calibrated. What is important is that the strongest laboratory lines of the element must be present and that the lines of medium intensity should be present down to a certain laboratory intensity. Despite these precautions there always exists a certain possibility of misidentification, the more so if the element is represented only by a few lines.

The preceding discussion may surprise many readers because most astronomers take for granted that these problems have been solved a long time ago. Nevertheless, it is true that, after so many years of spectroscopy, we still have to discuss identification problems, simply because the laboratory work on the elements is not yet finished.

Most users of laboratory spectroscopy need just a list of the strongest lines of each element for their work. Astronomers, on the other hand, need complete line lists for each element. The need arises because the astronomer has to deal in a stellar spectrum with a mixture of lines from many elements. In consequence he or she does not know whether a given line corresponds to the strongest line of an exotic element (for instance Hf) or to a weak line of a fairly common element (for instance Fe). Such a situation happens very seldom in laboratory work.

In what follows we shall discuss briefly two other topics that are of importance for line identification in certain objects, namely magnetic fields and stellar rotation.

Magnetic fields are present in many types of cosmic sources, with a wide variety of field strengths – from a few gauss in solar-type stars to 10^{10} G in *neutron stars*. (10^4 G = 1 Tesla.)

Magnetic fields are important for identification if they cause known spectral lines to appear at wavelengths that are unexpected and defy identification. This has happened in *degenerates* for the hydrogen lines (see the discussion on hydrogen). For a long time a number of lines in the vicinity of known Balmer lines defied identification until it was established that they were the Balmer lines split into components due to strong magnetic fields. Similar effects also exist for lines of other elements. For a review of the physical aspects of this problem, see Garstang (1977), and Chanmugam (1992) for a review of the astronomical aspects. For stars on or near the *main sequence*, magnetic fields are usually small, being of the order of at most a few kilogauss. Their net effect is to broaden those lines that are sensitive to the Zeeman effect, producing a kind of (pseudo)turbulence. An empirical rule states that this pseudo-turbulence (measured in km s^{-1}) is roughly equal to the strength of the magnetic field (measured in kilogauss) (Sadakane 1992).

It is clear that this effect must be taken into account when deriving abundances.

On the other hand, not all lines of a given element are affected in the same way by the Zeeman effect. This permits one to use the ratio of the equivalent widths of selected lines to establish the presence of magnetic fields. This was done for instance by Takeda (1991), who computed theoretical line profiles of the Fe II lines at 4385.38 and 4416.82. When no magnetic field is present, the equivalent widths are equal. When a magnetic field is present, the difference in equivalent widths grows to 0.20 Å for a field of 5 kG.

Stellar rotation manifests itself by a broadening of all the lines of the spectrum. Since the equivalent width of the line is preserved, this implies that the line becomes shallower with faster rotation. If the rotation is very fast, then faint lines become practically invisible. For this reason, most of the analysis of stellar atmospheres was done on slowly rotating stars, if possible with $V \sin i < 10$ km s^{-1}. This is clearly advantageous for definition of the lines and thus for ease of measurement of equivalent widths, positions and line profiles. On the other hand, it is also true that it may introduce a considerable bias, because one ignores the question of whether slowly and rapidly rotating stars have the same composition. One could, for instance, think that fast rotation produces internal circulation of matter and thus causes a modification of the abundance pattern that is observed in slowly rotating stars. More analysis is needed to solve this problem.

4 EQUIVALENT WIDTHS

4.1 Introduction

The strength of a line is characterized by its equivalent width (see figure 4). The continuous background – or continuous spectrum – being taken as unity, the equivalent width is the surface enclosed by the line profile. The abscissa can be expressed either in terms of wavelength or on a frequency (or wavenumber) scale. The equivalent width is given in the same units as the abscissa and can thus be expressed in ångström units (1 Å$=10^{-8}$ cm), or in reciprocal centimeters. Habitually one speaks of a strong line as a line having an equivalent width ($W>1$ Å and of a weak line as one having $W<0.1$ Å.

The difficulties in measuring equivalent widths usually come from the tracing of the continuum. It can happen that there are too many lines in the region that one is studying, so that the region is so crowded that the background becomes invisible. This happens in all late type stars because of the number of lines present in their spectra. The larger the plate factor is (in Å mm^{-1}), the more the lines are compressed and one cannot any longer see a stretch of unperturbed continuum. A partial solution is to use lower plate factors, which may allow one to find (narrow) windows of the true continuum.

In early type stars fewer lines are present and therefore one might think that the true continuum is easy to trace; but among these stars one finds a large number of rapid rotators. If rotation is very fast it draws out the line profiles, increasing superposition of neighboring lines and again the continuous spectrum is perturbed. There is no easy solution to overcome this difficulty, except for a deconvolution of the line profiles, which is feasible only for spectra having very high signal/noise ratios.

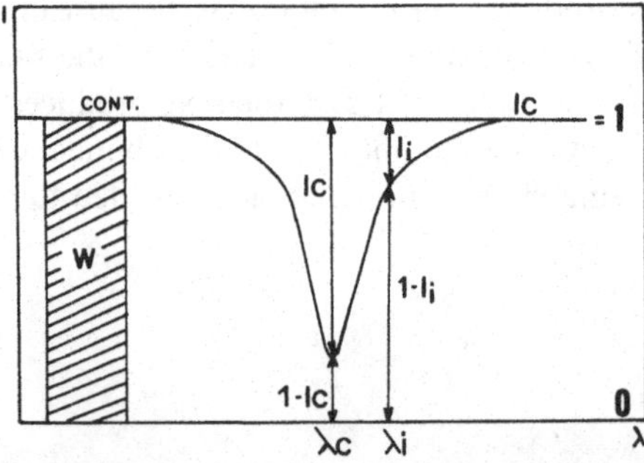

Fig. 4. Equivalent width. Abscissa: wavelength. Ordinate: intensity. I_c is the intensity of the continuum, set equal to unity. I_i is the intensity of the line at wavelength λ. $_i$. I_c is the intensity of the line centre at wavelength λ_c.1$-$ I_c is the residual intensity of the line center. W is the equivalent width of the line. The integral of the line profile is equal to the width of the rectangle of height 1, marked 'W'.

A further difficulty is that strong lines have extended wings, which may extend to large distances from the line center. This means that, if one measures lines in the vicinity of a strong line, then all possible precautions must be taken to recover the true continuum.

As a result of the difficulties with the tracing of the continuum, equivalent widths always have a rather large error attached. Perhaps the best illustration is given in the workshop on 'Element abundance analysis', where different teams were provided with the same tracings of two early type sharp-lined stars, to measure the equivalent widths of a set of pre-specified lines. For a line of 0.150 Å the measurements scattered between 0.130 and 0.162 in the worst cases (i.e. ±10%).

Problems also exist for very strong lines, for instance the hydrogen lines. As we have already remarked, such lines always have extended wings and a small error in fixation of the continuum has a large influence upon W. As a consequence the values of W are also beset by errors of at least ±10%.

To avoid such large errors, many authors have preferred to use photoelectric measurements. One uses in this case one filter centered on the line, plus one or two filters placed at a safe distance from the line center to obtain an undisturbed measurement of the continuum. Such a procedure is valid provided that one can prove that no other spectral lines fall within the filters. This happens only in theory and never in practice and as a result photoelectric measurements may have smaller random errors but may also have sizable systematic errors.

Another question of interest concerns the faintest line measurable on a given spectroscopic material. Seitter, in a private communication, says that the equivalent width (in milli-ångström units) of the faintest visible feature in a spectrum is approximately equal to the plate factor. So a plate factor of 5 Å mm^{-1} allows one to recognize features of 0.005 Å (5 mÅ).

As for the errors, we have adopted in this book the rule that errors are at least twice as large as the figure given before, i.e. ±10 mÅ for a 5 Å mm^{-1} plate factor.

It should be added, to be fair, that the errors that we have adopted apply mainly to measurements based upon photographic spectra. Measures based upon spectrograms obtained with solid state receivers can have a much higher signal/noise ratio and equivalent widths can have smaller errors. Probably a precision of the order of 1 or 2 mÅ is within the reach of present measuring technology, but it is not certain that the continuum may be fixed with a much higher precision. In the end the errors have diminished with new techniques, by a factor of 2–3.

A last point must be mentioned with regard to measurements of molecular band strengths. Since molecules occur only in late type stars, equivalent widths have seldom been measured. Most authors prefer instead to quote residual intensities. Such 'residual intensities' (RI) are defined as the ratio between the intensity of the spectrum at the band head and the intensity of a nearby point unaffected by the band (see figure 5). It would clearly be desirable that the point chosen for comparison should be entirely free of line absorptions, but this almost never happens in practice. With our definition, it is clear that large values of the

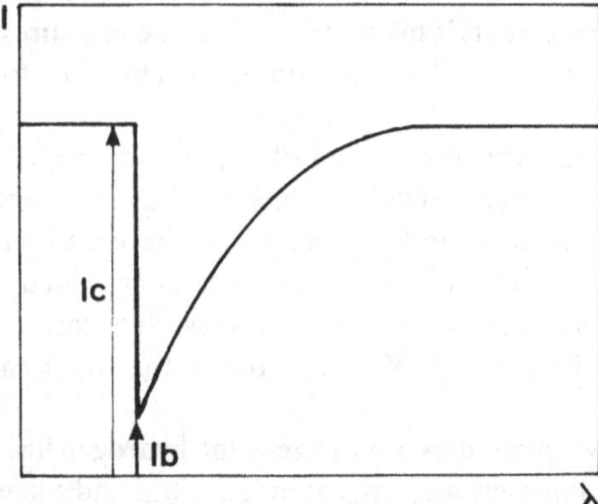

Fig. 5. Residual intensity. Abscissa: wavelength. Ordinate: intensity I_c, the intensity of the continuum, and I_b, the intensity at the band head. The ratio I_b/I_c is called the residual intensity.

residual intensity correspond to a weak band and small residual intensities to a strong band. Measures of residual intensities allow one to quantify the band intensities, but are not directly convertible into meaningful physical information.

4.2 Sources of equivalent widths

The equivalent widths for normal stars were taken from a variety of sources and the errors are therefore larger than the few milli-ångström units quoted above. This is a fact that the reader may easily check by comparing the scatter of the points around the average curve in all the plots for W against spectral type.

Because of the large number of measurements used in this book, it is impractical to provide the bibliographic source for each individual measurement.

To permit the reader to retrace the sources, we have compiled table 2, which gives the authors and bibliographic references of the papers from which the data on stars were taken. It also provides the spectral types and luminosity classes of the stars. Since all our stars are MK standards, we have not given the name of the star, but only its classification. The last column lists the plate factors of the spectrograms used, coded as follows:

A stands for plate factors less than 5 Å mm^{-1},
B for 5–10 Å mm^{-1},
C for 10–20 Å mm^{-1},
D for values larger than 20 Å mm^{-1} and
E for 15–35 Å mm^{-1}.

We have used W values of qualities D and E only when nothing else was available.

Table 2. *Sources of the equivalent widths*

Source	Types of stars	Code
Adelman (1973a)	B 9V	A
Adelman (1988)	B 5V, A 1V, A 2V	A
Adelman (1991)	B 7V, B 9.5V, B 8III	A
Adelman and Philip (1992b)	A 0IV	B
Aydin (1972)	A 0–A 3Ia, Ib	C
Boyartchuk (1957)	B 1Ia, B 3, B 5, B 7V, B 8Ia	D
Buscombe (1969)	O 9 to B 5 supergiants	B
Cayrel de Strobel (1968)	F 8V	A, B, C
Cayrel de Strobel (1968)	K 2V	B
Conti and Strom (1968)	A 1V, A 2V, A 7V	A, B
Conti (1974)	O-type	C
Conti and Leep (1974)	O-type	C
Conti *et al.* (1966)	A 1V, A 7V, F 4V, F 6V, F 8V, G 0V, G 1V, G 5V, K 2V	C
Crivellari *et al.* (1981)	B 7V, B 9.5V	B, C
Fujita *et al.* (1963)	M 0III, M 3III, M 4III, M 5III, M 7III, M 5II, M 5Ib, M 2Ia	E
Galkina and Kopylov (1962)	A 2, F 0Ia, F 2Ib	D
Gehren *et al.* (1978)	G 0V	A, B
Greenstein (1942)	F 0II	A
Greenstein (1948)	F 5V, F 5Ib	A
Grigsby *et al.* (1992)	O 9V, B 0V, B 1V, B 2III	A, B, C
Haenni and Kipper (1975)	M 0, M 2.5III	A, B
Hardorp and Scholz (1970)	B 0V	A, B
Helfer and Wallerstein (1968)	G 8III, G 9III, K 0III, K 1III, K 3III	B
Hollandsky and Kopylov (1962)	B 0Ib, B 1Ib, B 2Ib, B 3Ia, B 5Ib	D
Kilian and Nissen (1989)	O 9V, B 0V, B 2V, B 2III	B
Kodaira and Scholz (1970)	B 3V	A
Koelblod and van Paradijs (1975)	G 9III, K 0III	B
Kollatschny (1980)	K 5V	A, B
Lamers (1972)	B 0Ia	A
Luck (1992)	K 3II, G 0Ib, G 2Ib, G 8Ib,	A, B

Table 2. (*cont.*)

Source	Types of stars	Code
	K 1.5 Ib, K 3 Ib, K 4 Ib, M 2 Ib, F 5 Ia, G 0 Ia, G 4 Ia, K 0 Ia, K 1 Ia, M 1.5 Ia	
Maeckle *et al.* (1975b)	K 1.5 III	A
Peterson and Scholz (1971)	O 6–O 9	A
Powell (1970)	G 0V	B
Provost and van t'Veer-Menneret (1969)	F 0V	B
Ruland *et al.* (1980)	K 0III	A
Simon *et al.* (1983)	O 3	C
Strohbach (1970)	K 0V, K 5V	B
Underhill (1973)	B 6III, B 6V	A
Underhill and Fahey (1973)	B 5Ia	A
van Helden (1972)	B 3Ia	A, C
van Paradijs (1973)	G 0–K 5Ib	B
Vitrichenko and Kopylov (1962)	B 8–A 0Ia	D
Wallerstein and Conti (1964)	G 8–K 0III, G 3–G 6II, G 3Ib	C
Wright (1947)	F 5Ib, F 8Ib, F 5V	A, B
Wright (1963)	G 0V, G 5V	B, C
Wolf (1972)	A 3I–O	C

4.3 Lines selected as being representative for each species

As explained in the preface, we have tried to provide for each element and each ionization stage the equivalent widths of at least one line (or more, if available) to get a quantitative idea of the behavior of the species in stars.

The lines selected were chosen according to the following criteria:

1 to be either the resonance line or at least a strong line,
2 to lie in an accessible wavelength region,
3 to be free of important blends over a large spectral type interval and
4 to possess measurements for stars of different spectral types and luminosities.

It is obvious that all four criteria can seldom be satisfied simultaneously. Very often they are conflicting and we have been obliged to make compromises. Since we are trying to illustrate how a species behaves in the HR diagram, we have given more weight to criterion 4.

To illustrate the behavior of the species with spectral type, we give both the individual values (listed in tables) and the general tendency, illustrated in the figures. If, for a given point (spectral type and luminosity class), two values were available, we have averaged them if they were not too discrepant. Otherwise both are quoted. In the figures we have usually drawn the curves for luminosity classes V and I. If values were available, class III was added. For class I the reader is advised to consult the tables, which accompany the figures, to find out whether I stands for Ia or Ib – homogeneity could not be achieved for all species. If the curve is drawn as a dashed line, this means either that it is based on too few points or that it is an extrapolation.

It should be added that the qualitative behavior of many of the stronger lines of the elements can be followed in the spectral atlases listed in table 3.

Table 3. *List of spectral atlases in different wavelength regions*

Ultraviolet

'An atlas of ultraviolet stellar spectra' by Code and Meade (1979). (Range 1200–1850 Å at 12 Å resolution and 1850–3600 Å at 22 Å resolution.) A second part of the atlas was published by Meade and Code (1980).

The IUE Low Dispersion Reference Atlas by Heck *et al.* (1984). (Range 1350–2740 Å at 2 Å resolution.)

'The IUE ultraviolet spectral atlas' by Wu *et al.* (1983). (Range 1150–3200 Å at 7 Å resolution.)

'A spectrophotometric atlas of white dwarfs compiled from the IUE archives' by Wegner and Swanson (1991). (Range 1200–3200 Å at 7 Å resolution.)

'IUE atlas of O-type spectra from 1200–1900 Å by Walborn *et al.* (1985).

'Atlas of high resolution IUE spectra of late type – 2400–2778 Å by Wing *et al.* (1983). (0.3 Å resolution.)

The photographic region

Revised MK Spectral Atlas for Stars Earlier than the Sun by Morgan *et al.* (1978) (plate factor 118 Å mm^{-1}).

An Atlas of Spectra of the Cooler Stars. Types G, K, M, S and C, by Keenan and McNeil (1976) (plate factor 75–85 Å mm^{-1}).

An Atlas of Representative Stellar Spectra by Yamashita *et al.* (1977) (plate factor 73 Å mm^{-1}).

A Second Atlas of Objective Prism Spectra by Houk and Newbery (1984) (plate factor 108 Å mm^{-1}).

Atlas of Stellar Spectra by Ginestet *et al.* (1992) (plate factor 40 Å mm^{-1}).

'An atlas of yellow–red OB spectra' by Walborn (1980) (5400–6500 Å).

Tracings

An Atlas of Digital Spectra of Cool Stars. Types G, K, M, S and C, by Turnshek *et al.* (1985) (range 4200–7900 or 4500–8200 Å; 8–12 Å resolution).

Atlas of Tracings of Spectra by Goy *et al.* (1995).

Table 3. (*cont.*)

'An atlas of optical spectra of DZ white dwarfs and related objects' by Sion *et al.* (1990) (3700–4650 Å).

Infrared

'An atlas of stellar spectra' by Johnson (1977) (17 stars at 4–10 Å resolution, range 4000–10 000 Å).

'Atlas of M2–M9 dwarfs in the region 0.6–1.5 microns' by Kirkpatrick *et al.* (1993) (resolution 48 Å).

'Spectra of late type standard stars in the region 2.0–2.5 microns' by Kleinmann and Hall (1986) (types F to M, resolution of 1.6 cm^{-1}).

'An atlas of stellar spectra between 2.00 and 2.45 μm' by Arnaud *et al.* (1989) (types F to M, resolution of 0.02 μm).

'An atlas of late type stellar spectra 2400–2778 inverse centimeters' by Ridgway *et al.* (1984).

'An atlas of infrared spectra' by Jaschek *et al.* (1995, in press).

'A digital atlas of O and B type spectra' by Walborn and Fitzpatrick (1990).

5 ABUNDANCES

Conversion of the measured continuum, of the equivalent widths and of the line profiles into physically significant parameters, such as temperature, gravity, turbulence and chemical abundances of elements is the domain of stellar atmosphere studies, for which a number of good textbooks exist, for example

The Observation and Analysis of Stellar Photospheres by D. Gray, second edition, 1992, Cambridge University Press and

Stellar Atmospheres by D. Mihalas, second edition, 1978, Freeman.

Analysis of stellar atmospheres falls outside the scope of this book. Nevertheless, some comments on the precision of the abundances, estimates of abundances and catalogs are necessary.

5.1 The precision of the abundances

Abundances are quoted in dex and usually they are given to tenths of a dex (see part two, chapter 2). It is thus often assumed that the precision of the abundances is 0.1 dex. Such a precision is, however, not attainable except in very special cases. A good illustration of the problems that are encountered is given in the proceedings of the workshop on 'Elemental abundance analysis' (Adelman and

Lanz 1987). At this meeting the same initial data of two sharp-lined Ap stars were provided to several groups of astronomers. The abundances obtained by the different groups show that uncertainties still exist at the level of 0.1–0.2 dex, due to different physical parameters used, differences in computer codes and so on.

If this is true for stars for which the observational data were strictly the same, it is clear that, for stars observed by different observers with different equipment, using different sets of atomic data and different computer codes, precision can only decrease. A comparison of the results of different authors who analyzed the same object (under the conditions described above) leads to the conclusion that the precision of the abundances is of the order of ±0.3 dex (within a factor of two).

A precision of this order is compatible with Luck (1991b), who discussed determination of different parameters (temperature, log$g$ and [Fe/H] in stars with weak metal lines. For one object he was able to list 20 different analyses; their statistics show that [Fe/H] has a spread of 0.6 dex, with a dispersion of ±0.16 dex. Luck showed that the spread is due to the fact that [Fe/H] is closely linked to other parameters, which have to be determined simultaneously. What remains is the fact that an abundance determination taken from the literature has an uncertainty of a factor of two.

With the introduction in recent years of better instrumentation (CCDs replacing photographic plates) and more realistic physical theories, the determinations made by different authors of the elemental abundances for the same star now tend to converge to the same value. For a critical appraisal of the abundances see Taylor (1991) and for a detailed discussion, Wheeler *et al.* (1989). A more optimistic view is held by Gustafsson (1992).

Obviously the abundances depend furthermore upon the number and quality of the lines used. For elements having, for instance, only two lines, the accuracy is down by (at least) another factor of two.

To have an idea about the relation between abundances and equivalent widths, we repeat what we said before, namely that an underabundance by a factor of ten diminishes the equivalent widths by a factor of about three. An underabundance of a factor of 100 times diminishes the equivalent widths by a factor of ten. It should be added that these rules are not strictly applicable to strong lines, but they give a useful order of magnitude estimate.

The preceding considerations apply to stars, with the exception of the sun. For derivation of the solar abundances excellent observations and detailed atmospheric theories are available, which allow one to derive abundances that are probably accurate to within ±0.05 dex for elements with a reasonable number of lines.

5.2 Estimates of abundances

Since detailed atmospheric analyses are always very time-consuming, it is quite natural that astronomers have looked for quicker methods of detecting stars that have very large under- or overabundances. Such methods are the spectral or photometric classification methods, or, in general, stellar classification, which has been discussed at length in our book '*The Classification of Stars*'. To illustrate some of the results obtained with these methods we shall quote two papers.

The first paper is by Jaschek *et al.* (1989) and the second by Fernandez-Villacañas *et al.* (1990).

Jaschek *et al.* (1989) have classified visually a large number of F-type stars at 80 Å mm^{-1} and have separated out all those objects in which metal lines are weakened. A detailed comparison with abundances derived from atmospheric analysis shows that the visual procedure is capable of picking out all stars in which the underabundances are larger than a factor of four (i.e. -0.6 dex) and half of those with 0.3 dex underabundance. This implies that large underabundances can be picked out fairly easily with spectral classification. The same result can be obtained with photometric techniques, using a multicolor photometry like the Strömgren or Geneva system photometry (see Hauck *et al.* (1991)). The only restriction is that, before applying photometric methods, one must assure oneself that the star is a normal object – i.e. that it does not have a 'peculiar' spectrum.

In the case that we have just considered – stars with weak metals – the weakening affects many lines. Let us now illustrate the opposite case – when only one spectral line is concerned – given in the paper by Fernandez-Villacañas *et al.* (1990) concerning Ba stars. These authors compared estimated Ba line strengths with measured equivalent widths of the 4554 Ba II line. They found that estimates of the Ba line strengths do not correlate very closely with the equivalent widths, except for very strong lines. Overabundances of 1.0 dex (i.e. a factor of ten) can be picked out easily by visual estimates.

The conclusion from the two papers quoted can be summarized by saying that, if one wants quick estimates of (large) underabundances, or a rapid selection of non-normal objects, then one can certainly use (spectral or photometric) classification methods. Application of such methods is a necessary first step and a detailed analysis of an object should only be undertaken afterwards. This is an elementary consideration, which may save a lot of time.

Many astronomers do not realize the fact that most objects of abnormal composition that we know today were discovered by classification techniques and not by atmospheric analysis. This can be demonstrated easily by examining the history of astronomy. The first star that was analyzed for composition was the sun, and this was done in the late 1920s. However, at that time there already existed a long list of peculiar objects, such as the WR, Be, Ap and carbon stars, all discovered before this first atmospheric analysis had been performed.

When, in the 1950s, nuclear physicists started solving the problem of the production of chemical elements in stars, they could check their predictions of nuclear processes against a long list of stellar groups of anomalous composition. Again, all these anomalies had been detected by classification techniques.

5.3 Catalogs

At present there does not exist a data bank providing the abundances of the different chemical elements in individual stars. This is rather surprising after 50 years of studies of stellar atmospheres. The only catalogs that exist are those of the [Fe/H] ratio in stars. The most recent catalog is by Cayrel de Strobel *et al.*, which has had successive re-editions. The latest edition (Cayrel de Strobel *et al.* 1992) contains about 1700 stars with 3250 determinations. This is a bibliographic catalog compiling the data from the literature. A critical compilation catalog is the one by Koeppen (1988), who provides elements by which to judge the quality of the [Fe/H] values.

6 AFTERTHOUGHTS

Let us conclude the book with some general reflections.

After reading this book we hope that readers will conclude that our knowledge of the behavior of the chemical elements in stars has advanced significantly since the time of Merrill's (1954) book *The Behavior of Chemical Elements in Stars*. In this book, which is an excellent summary of the state of knowledge in the early 1950s, Merrill was able to report on 41 elements, whereas in this book we now have some information for 85 elements. In Merrill's time, studies were based upon the optical wavelength region (approximately 3000–8000 Å), but now we are able to provide results from the ultraviolet to the extreme infrared and the radio region. This has helped very much in the detection of molecules – whereas Merrill quoted some 40 molecules, we are able to list 89.

Progress is clearly visible in the production of papers on stellar spectroscopy. In Merrill's time, the annual production was of the order of 100 papers per year, whereas we now find about 1000 papers annually. It is thus legitimate to conclude that knowledge of the subject has advanced significantly.

However, if one regards the subject in more detail, a number of black spots appear on this rosy picture. A perusal of this book shows for instance that, for many of the elements discovered, we still do not know how they behave in stars of the main sequence, not to speak of their behavior in other regions of the HR diagram. For some elements, we know of their existence in a few stars but we are ignorant of whether the stars in which the element was detected are representative of the group to which they belong. For other elements we do not know whether they exist in stars. Do they really not exist in stars, or did we not look hard enough?

On the other hand, one can state the rule that, whenever an element has been studied in detail, one always finds a group of stars in which it behaves peculiarly, being either unexpectedly weak or overenhanced. We have stars with H, He, C, N, O etc. either strong or weak, and the same applies to rare elements like Ho or

Hg. Apparent exceptions to the rule (i.e. elements that behave normally in all stars) simply show a lack of studies. For instance we thought that we had found one exception with Ne, which is cosmically very abundant and yet little studied. Up to 1991 this was true, since we did not know of any Ne-anomalous stars, but then Smits (1991) called attention to certain novae that are Ne-rich and we thus find another confirmation of the rule.

If we assume that the extent of the different sections on the elements reflects the extent of our knowledge, then we can divide the elements into three groups: those that are well studied, those that are reasonably well studied and the poorly known ones. In the first category we find nine (mostly light) elements, namely H, He, C, N, O, Mg, Si, Ca and Fe. In the second group we find 24 elements, mostly metals and rare earths: Li, Ne, Na, Al, S, Sc, Ti, V, Cr, Mn, Co, Ni, Cu, Zn, Y, Zr, La, Ce, Pr, Nd, Sm, Eu, Gd and Dy. In the third group, finally, we find the largest number of elements – 50 – about which we know next to nothing, except that they exist in some stars.

We believe that these figures illustrate very nicely the ocean of ignorance in which we are still swimming.

A perusal of the bibliography quoted in this book brings out the curious fact that many elements were searched for and found in the 1970s, but that the next decades did not add many new elements. Why is that so? Sceptics will say that this is simply due to a passing fashion, but it may also be the sign of some deeper change. It seems to us that, at present, astrophysicists are too preoccupied with the immediate physical explanation of phenomena. Comparatively little interest is shown in the purely descriptive side of the phenomena. If something cannot be explained by stellar atmospheric theories or by nuclear physics then it is considered either unworthy of interest or directly wrong. We had for instance the experience of a paper on He I emission lines (which were theoretically unexpected in a certain type of stars), which was rather severely criticized by the referee. Similarly Ga 'cannot be present in stars' because it is cosmically underabundant and its discovery in stars was considered by some theoreticians as being incredible. One could go on with such quotes, but these two examples may suffice. Let us remember, however, that astronomy is a natural science – like botany – so that, despite all theory, there is still room and need for descriptions and for discoveries of things unforeseen by theory. The spectroscopists of the past ignored current physical theories, but they were able to describe and classify phenomena that became understood only afterwards. When first Saha's theory of ionization and later the nuclear theories came up, they could be checked against a rich background of observed facts, which had been discovered long before theory was capable of interpreting them.

This attitude toward immediate explanation has also produced curious neglects. It is fashionable to discuss the behavior of metals in metal-weak stars, but metals are reduced to three or four easy elements like iron, chromium and so on, leaving aside the less common metals. This is definitely not due to a lack of possibilities because our technology for detection of faint lines has improved – it is a consequence of the belief that the other metals are less important, because a first-order theory says so.

Comments of similar type can be made on line identification in stars. Decades ago this was a good subject for research and as a consequence we have a wealth of line identifications for all types of stars, upon which much of our present knowledge is based. A perusal of the recent literature shows that line identification papers are now rare. Apparently astronomers think that line identification problems do not exist anymore. We hope that this book shows a sufficient number of gaps to encourage colleagues to reconsider these problems, especially for late type stars and for the new wavelength regions.

The line identification problem is especially acute in the 'new' spectral ranges – that is, outside the 3500–6500 Å interval. We have very few detailed line identifications in the 1200–3000 Å range and the problem is even worse for equivalent widths. Despite a decade of studies with the IUE satellite we have no systematic studies of the behavior of lines with spectral type, if a few specific lines are excepted, like those of Mg II.* The same is true for the near infrared and far infrared spectral regions. Work in progress at Strasbourg, Montpellier and Toulouse will provide at least equivalent width measurements of the stronger lines of the fairly abundant elements like H, He, O, N, C, Mg, Si, Fe and Cr. As far as we know, nothing is being done regarding the lines of medium and weak intensity in this region.

Beyond 1.0 μm one can say that there is a 'white spot' (or a black hole) on the map. The region is open for work, for many discoveries and probably also for surprises.

With regard to the behavior of molecules in stars, our knowledge is definitely less complete than for atoms. Many molecules were just observed in one star and their existence was then generously extended to all stars of the group. This ignores completely the nasty possibility that there might exist stars that are anomalous in molecules, just as there exist stars that have an anomalous atomic composition.

Observe for instance that most of the complex carbonated molecules were discovered in IRC + 10216 – if we were to forget about this single object we would have to omit 20 molecules out of a total of 89 molecules identified in stars!

We have, furthermore, some (curiously ignored) information on matters that do not fit into present day explanations, like variable stars that during their cycle change from type S to type C. Now an S-type object has, according to theory, more oxygen atoms than carbon atoms, whereas a C star has more carbon atoms than oxygen atoms. How can a star then change periodically between S and C? The easy answer that the oxygen and carbon anomalies are superficial anomalies (i.e. that they do not come from the deep interior) runs into the difficulty that, if

* Note added in proof. Slettebak (1994) *Ap. J. Suppl.* **94**, 163 has recently measured equivalent widths of a number of lines of different elements in the spectral range B 1–B 9.

this is true, then we should find stars where the C or S type is only a superficial feature, not something really connected with the deep interior. Such little-studied objects may hold important clues for new interpretations.

Concern can be expressed over the small amount of interaction between stellar atmospherists and observers. Apparently both think that the situation is satisfactory and that all problems have been solved. If one looks in detail, the situation is rather unsatisfactory. Take for instance a well-known element like helium. One might expect that at least for this element theory could predict the variation of the observed equivalent width with temperature and gravity. However, this is not so. We found recently (Jaschek *et al.* 1994) that the predictions made by Auer and Mihalas (1972, 1973) for some of the near infrared lines are good, whereas for others they are poor or predict luminosity effects that are not observed. We detected also a couple of new helium lines in early type stars, although at first we could not believe that such lines had not been detected earlier. If this happens with helium, one can surmise that the situation is not much better for other elements, and this is certainly true. (It would be easy to quote more examples.) Generally, first-order theories are available, which explain the gross behavior of the lines with temperature and gravity, but many lines require more realistic theories (for instance implying non-local thermodynamic equilibrium) to interpret the observations.

On the side of the observers one finds a growing separation from laboratory physicists. The mildest form of disregard is to continue to use for line identification work Miss Moore's tables – which are correct but incomplete. Since the Moore tables date from 1945, this means that one disregards the wealth of information produced by laboratory physicists afterwards – almost 50 years of work! This is very regrettable and matters were definitely not so at the time of Miss Moore. The spectroscopists of those days – Struve, Merrill, Bowen, Joy, Sanford (to name but a few) – were very well informed about progress in laboratory spectroscopy.

Just for completeness we mention another sore point, namely the preservation of what is known, that is archiving. We have already discussed some of this in the chapters on line identification and equivalent widths so we shall only repeat some of the essential points. At the time of photographic plates, information was stored safely on the plate and everybody knows that plates last for many decades. Nowadays spectra are read from solid state receivers on to magnetic tapes and nobody knows what happens to a particular magnetic tape – usually it is erased and re-used. The information on wavelengths and equivalent widths is definitely lost, except if the author was able to convince an editor of a scientific journal to publish the data. The latter rarely happens because space in journals costs money and as a result editors are not particularly fond of publishing long tables. In essence, very often the information collected is simply thrown away. Let us recall that some of this information is irreplaceable, since the spectrum of many (if not all) stars may change with time.

The whole issue of archiving spectroscopic data must urgently be solved. If astronomy is an observational science, how can you throw away the information upon which it is built? Years ago archiving was difficult, but nowadays with mass storage media there exists no valid excuse for not archiving!

We have already mentioned the happy initiative of Luck to hand over his equivalent width measurements to data banks. This is a very good solution, because astronomical data banks were created just for such a purpose. It is to be hoped that Luck's initiative shall not remain an isolated case.

Let us stop here. We do not want to convey the impression that everything is incomplete or in bad shape; we have tried simply to call attention to some of the major problems encountered when writing this book.

We hope that this book may help colleagues to rediscover the fascinating area of research that is called broadly 'the behavior of chemical elements in stars'.

Part Four

Content description

This part contains auxiliary tables, which facilitate reading of the book.

1 the periodic table of chemical elements
2 a table of chemical elements ordered by name
3 a table of chemical elements ordered by formula (chemical symbol)
4 a table of chemical elements ordered by atomic number
5 a table of cosmic abundances of the elements.
6 a table of surface gravity as a function of spectral type and luminosity class
7 a table of effective temperatures as a function of spectral type and luminosity class.

Table 1. *The periodic table of elements.*

In each case the atomic number and the symbol of the element are indicated. The designations indicated at the top of the different columns are those used for ordering the molecules.

1a	2a	3b		4b	5b	6b	7b	8b			1b	2b	3a	4a	5a	6a	7a	8a
																	1 H	2 He
3 Li	4 Be												5 B	6 C	7 N	8 O	9 F	10 Ne
11 Na	12 Mg												13 Al	14 Si	15 P	16 S	17 Cl	18 Ar
19 K	20 Ca		21 Sc	22 Ti	23 V	24 Cr	25 Mn	26 Fe	27 Co	28 Ni	29 Cu	30 Zn	31 Ga	32 Ge	33 As	34 Se	35 Br	36 Kr
37 Rb	38 Sr		39 Y	40 Zr	41 Nb	42 Mo	43 Tc	44 Ru	45 Rh	46 Pd	47 Ag	48 Cd	49 In	50 Sn	51 Sb	52 Te	53 I	54 Xe
55 Cs	56 Ba		71 Lu	72 Hf	73 Ta	74 W	75 Re	76 Os	77 Ir	78 Pt	79 Au	80 Hg	81 Tl	82 Pb	83 Bi	84 Po	85 At	86 Rn
87 Fr	88 Ra		103 Lr															

57 La	58 Ce	59 Pr	60 Nd	61 Pm	62 Sm	63 Eu	64 Gd	65 Tb	66 Dy	67 Ho	68 Er	69 Tm	70 Yb
89 Ac	90 Th	91 Pa	92 U	93 Np	94 Pu	95 Am	96 Cm	97 Bk	98 Cf	99 Es	100 Fm	101 Md	102 No

Table 2. *Elements in alphabetical order of names with formula and atomic numbers.*

The first and fourth columns provide the name, the second and fifth the formula (abbreviation) and the third and sixth the atomic number (*Z*)

Actinium	Ac	89	Hydrogen	H	1
Aluminum	Al	13	Indium	In	49
Americium	Am	95	Iodine	I	53
Antimony	Sb	51	Iridium	Ir	77
Argon	Ar	18	Iron	Fe	26
Arsenic	As	33	Krypton	Kr	36
Astatine	At	85	Lanthanum	La	57
Barium	Ba	56	Lawrencium	Lr	103
Berkelium	Bk	97	Lead	Pb	82
Beryllium	Be	4	Lithium	Li	3
Bismuth	Bi	83	Lutetium	Lu	71
Boron	B	5	Magnesium	Mg	12
Bromine	Br	35	Manganese	Mn	25
Cadmium	Cd	48	Mendelevium	Md	101
Calcium	Ca	20	Mercury	Hg	80
Californium	Cf	98	Molybdenum	Mo	42
Carbon	C	6	Neodymium	Nd	60
Cerium	Ce	58	Neon	Ne	10
Cesium	Cs	55	Neptunium	Np	93
Chlorine	Cl	17	Nickel	Ni	28
Chromium	Cr	24	Niobium	Nb	41
Cobalt	Co	27	Nitrogen	N	7
Copper	Cu	29	Nobelium	No	102
Curium	Cm	96	Osmium	Os	76
Dysprosium	Dy	66	Oxygen	O	8
Einsteinium	Es	99	Palladium	Pd	46
Erbium	Er	68	Phosphorus	P	15
Europium	Eu	63	Platinum	Pt	78
Fermium	Fm	100	Plutonium	Pu	94
Fluorine	F	9	Polonium	Po	84
Francium	Fr	87	Potassium	K	19
Gadolinium	Gd	64	Praseodymium	Pr	59
Gallium	Ga	31	Promethium	Pm	61
Germanium	Ge	32	Protactinium	Pa	91
Gold	Au	79	Radium	Ra	88
Hafnium	Hf	72	Radon	Rn	86
Helium	He	2	Rhenium	Re	75
Holmium	Ho	67	Rhodium	Rh	45

Table 2. (*cont.*)

The first and fourth columns provide the name, the second and fifth the formula (abbreviation) and the third and sixth the atomic number (*Z*)

Rubidium	Rb	37	Thallium	Tl	81
Ruthenium	Ru	44	Thorium	Th	90
Samarium	Sm	62	Thulium	Tm	69
Scandium	Sc	21	Tin	Sn	50
Selenium	Se	34	Titanium	Ti	22
Silicon	Si	14	Tungsten	W	74
Silver	Ag	47	Uranium	U	92
Sodium	Na	11	Vanadium	V	23
Strontium	Sr	38	Xenon	Xe	54
Sulfur	S	16	Ytterbium	Yb	70
Tantalum	Ta	73	Yttrium	Y	39
Technetium	Tc	43	Zinc	Zn	30
Tellurium	Te	52	Zirconium	Zr	40
Terbium	Tb	65			

Table 3. *Elements in alphabetical order of formula with names and atomic numbers.*

The first and fourth columns provide the formula (chemical symbol), the second and fifth the complete name and the third and sixth the atomic number (*Z*)

Ac	Actinium	89	He	Helium	2
Ag	Silver	47	Hf	Hafnium	72
Al	Aluminum	13	Hg	Mercury	80
Am	Americium	95	Ho	Holmium	67
Ar	Argon	18	I	Iodine	53
As	Arsenic	33	In	Indium	49
At	Astatine	85	Ir	Iridium	77
Au	Gold	79	K	Potassium	19
B	Boron	5	Kr	Krypton	36
Ba	Barium	56	La	Lanthanum	57
Be	Beryllium	4	Li	Lithium	3
Bi	Bismuth	83	Lr	Lawrencium	103
Bk	Berkelium	97	Lu	Lutetium	71
Br	Bromine	35	Md	Mendelevium	101
C	Carbon	6	Mg	Magnesium	12
Ca	Calcium	20	Mn	Manganese	25
Cd	Cadmium	48	Mo	Molybdenum	42
Ce	Cerium	58	N	Nitrogen	7
Cf	Californium	98	Na	Sodium	11
Cl	Chlorine	17	Nb	Niobium	41
Cm	Curium	96	Nd	Neodymium	60
Co	Cobalt	27	Ne	Neon	10
Cr	Chromium	24	Ni	Nickel	28
Cs	Cesium	55	No	Nobelium	102
Cu	Copper	29	Np	Neptunium	93
Dy	Dysprosium	66	O	Oxygen	8
Er	Erbium	68	Os	Osmium	76
Es	Einsteinium	99	P	Phosphorus	15
Eu	Europium	63	Pa	Protactinium	91
F	Fluorine	9	Pb	Lead	82
Fe	Iron	26	Pd	Palladium	46
Fm	Fermium	100	Pm	Promethium	61
Fr	Francium	87	Po	Polonium	84
Ga	Gallium	31	Pr	Praseodymium	59
Gd	Gadolinium	64	Pt	Platinum	78
Ge	Germanium	32	Pu	Plutonium	94
H	Hydrogen	1	Ra	Radium	88

Table 3. (*cont.*)

The first and fourth columns provide the formula (chemical symbol), the second and fifth the complete name and the third and sixth the atomic number (*Z*)

Rb	Rubidium	37	Tc	Technetium	43
Re	Rhenium	75	Te	Tellurium	52
Rh	Rhodium	45	Th	Thorium	90
Rn	Radon	86	Ti	Titanium	22
Ru	Ruthenium	44	Tl	Thallium	81
S	Sulfur	16	Tm	Thulium	69
Sb	Antimony	51	U	Uranium	92
Sc	Scandium	21	V	Vanadium	23
Se	Selenium	34	W	Tungsten	74
Si	Silicon	14	Xe	Xenon	54
Sm	Samarium	62	Y	Yttrium	39
Sn	Tin	50	Yb	Ytterbium	70
Sr	Strontium	38	Zn	Zinc	30
Ta	Tantalum	73	Zr	Zirconium	40
Tb	Terbium	65			

Table 4. *Elements ordered by atomic number, with formula and atomic relative weights.*

The first column provides the atomic number (*Z*), the second the formula and the third the relative atomic weight, on the basis of $C^{12} = 12$. The following columns are arranged similarly. Values in parentheses are for radioactive elements.

1	H	1.0079	36	Kr	83.80
2	He	4.00260	37	Rb	85.4678
3	Li	6.941	38	Sr	87.62
4	Be	9.01218	39	Y	88.9059
5	B	10.81	40	Zr	91.22
6	C	12.011	41	Nb	92.9064
7	N	14.0067	42	Mo	95.94
8	O	15.9994	43	Tc	(97)
9	F	18.998403	44	Ru	101.17
10	Ne	20.179	45	Rh	102.9055
11	Na	22.98977	46	Pd	106.4
12	Mg	24.305	47	Ag	107.868
13	Al	26.98154	48	Cd	112.41
14	Si	28.0855	49	In	114.82
15	P	30.97376	50	Sn	118.69
16	S	32.06	51	Sb	121.75
17	Cl	35.453	52	Te	127.60
18	Ar	39.948	53	I	126.9045
19	K	39.0983	54	Xe	131.30
20	Ca	40.08	55	Cs	132.9054
21	Sc	44.9559	56	Ba	137.33
22	Ti	47.90	57	La	138.9055
23	V	50.9415	58	Ce	140.12
24	Cr	51.996	59	Pr	140.9077
25	Mn	54.9380	60	Nd	144.24
26	Fe	55.847	61	Pm	(145)
27	Co	58.9332	62	Sm	150.4
28	Ni	58.70	63	Eu	151.96
29	Cu	63.546	64	Gd	157.25
30	Zn	65.38	65	Tb	158.9254
31	Ga	69.78	66	Dy	162.50
32	Ge	72.59	67	Ho	164.9304
33	As	74.9216	68	Er	167.26
34	Se	78.96	69	Tm	168.9342
35	Br	79.904	70	Yb	173.04

Table 4. (*cont.*)

The first column provides the atomic number (Z), the second the formula and the third the relative atomic weight, on the basis of $C^{12} = 12$. The following columns are arranged similarly. Values in parentheses are for radioactive elements.

71	Lu	174.967	88	Ra	226.0254
72	Hf	178.49	89	Ac	227.08
73	Ta	180.9479	90	Th	232.0381
74	W	183.85	91	Pa	231.0359
75	Re	186.207	92	U	238.029
76	Os	190.2	93	Np	237.0482
77	Ir	192.22	94	Pu	(244)
78	Pt	195.09	95	Am	(243)
79	Au	196.9665	96	Cm	(247)
80	Hg	200.59	97	Bk	(247)
81	Tl	204.37	98	Cf	(251)
82	Pb	207.2	99	Es	(254)
83	Bi	208.9804	100	Fm	(257)
84	Po	(209)	101	Md	(258)
85	At	(210)	102	No	(259)
86	Rn	(222)	103	Lr	(260)
87	Fr	(223)			

Source: The values were taken from the *CRC Handbook*, 63rd Edition (1982).

Table 5. *Abundances of chemical elements.*

The first column provides the atomic number and formula of the element. The second provides the abundance in the solar photosphere, with uncertain values placed in parentheses. The third provides the abundance in meteorites. Values signaled by an † are based on coronal or stellar data.

1 H	12.00		37 Rb	2.60	2.40
2 He	10.99†		38 Sr	2.90	2.93
3 Li	1.16	3.31	39 Y	2.24	2.22
4 Be	1.15*	1.42	40 Zr	2.60	2.61
5 B	(2.6)	2.88	41 Nb	1.42	1.40
6 C	8.55*		42 Mo	1.92	1.96
7 N	7.97*		44 Ru	1.84	1.82
8 O	8.87*		45 Rh	1.12	1.09
9 F	(4.56)	4.48	46 Pd	1.69	1.70
10 Ne	8.08*†		47 Ag	(0.94)	1.24
11 Na	6.33	6.31	48 Cd	1.77*	1.76
12 Mg	7.58	7.58	49 In	(1.66)	0.82
13 Al	6.47	6.48	50 Sn	2.0	2.14
14 Si	7.55	7.55	51 Sb	1.0	1.04
15 P	5.45	5.57	52 Te		2.24
16 S	7.21	7.27	53 I		1.51
17 Cl	(5.5)	5.27	54 Xe		2.23
18 Ar	6.52*†		55 Cs		1.12
19 K	5.12	5.13	56 Ba	2.13	2.21
20 Ca	6.36	6.34	57 La	1.22	1.20
21 Sc	3.17*	3.09	58 Ce	1.55	1.61
22 Ti	5.02*	4.93	59 Pr	0.71	0.78
23 V	4.00	4.02	60 Nd	1.50	1.47
24 Cr	5.67	5.68	62 Sm	1.01*	0.97
25 Mn	5.39	5.53	63 Eu	0.51	0.54
26 Fe	7.50*	7.51	64 Gd	1.12	1.07
27 Co	4.92	4.91	65 Tb	(0.1)*	0.33
28 Ni	6.25	6.25	66 Dy	1.14	1.15
29 Cu	4.21	4.27	67 Ho	(0.26)	0.50
30 Zn	4.60	4.65	68 Er	0.93	0.95
31 Ga	2.88	3.13	69 Tm	(0.0)	0.13
32 Ge	3.41	3.63	70 Yb	1.08	0.95
33 As		2.37	71 Lu	(0.76)	0.12
34 Se		3.35	72 Hf	0.88	0.73
35 Br		2.63	73 Ta		0.13
36 Kr		3.23	74 W	(1.11)	0.68

Table 5. (*cont.*)

The first column provides the atomic number and formula of the element. The second provides the abundance in the solar photosphere, with uncertain values placed in parentheses. The third provides the abundance in meteorites. Values signaled by an † are based on coronal or stellar data.

75 Re		0.27	81 Tl	(0.9)	0.82
76 Os	1.45	1.38	82 Pb	1.95*	2.05
77 Ir	1.35	1.37	83 Bi		0.71
78 Pt	1.8	1.68	90 Th	(0.08)*	0.08*
79 Au	(1.01)	0.83	92 U		0.49
80 Hg		1.09			

Source: All data are from Anders and Grevesse (1989), except those signaled by an *, which come from an unpublished compilation by N. Grevesse and A. Noels (1993), which those authors have kindly made available.

Table 6. *Surface gravity (g) as a function of spectral type and luminosity class.*

The table provides the values expressed as $\log(g/g_0)$, g_0 = solar surface gravity.

Spectral type	Dwarfs	Giants	Supergiants
O 3	−0.3		
O 5	−0.4		−1.1
O 6	−0.45		−1.2
O 8	−0.5		
B 0	−0.5	−1.1	−1.6
B 3	−0.5		
B 5	−0.4	−0.85	−2.0
B 8	−0.4		
A 0	−0.3		−2.3
A 5	−0.15		−2.4
F 0	−0.1		−2.7
F 5	−0.1		−3.0
G 0	−0.05	−1.5	−3.1
G 5	0.05	−1.9	−3.3
K 0	0.05	−2.3	−3.5
K 5	0.1	−2.7	−4.1
M 0	0.15	−3.1	−4.3
M 2	0.2		−4.5
M 5	0.5		

Source: The values given are taken from Schmidt-Kaler (1982).

Table 7. *Effective temperature as a function of spectral type and luminosity class.*

The table provides the effective temperatures in kelvin. Supergiants refer to luminosity class Iab

Spectral type	Dwarfs	Giants	Supergiants
O 3	52500	50000	47300
O 5	44500	42500	40300
O 8	35800	34700	34200
B 0	30000	29000	26000
B 3	18700	17100	16200
B 5	15400	15000	13600
B 8	11900	12400	11200
A 0	9520	10100	9730
A 5	8200	8100	8510
F 0	7200	7150	7700
F 5	6440	6470	6900
G 0	6030	5850	6550
G 5	5770	5150	4850
K 0	5250	4750	4420
K 5	4350	3950	3850
M 0	3850	3800	3650
M 2	3580	3620	3450
M 4	3370	3430	2980
M 6	3050	3240	2600
M 8	2640		

Source: Data are from Lang (1992).

References

The following short journal abbreviations have been used

AA Astronomy and Astrophysics
AJ Astronomical Journal
Ap. J. Astrophysical Journal
Suppl. Supplement
MN Monthly Notices of the Royal Astronomical Society
PASP Publications of the Astronomical Society of the Pacific
PAS Japan Publication of the Astronomical Society of Japan
Quart. J. RAS Quarterly Journal of the Royal Astronomical Society
ARAA Annual Review of Astronomy and Astrophysics
Ann. d'Astroph Annales d'Astrophysique
Z. Astroph. Zeitschrift für Astrophysik
Publ. Dominion Obs. Publications of the Dominion Observatory Victoria

The abbreviations for meetings are given at the end.

Abbott D. C., Bohlin R. C. and Savage B. D. (1982) *Ap. J. Suppl.* **48**, 369
Abia C., Boffin H. M. J., Isern J., and Rebolo R. (1991) *AA* **245**, Ll
Abt H. A. (1984) *MK Process and Stellar Classification* p. 340 ed. R. Garrison, Toronto
Acker A., Koeppen J., Stenholm B. and Raytchev B. (1991) *AA Suppl.* **89**, 237
Adams W. S. and Pease F. G. (1914) *PASP* **26**, 258
Adelman S. (1973a) *Ap. J.* **182**, 531
Adelman S. (1973b) *Ap. J. Suppl.* **26**, 1
Adelman S. J. (1974) *Ap. J. Suppl.* **28**, 51
Adelman S., Bidelman W. P. and Pyper D. (1979) *Ap. J. Suppl.* **40**, 371
Adelman S. J. (1987) *MN* **228**, 573
Adelman S. J. and Hill G. (1987) *MN* **226**, 581
Adelman S. J. and Lanz T. (1987) ed. *Elemental Abundance Analysis*, Institute d'astronomie de l'Université de Lausanne
Adelman S. (1988) *MN* **230**, 671
Adelman S. (1989) *MN* **239**, 487
Adelman S. J. and Philip A. G. Davis (1990) *MN* **247**, 132
Adelman S. J. (1991) *MN* **252**, 116
Adelman S. J. (1992) *MN* **258**, 167
Adelman S. J. and Philip A. G. Davis (1992a) *MN* **254**, 539
Adelman S. J. and Philip A. G. Davis (1992b) *PASP* **104**, 316
Aikman G. L., Cowley C. R. and Crosswhite H. (1979) *Ap. J.* **232**, 812

Alcolea J. and Bujarrabal V. (1992) *AA* **253**, 475
Allen C. W. (1973) *Astrophysical Quantities*, third edition (Athlone)
Allen D. A. and Swings J. P. (1976) *AA* **47**, 293
Allen D. (1988) *IAU Coll.* **103**, 3
Aller M. F. and Cowley C. R. (1970) *Ap. J.* **162**, L145
Anders F. and Grevesse N. (1989) *Geochimica et Cosmochimica Acta* **53**, 197
Andersen J., Jaschek M. and Cowley C. R. (1984) *AA* **132**, 354
Andreae J. (1993) *Reviews in Modern Astronomy* vol. 5, p. 58, ed. G. Klare (Springer)
Andrillat, Y. and Swings J. P. (1976) *Ap. J.* **204**, L123
Andrillat Y. and Vreux J. M. (1979) *AA* **76**, 221
Andrillat Y. and Fehrenbach Ch. (1982) *AA Suppl.* **48**, 98
Andrillat Y., Vreux J. M. and Dennefeld M. (1982) *IAU Symp.* **98**, 229
Andrillat Y., Jaschek M. and Jaschek C. (1990) *AA Suppl.* **84**, 11
Andrillat Y. and Vreux J. M. (1991) *Montpellier Coll.*, 127
Andrillat Y., Jaschek M. and Jaschek C. (1993) *AA Suppl.* **97**, 781
Andrillat Y., Jaschek M. and Jaschek C. (1994) *AA Suppl.* **103**,135
Angel J. R. P., Liebert J. and Stockman H. S. (1985) *Ap. J.* **292**, 260
Anthony-Twarog B. J., Shawl S. J. and Twarog B. A. (1992) *AJ* **104**, 2229
Appenzeller I., Chavarria C., Krautter J., Mundt R. and Wolf B. (1980) *AA* **90**, 184
Arellano Ferro A., Giridhar S. and Goswami A. (1991) *MN* **250**, 1
Arimoto N. and Cayrel de Strobel G (1988) *IAU Symp.* **132**, 453
Arnaud K. A., Gilmore G. and Cameron A. C. (1989) *MN* **237**, 495
Arnett W. D., Bahcall J. N., Kirshner R. P. and Woosley S. E. (1989) *ARAA* **27**, 629
Artru M. C., Borsenberger J. and Lanz T. (1989) *AA Suppl.* **80**, 17
Aslanov I. A., Rustamov Yu. S. and Kowalski M. (1975) *IAU Coll.* **32**, 311
Auer L. H. and Mihalas D. (1972) *Ap. J. Suppl.* **24**, 193
Auer L. H. and Mihalas D. (1973) *Ap. J. Suppl.* **25**, 433
Avery L. W., Amano T., Bell M. B., Feldman P. A., Johns J. W. C., MacLeod J. M., Matthews H. E., Morton D. C., Watson J. K. G., Turner B. E., Hayahsi S. S., Watt G. D. and Webster A. S. (1992) *Ap. J. Suppl.* **83**, 363
Aydin C. (1972) *AA Suppl.* **7**, 331
Ayres, T. R., Linsky J. L., Simon T., Jordan C. and Brown A. (1983) *Ap. J.* **274**, 784
Ayres, T. R., Marstad N. C. and Linsky J. L. (1981a) *Ap. J.* **247**, 545
Ayres T. R., Moos H. W. and Linsky J. L. (1981b) *Ap. J.* **248**, L137

Babcock H. D. (1945) *Ap. J.* **102**, 154
Balachandran S. (1990) *Ap. J.* **354**, 310
Baldwin J. R., Frogel J. A. and Persson S. E. (1973) *Ap. J.* **184**, 427
Baratta G. B., Damineli Neto A., Rossi C. and Viotti R. (1991) *AA* **251**, 75
Barba R., Brandi E., García L. and Ferrer O. (1992) *PASP* **104**, 330
Barbier D. and Chalonge D. (1941) *Ann. d'Astroph.* **4**, 293
Barbuy B. (1988) *AA* **191**, 121
Barbuy B. (1992) *IAU Symp.* **149**, 143
Barbuy R., Spite M., Spite F. and Milone A. (1981) *AA* **247**, 15
Barlow M. J. and Hummer D. G. (1982) *IAU Symp.* **99**, 387
Barnbaum C., Morris M., Lickel L. and Kastner J. H. (1991) *AA* **251**, 79
Barnett E. W. and McKeith C. D. (1988) *MN* **234**, 325
Baschek B. and Norris J. (1970) *Ap. J. Suppl.* **19**, 327

Baschek B., Hoeflich P. and Scholz M. (1982) *AA* **112**, 76
Baschek B., Heck A., Jaschek C., Jaschek M., Koeppen J., Scholz M. and Wehrse R. (1984) *AA* **131**, 1984
Basri G. S. and Linsky J. L. (1979) *Ap. J.* **234**, 1023
Basri G. (1987) in *Cool Stars, Stellar Systems and the Sun* (Springer)
Basri G., Martin E. L. and Bertout C. (1991) *AA* **252**, 625
Baxandall F. E. (1928) *MN* **88**, 679
Beals C. S. (1951) *Publ. Dominion Obs.* **9**, 1
Beers T. C., Preston G. W., Shectman S. A. and Kage J. (1990) *AJ* **100**, 849
Bell M. B., Feldman P. A., Sun Kwok and Matthews H. E. (1982) *Nature* **295**, 389
Bennett P. D. and Johnson H. R. (1985) *Str. Coll.*, p. 249
Benson P. J. and Little-Marenin I. R. (1987) *Ap. J.* **316**, L39
Benson P. J., Little-Marenin I. R., Woods T. C., Attridge J. M., Blais K. A., Rudolph D. B., Rubiera M. E. and Keefe H. L. (1990) *Ap. J. Suppl.* **74**, 911
Bergeron P., Ruiz M. T. and Leggett S. K. (1992) *Ap. J.* **400**, 315
Bernath P. F., Hinkle K. H. and Keady J. J. (1989) *Science* **244**, 562
Berthet S. (1990) *AA* **227**, 156
Berthet S. (1991) *AA* **251**, 171
Bertout C. (1989) *ARAA* **27**, 351
Betz A. L., McLaren R. A. and Spears D. L. (1979) *Ap. J.* **229**, L97
Betz A. L. and McLaren R. A. (1980) *IAU Symp.* **87**, 503
Betz A. L. (1981) *Ap. J.* **244**, L103
Bianchi I. and Grewing M. (1987) *AA* **181**, 85
Bidelman W. P. and Keenan P. C. (1951) *Ap. J.* **114**, 473
Bidelman W. P. (1953) *Ap. J.* **117**, 377
Bidelman W. P. (1954) *Ap. J. Suppl.* **1**, 214
Bidelman W. P. (1960) *PASP* **72**, 24
Bidelman W. P. (1962a) *Sky and Telescope* **23**, 140
Bidelman W. P. (1962b) *Ap. J.* **135**, 651
Bidelman W. P. (1962c) *AJ* **67**, 111
Bidelman W. P. (1966) *IAU Symp.* **26**, 229
Bidelman W. P. and Corliss C. H. (1966) *Ap. J.* **135**, 968
Bidelman W. P. and Irvine C. E. (1974) *Bull. AAS* **6**, 365
Bidelman W. P. (1979) *Bull. AAS* **11**, 364
Boesgaard A. M. (1976a) *PASP* **88**, 353
Boesgaard A. M. (1976b) *Ap. J.* **210**, 466
Boesgaard A. M. and Heacox W. D. (1978) *Ap. J.* **226**, 888
Boesgaard A. M. and Lavery R. J. (1986) *Ap. J.* **309**, 762
Boesgaard A. M. and Tripicco M. J. (1986) *Ap. J.* L49
Boesgaard A. M. and Budge K. G. (1989) *Ap. J.* **338**, 875
Boesgaard A. M. and Friel E. D. (1990) *Ap. J.* **351**, 467
Bond H. E. and Luck R. E. (1988) *IAU Symp.* **132**, 477
Bond H. E. (1991) *IAU Symp.* **145**, 341
Bouigue R. (1954) *Ann d'Astroph.* **17**, 97
Bowen I. S. and Swings P. (1947) *Ap. J.* **105**, 92
Boyartchuk A. A. (1557) *Isv. Crimea* **17**, 89
Boyartchuk A. A. and Savonov I. S. (1986) *IAU Coll.* **90**, 433
Boyartchuk A. A. and Snow T. P. (1978) *Ap. J.* **219**, 519

Boyartchuk M. E. (1969) *Isv. Crimea* **39**, 114
Branch D. and Peery B. F. (1970) *PASP* **82**, 1060
Branch D. (1990) in *Supernovae* ed. A. G. Petschek (Springer)
Brandi E. and Jaschek M. (1970) *PASP* **82**, 847
Bregman J.-D., Goebel J. H. and Strecker D. W. (1978) *Ap. J.* **223**, L45
Bretz M. C. (1966) *Trieste Coll.*, 166
Broten *et al.* (1978)
Brown A., de Ferraz M. C. and Jordan C. (1984) *MN* **207**, 831
Brown J. A., Sneden C., Lambert D. L. and Dutchover E. (1989) *Ap. J. Suppl.* **71**, 293
Brown, J., Smith V. V., Lambert D. L., Dutchover E., Hinkle K. and Johnson H. R. (1990) *AJ* **99**, 1930
Brown J. A., Wallerstein G., Cunha K. and Smith V. V. (1991) *AA* **249**, L13
Bruhweiler F. C., Kondo Y. and McCluskey G. E. (1981) *Ap. J. Suppl.* **46**, 255
Bujarrabal V. and Planesas P. A. del Romero (1987) *AA* **175**, 164
Burbidge E. M., Burbidge G. R., Fowler W. A. and Hoyle F. (1957) *Rev. Modern Physics* **29**, 547
Burbidge G. R. and Burbidge E. M. (1955) *Ap. J. Suppl.* **1**, 431
Burkhart C. and Coupry M. F. (1991) *AA* **249**, 205
Burwell C. G. (1930) *PASP* **42**, 351
Buscombe W. and Merrill P. W. (1952) *Ap. J.* **116**, 525
Buscombe W. (1969) *MN* **144**, 1
Butcher H. (1988) *The Messenger* **51**, 12
Byrne P. B. and Doyle J. G. (1990) *AA* **238**, 221

Carbon D. F., Barbuy B., Kraft R., Friel E. and Suntzeff N. (1987) *PASP* **99**, 335
Carpenter K. G. and Wing R. F. (1985) *Ap. J. Suppl.* **57**, 405
Carpenter K. G. (1992) *IAU Symp.* **151**, 51
Carroll P. K., McCormack and O'Connor S. (1976) *AJ* **208**, 903
Carter B. D. and O'Mara B. J. (1991) *MN* **253**, 47
Cassatella A. and Gonzalez Riestra R. (1988) *IAU Coll.* **94**, 115
Catchpole R. M. (1975) *PASP* **87**, 397
Catchpole R. M. and Feast M. W. (1976) *MN* **175**, 501
Cayrel de Strobel G. (1966) *Ann. d'Astroph.* **29**, 413
Cayrel de Strobel G. (1968) *Ann. d'Astroph.* **31**, 43
Cayrel de Strobel G. (1983) *PASP* **95**, 111
Cayrel de Strobel G. (1990) *Mem. S. A. Ital.* **61**, 613
Cayrel de Strobel G. (1991) *Rome Coll.*, 27
Cayrel de Strobel G. and Cayrel R. (1989) *AA* **218**, L9
Cayrel de Strobel G., Hauck B., Francois P., Thévenin, F., Friel E., Mermilliod M. and Borde S. (1992) *AA Suppl.* **95**, 273
Cernicharo J., Kahane C., Gomez-Gonzalez J. and Guelin M. (1986a) *AA* **164**, L1
Cernicharo J., Kahane C., Gomez-Gonzalez J. and Guelin M. (1986b) *AA* **167**, L5
Cernicharo J. and Guelin M. (1987) *AA* **183**, L10
Cernicharo J., Guelin M., Hein H. and Kahane C. (1987) *AA* **181**, L9
Cernicharo J., Gottlieb C. A., Guelin M., Thaddeus P. and Vrtilek J. M. (1989) *Ap. J.* **341**, L25
Cernicharo J., Gottlieb C. A., Guelin M., Killian T. C., Paubert G., Thaddeus P. and Vrtilek J. M. (1991a) *Ap. J.* **368**, L39

Cernicharo J., Gottlieb C. A., Guelin M., Killian T. C., Thaddeus P. and Vrtilek J. M. (1991b) *Ap. J.* **368**, L43
Cernicharo J., Guelin M., Kahane C., Bogey M., Demuynck C. and Destombes J. L. (1991c) *AA* **246**, 213
Chanmugam G. (1992) *ARAA* **30**, 143
Chen Peisheng, Gao Heng and Bao Mengxian (1992) *Chinese Astr. Astroph.* **16**, 153
Ciatti F., d'Odorico S. and Mammano A. (1974) *AA* **34**, 181
Clayton G. C., Whitney B. A., Stanford S. A., Drilling J. S. and Judge P. G. (1992) *Ap. J.* **384**, L23
Clegg R. E. S. and Lambert D. L. (1978) *Ap. J.* **226**, 931
Clegg R. E. S., Lambert D. L. and Bell R. A. (1979) *Ap. J.* **234**, 188
Clegg R. and Wyckoff S. (1979) *MN* **179**, 417
Clegg R. E. S., Hinkle K. H. and Lambert D. L. (1982) *MN* **201**, 95
Code A. D. and Meade M. R. (1979) *Ap. J. Suppl.* **39**, 195
Cohen J. G., Deutsch A. J. and Greenstein J. L. (1969) *Ap. J.* **156**, 629
Conti P. S. (1965) *Ap. J.* **142**, 1594
Conti P. S., Wallerstein G. and Wing R. F. (1966) *Ap. J.* **142**, 999
Conti, P. S. and Strom S. E. (1968) *Ap. J.* **152**, 483
Conti P. S. (1974) *Ap. J.* **187**, 539
Conti P. S. and Leep E. M. (1974) *Ap. J.* **193**, 113
Conti P. S. and Massey P. (1989) *Ap. J.* **337**, 251
Conti P. S., Massey P. and Vruex J. M. (1990) *Ap. J.* **354**, 359
Cottrell P. L. and Norris J. (1978) *Ap. J.* **221**, 893
Cottrell P. L. and Lambert D. L. (1982) *Ap. J.* **261**, 595
Cowley C. R., Hartoog M. and Cowley A. P. (1974) *Ap. J.* **194**, 343
Cowley C. R. (1976) *Ap. J. Suppl.* **32**, 631
Cowley C. R., Aikman G. C. L. and Hartoog M. R. (1976) *Ap. J.* **206**, 196
Cowley, C. R., Aikman G. C. L. and Fisher W. A. (1977) *Publ. Dominion Obs.* **15**, 37
Cowley C. R. and Arnold C. N. (1978) *Ap. J.* **226**, 420
Cowley C. R. and Crosswhite H. M. (1978) *PASP* **90**, 108
Cowley C. R. (1979) *Les éléments et leur isotopes dans l'Univers, Liège XXII Coll.* 373
Cowley C. R. and Henry R. (1979) *Ap. J.* **233**, 633
Cowley C. R. (1980) *Vistas in Astronomy* **24**, 245
Cowley C. R. (1981) *23rd Liège Astroph. Coll.*, 169
Cowley C. R. (1984) *Physica Scripta* T8, 28
Cowley C. R. and Dempsey R. C. (1984) *Str. Coll.*, 225
Cowley C. R. (1987) *Obs.* **107**, 188
Cowley C. R. and Greenberg M. (1987) *PASP* **99**, 1201
Cowley C. (1991) *IAU Symp.* **145**, 183
Crivellari L., Flora U, Mercanti M., Morossi C., Rusconi L. and Sedmak G. (1981) *Astroph. Space Sci.* **80**, 425
Curiel S. (1992) *IAU Symp.* **150**, 373

Dahari O. and Osterbrock D. E. (1984) *Ap. J.* **277**, 648
Dalgarno A. and Layzer D. L. (1987) ed. *Spectroscopy of Astrophysical Plasmas* (Cambridge University Press)

Danziger I. J. (1965) *MN* **131**, 51
Davis D. N. (1940) *Publ. Amer. Astr. Soc.* **10**, 48
Davis D. N. (1947) *Ap. J.* **106**, 28
Davis D. N. (1984) *Str. Coll.* 261
Davis S.-P. and Hammer P. D. (1981) *Ap. J.* **250**, 805
Dean C. A. and Bruhweiler F. C. (1985) *Ap. J. Suppl.* **57**, 133
Dempsey R. C., Bopp B. W., Henry G. W. and Hall D. S. (1993) *Ap. J. Suppl.* **86**, 293
Denn G. R., Luck R. E. and Lambert D. L. (1991) *Ap. J.* **377**, 657
Deutsch A. (1956) *PASP* **68**, 92
Deutsch A. and Merrill P. W. (1959) *Ap. J.* **130**, 570
Diaz A. J., Terlevich E. and Terlevich R. (1989) *MN* **239**, 325
Didelon P. (1982) *AA Suppl.* **50**, 199
Didelon P. (1986) *Rev. Méx. Astr. Astrof.* **12**, 163
Doherty, L. R. (1985) *MN* **217**, 41
Dominy J. F. (1984) *Ap. J. Suppl.* **55**, 27
Dominy J. F. (1985) *PASP* **97**, 1104
Doschek G. A., Feldman U., van Hoosler M. E. and Bartoe J. D. (1976) *Ap. J. Suppl.* **312**, 417
Doyle J. G., Panagi P. and Byrne P. B. (1990) *AA* **228**, 443
Doyle J. G. (1991) *Irish AJ* **20**, 47
Duncan D. K., Vaughan A. H., Wilson O. C., Preston G. W., Frazer J., Lanning H., Misch A., Mueller J., Soyumer D., Woodard L., Baliunas S., Noyes R. W., Hartmann L. W., Porter A., Zwaan C., Middelkoop F. and Rutten R. G. M. (1991) *Ap. J. Suppl.* **76**, 383
Duncan D. K., Lambert D. L. and Lemke M. (1993) *Ap. J.* **401**, 584
Dupree A. K. (1986) *ARAA* **24**, 377
Dupree A. K., Hartmann L. and Smith G. H. (1990) *Ap. J.* **353**, 623
Dworetsky M. M. (1969) *Ap. J.* **156**, 1969
Dworetsky M. M. (1985) *IAU Coll.* **90**, 397

Eaton J. A., Johnson H. R., O'Brien G. T. and Baumert J. H. (1985) *Ap. J.* **290**, 276
Edlen B. (1942) *Arkiv Mat. Astr. Fys.* B28, N.1
Eenens P. R. J., Williams P. M. and Wade R. (1991) *MN* **252**, 300
Eenens P. R. and Williams P. M. (1991) *Montpellier Coll.* 158
Eidelsberg M., Crifo-Magnant F. and Zeippen C. J. (1981) *AA Suppl.* **43**, 455
Elgoroy O. (1988) *AA* **204**, 147
Elmsley J. (1989) *The Elements* (Clarendon)
Engels D. (1987) *Calgary Conference*, 87
Engvold O. (1973) *AA Suppl.* **10**, 11
Engvold O., Woehl H. and Brault J. W. (1980) *AA Suppl.* **42**, 209
Evans A., Whittet D. C. B., Davies J. K., Kilkenny D. and Bode M. F. (1985) *MN* **217**, 767
Evans A. (1989) in *Classical Novae* ed. Bode M. F. and Evans A. (Wiley)

Fanelli M. N., O'Connell R. W., Burstein D. and Wu C. C. (1990) *Ap. J.* **364**, 272
Faraggiana R., Gerbaldi M., Castelli F. and Floquet M. (1986) *AA* **158**, 200
Faraggiana R., Gerbaldi M., van t'Veer C. and Floquet M. (1988) *AA* **201**, 259
Fay T. D., Fredrick L. W. and Johnson H. R. (1968) *Ap. J.* **152**, 151

Fehrenbach Ch. (1956) *Handbuch der Physik* vol. 50, p. 1 (Springer)
Feibelman W. A., Bruhweiler F. C. and Johansson S. (1991) *Ap. J.* **373**, 649
Feldman U. and Doschek G. A. (1991) *Ap. J. Suppl.* **75**, 925
Ferlet R., Dennefeld M. and Spite M. (1983) *AA* **124**, 172
Fernandez-Villacañas J.-L., Rego M. and Cornide M. (1990) *AJ* **99**, 1961
Fillipenko A. V. (1988) *AJ* **96**, 1941
Firth J. G., Freeman F. F., Gabriel A. H., Jones B. B., Jordan C., Negus C. R., Shenton D. B. and Turner R. F. (1974) *MN* **166**, 543
Fix J. D. and Cobb M. L. (1987) *Ap. J.* **312**, 290
Francois P. (1991) *AA* **247**, 56
Francois P. (1992) *IAU Symp.* **149**, 417
Freil E. D. and Boesgaard A. M. (1990) *Ap. J.* **351**, 480
Freire Ferrero R., Catalano S., Marilli E., Wonnacot D., Gouttebroze P., Bruhweiler F. and Talavera A. (1992) *Vaiana Memorial Symp. Advances in Stellar and Solar Coronal Physics*
Freitas Pacheco J. A. and Costa R. D. D. (1992) *AA* **257**, 619
Frost E. B. (1895) *PASP* **7**, 317
Frost S. A. and Conti P. S. (1976) *IAU Symp.* **70**, 139
Fuhrmann K. (1989) *AA Suppl.* **77**, 345
Fujita Y., Yamashita Y., Kamijo F., Tsuji T. and Utsumi K. (1963) *Publ. Dominion Obs.* **12**, 293
Fujita Y., Tsuji T. and Maehara H. (1966) *Trieste Coll.* 75
Fujita Y. (1990) *Proc. Japan Acad.* **66**, N.5, 1
Fujita Y. (1992) *Proc. Japan Acad.* **68**, ser. B, 1

Gahm G. F. (1974) *AA Suppl.* **18**, 259
Gahm G. F., Lindgren B. and Lindroos K. P. (1977) *AA Suppl.* **27**, 277
Galkina T. S. and Kopylov I. M. (1962) *Isv. Crimea* **28**, 35
García López R. J., Rebolo R., Beckman J. E. and McKeith C. D. (1993) *AA* **273**, 482
Garstang R. H. (1977) *Rep. Prog. Physics* **40**, 105
Gehren, R., Reimers D., Berthold L., Berthold J. and Hennig R. (1978) *AA Suppl.* **31**, 297
Gerbaldi M., Faraggiana R. and Molaro P. (1986) *ESA SP-263*, p. 49
Gerbaldi M., Floquet M., Faraggiana R. and van t'Veer-Menneret C. (1989) *AA Suppl.* **81**, 127
Gillet D. (1988) *IAU Symp.* **132**, 143
Gilroy K. K., Sneden C., Pilachowski C. and Cowan J. J. (1988) *Ap. J.* **327**, 298
Ginestet N., Carquillat J. M., Jaschek M., Jaschek C., Pedoussaut A., Rochette J. (1992) *Atlas of Stellar Spectra* (Toulouse Observatory)
Glebocki R., Musielak G. and Stawikowski (1980) *Acta Astron.* **30**, N.4
Goebel J. H., Bregman J. D., Strecker D. W., Witteborn F. C. and Erickson E. F. (1978) *Ap. J.* **222**, L129
Goebel J. H., Bregman J. D., Witteborn F. C., Taylor B. J. and Willner S. P. (1981) *Ap. J.* **246**, 455
Goebel J. H., Bregman J. D., Cooper D. M., Goorvitch D., Langhoff S. R. and Witteborn F. C. (1983) *Ap. J.* **270**, 190
Golay M. (1974) *Introduction to Astronomical Photometry* (Reidel)

Goldberg L. and Mueller E. A. (1953) *Ap. J.* **118**, 397
Goldberg L., Parkinson W. H. and Reeves E. M. (1965) *Ap. J.* **141**, 1293
Goldhaber D. M., Betz A. L. and Ottusch J. J. (1987) *Ap. J.* **314**, 356
Gomez Y., Moran J. M. and Rodriguez L. F. (1991) *Third Haystack Conference* p. 403
Gonzalez G. and Wallerstein G. (1992) *MN* **254**, 343
Gopka V. F. and Komarov N. S. (1990) *Sov. Astr.* **34**, 610
Gopka V. F., Komarov N. S., Mishenina T. V. and Yuschchenko A. V. (1990) *Sov. Astr.* **34**, 606
Goy G., Jaschek C. and Jaschek M. (1995) *Atlas of Tracings of Spectra* (Geneva Observatory)
Gratton R. and Sneden C. (1987) *AA* **178**, 179
Gratton R. and Sneden C. (1988) *AA* **204**, 193
Gratton R. G. (1989) *AA* **208**, 171
Green P. J., Margon B. and McConnell J. (1991) *Ap. J.* **380** L31
Greene A. E. and Wing R. F. (1975) *Ap. J.* **200**, 688
Greenstein J. (1942) *Ap. J.* **92**, 161
Greenstein J. (1943) *Ap. J.* **97**, 252
Greenstein J. L. (1948) *Ap. J.* **107**, 151
Grevesse N. and Sauval A. J. (1971) *AA* **14**, 477
Grevesse N. and Sauval A. J. (1973) *AA* **27**, 29
Grevesse N. and Sauval A. J. (1991) *Montpellier Coll.* 215
Grigsby J. A., Morrison N. D. and Anderson L. S. (1992) *Ap. J. Suppl.* **78**, 205
Grotrian W. (1939) *Naturwissenschaften* **27**, 214
Grudzinska S. (1984) *Str. Coll.* 267
Guelin M. and Thaddeus P. (1977) *Ap. J.* **212**, L81
Guelin M., Cernicharo J., Kahane C., Gomez-Gonzalez J. and Walmsley C. M. (1987a) *AA* **175**, L5
Guelin M., Cernicharo J., Navarro S., Woodward D. R., Gottlieb C. A. and Thaddeus P. (1987b) *AA* **182**, L37
Guelin M., Cernicharo J., Paubert G. and Turner B. E. (1990) *AA* **230**, L9
Guelin M. and Cernicharo J. (1991) *AA* **244**, L21
Guilloteau S., Lucas R., Nguyen-Q-Rieu and Omont A. (1986) *AA* **165**, L1
Guinan E. F. and Smith G. H. (1984) *PASP* **96**, 354
Gurzadyan G.A. (1975) *PASP* **87**, 289
Gustafsson B. (1989) *ARAA* **27**, 701
Gustafsson B. (1992) *HC*, p. 78
Guthrie B. N. G. (1969) *Obs.* **89**, 224
Guthrie B. N. B. (1972) *Astroph. Space Sci.* **15**, 214
Guthrie B. N. G. (1984) *MN* **206**, 85

Hack M., Engin S. and Yilmaz N. (1989) *AA* **225**, 143
Haenni L. and Kipper M. (1975) *Publ. Tartu. Obs.* **43**, 51
Hagen Bauer W., Stencel R. E. and Neff D. H. (1991) *AA Suppl.* **90**, 175
Haisch B. M. (1987) *Cool Stars, Stellar Systems and the Sun* ed. Linsky J. L. and Stencel R. E. (Springer), p. 269
Hall D. N. B. and Noyes R. W. (1969) *Astroph. Lett.* **4**, 143

Hall D. N. B. and Noyes R. W. (1972) *Ap. J.* **175**, L95
Hall D. N. B. (1974) *An Atlas of Infrared Spectra of the Solar Photosphere and of Sunspot Umbrae* (Kitt Peak National Observatory)
Hall D. N. B. and Ridgway S. T. (1978) *Nature* **273**, 281
Hamann W. R., Duennebeil G., Koesterke L., Schmutz W. and Wessolowski U. (1991) *AA* **249**, 443
Hanuschik R. W., Kozok J. R. and Kaiser D. (1988) *AA* **189**, 147
Hardorp J. (1966) *Z. Astroph.* **63**, 137
Hardorp J. and Scholz M. (1970) *Ap. J. Suppl.* **19**, 193
Harper G. M. (1990) *MN* **243**, 381
Harris M. J., Lambert D. L. and Smith V. V. (1985a) *Ap. J.* **292**, 620
Harris M. J., Lambert D. L. and Smith V. V. (1985b) *Ap. J.* **299**, 375
Harris M. J., Lambert D. L., Hinkle K. H., Gustafsson B. and Eriksson K. (1987) *Ap. J.* **316**, 294
Harris M. J. and Lambert D. L. (1987) *Ap. J.* **318**, 868
Harris M. J., Lambert D. L. and Smith V. (1988) *Ap. J.* **325**, 768
Hartoog M. R., Cowley C. R. and Cowley A. P. (1973) *Ap. J.* **182**, 847
Hartoog M. R., Cowley C. R. and Adelman S. J. (1974) *Ap. J.* **187**, 551
Hartoog M. R. and Cowley A. P. (1979) *Ap. J.* **228**, 229
Harvey P. M. (1984) *PASP* **96**, 297
Hauck, B., Jaschek C., Jaschek M. and Andrillat Y. (1991) *AA* **252**, 260
Hauge O. (1971) *AA* **10**, 73
Heber U. (1987) *Mitt. Astron. Ges.* **70**, 71
Heber U. (1991) *IAU Symp.* **145**, 363
Heck, A., Egret D., Jaschek M. and Jaschek C. (1984) *ESA SP-1052*
Helfer H. L. and Wallerstein G. (1968) *Ap. J. Suppl.* **16**, 1
Henkel C., Matthews H. E., Morris M., Terebey S. and Fich M. (1985) *AA* **147**, 143
Herbig G. H. (1948) *PASP* **60**, 378
Herbig G. H. (1949) *Ap. J.* **110**, 143
Herbig G. H. (1956) *PASP* **68**, **204**
Herzberg G. (1945) *Atomic Spectra and Atomic Structure* (Dover)
Hill L. C. (1993) *Quart. J. RAS* **34**, 73
Hinkle K. H., Keady J. J. and Bernath P. F. (1988) *Science* **241**, 1319
Hinkle K. H., Lambert D. L. and Wing R. F. (1989a) *IAU Coll.* **106**, 61
Hinkle K. H., Lambert D. L. and Wing R. F. (1989b) *MN* **238**, 1365
Hjalmarson A. and Olofsson H. (1979) *Ap. J.* **234**, L199
Hobbs L. M. and Thorburn J. A. (1991) *Ap. J.* **375**, 116
Hollandsky O. P. and Kopylov I. M. (1962) *Isv. Crimea* **28**, 3
Holweger H., Gigas D. and Steffen M. (1986) *AA* **155**, 58
Holweger H. and Stuerenburg S. (1991) *AA* **252**, 255
Houk N. and Newbery M. V. (1984) *A Second Atlas of Objective Prism Spectra* (Department of Astronomy, University of Michigan)
Howarth I. D. and Prinja R. K. (1989) *Ap. J. Suppl.* **69**, 527
Humphreys R. M., Kudritzki R. P. and Groth H. G. (1991) *AA* **245**, 593
Hunger K., Gruschinske J., Kudritzki R. P. and Simon K. P. (1981) *AA* **95**, 244
Husfeld D., Butler K., Heber U. and Drilling J. (1989) *AA* **222**, 150
Hyland A. R. (1974) – *IAU Highlights of Astronomy*, 307

Jacobs J. M. and Dworetsky M. M. (1981) *23rd Liège Astroph. Coll.* 153
Jaschek C. and Jaschek M. (1987a) *The Classification of Stars* (Cambridge University Press)
Jaschek C. (1991) *Montpellier Coll.* 113
Jaschek M. and Jaschek C. (1962) *PASP* **74**, 154
Jaschek M. and Lopez García Z. (1966) *Z. Astroph.* **64**, 217
Jaschek M. and Lopez García Z. (1967) *PASP* **79**, 62
Jaschek M. and Malaroda S. (1970) *Nature* **225**, 246
Jaschek M. and Brandi E. (1972) *AA* **20**, 233
Jaschek M. and Jaschek C. (1974) *Vistas in Astronomy* **16**, 131
Jaschek M. and Jaschek C. (1987b) *AA*, **171**, 380
Jaschek M., Andrillat Y. and Jaschek C. (1989) *AA* **218**, 180
Jaschek M., Jaschek C., Andrillat Y. and Houziaux L. (1992) *MN* **254**, 413
Jaschek M., Jaschek C. and Andrillat Y. (1993) *AA Suppl.* **97**, 781
Jaschek M., Andrillat Y., Houziaux L. and Jaschek C. (1994) *AA* **282**, 911
Jefferies J. T., Orrall F. Q. and Zirker J. B. (1971) *Solar Physics* **16**, 103
Jefferts K. B., Penzias A. A. and Wilson R. W. (1970) *Ap. J.* **161**, L87
Jeffery C. S. and Heber U. (1993) *AA* **270**, 167
Jennings D. E., Boyle R. J., Wiedemann G. R. and Moseley S. H. (1993) *Ap. J.* **408**, 277
Johansson L. E. B., Andersson C., Ellder J., Friberg P., Hjalmarson A., Hoeglund B., Irving W. M., Olofsson H. and Rydbeck G. (1984) *AA* **130**, 227
Johansson S., Nave G., Learner R. C. M. and Thorne A. P. (1991) *Montpellier Coll.* 189
Johnson H. L. (1977) *Rev. Mexicana Astr. Astrof.* **2**, 71
Johnson H. R., Goebel J. H., Goorvitch D. and Ridgway S. T. (1983) *Ap. J.* **270**, L63
Johnson H. R. (1992) *IAU Symp.* **151**, 157
Jones D. H. P. (1973) *MN* **161**, 19P
Jones T. J. (1987) *Calgary Conference* 3
Jonsson J., Launila O. and Lindgren B. (1992) *MN* **258**, 49P
Joras P. (1986) *ESA SP-263* 129
Jordan C., Brueckner G. E., Bartoe J. D., Sandlin G. D. and van Hoosier M. E. (1977) *MN* **187** 473
Jordan C. (1988) *IAU Coll.* **94**, 223
Jorgenson U. F. (1991) *Montpellier Coll.* 165
Jorissen A. (1988) *IAU Symp.* **132**, 593
Jorissen A., Smith V. V. and Lambert D. L. (1992) *AA* **261**, 164
Joy A. H. (1926) *Ap. J.* **63**, 281
Joy A. H. (1945) *Ap. J.* **102**, 168
Joy A. H. and Swings P. (1945) *Ap. J.* **102**, 353
Joy A. H. (1954) *Ap. J. Suppl.* **1**, 39
Judge P. G. (1988) *IAU Symp.* **132**, 163
Judge P. G. (1989) *IAU Coll.* **106**, 303
Judge P. G. and Jordan C. (1991) *Ap. J. Suppl.* **77**, 75
Jura M. (1991) *Ap. J.* **372**, 208

Kahane G., Cernicharo J., Gomez-Gonzalez J. and Guelin M. (1992) *AA* **256**, 235

Kato K. (1987) *PAS Japan* **39**, 517

Kato K. and Sadakane K. (1986) *AA* **167**, 111

Kawaguchi K., Kagi E., Hirano T., Takana S. and Saito S. (1993) *Ap. J.* **406**, L39

Keenan F. P., Bates B., Dufton P. L., Holmgren D. E., Gilheany S. (1990) *Ap. J.* **348**, 322

Keenan P. C. and Hynek J. A. (1950) *Ap. J.* **111**, 1

Keenan P. C. (1957) *PASP* **69**, 5

Keenan P. C. and McNeil, R. C. (1976) *An Atlas of Spectra of the Cooler Stars* (Ohio State University)

Kelly R. L. (1987) *Atomic and Ionic Spectrum Lines Below 2000 Ångström, J. Phys. Chem. Ref. Data* **16**, suppl. 1

Kenyon S. J. and Fernandez-Castro T. (1987) *AJ* **93**, 938

Kerr and Trotman-Dickenson (1972) *Handbook of Chemistry and Physics* (CRC Press) 53rd edition F. 183

Kilian J. and Nissen P. E. (1989) *AA Suppl.* **80**, 255

Kilston S. (1975) *PASP* **87**, 189

Kipper T. (1976) *Publ. Tartu.* **44**, 91

Kipper T. A. and Kipper M. A. (1979) *Tartu Obs.* **47**, 222

Kipper T. and Wallerstein G. (1990) *PASP* **102**, 574

Kipper T. (1992a) *Baltic Astr.* **1**, 190

Kipper T. (1992b) *Baltic Astr.* **1**, 181

Kirkpatrick J. D., Henry T. J. and McCarthy D. W. (1991) *Ap. J. Suppl.* **77**, 417

Kirkpatrick J. D., Kelly D. M., Rieke G. H., Liebert J., Allard F. and Wehrse R. (1993) *Ap. J.* **402**, 643

Kleinmann S. G. and Hall D. N. B. (1986) *Ap. J. Suppl.* **62**, 501

Knapp G. R. and Morris M. (1985) *Ap. J.* **292**, 640

Kodaira K. and Scholz M. (1970) *AA* **6**, 93

Kodaira K. and Takada M. (1978) *Ann. Tokyo Obs.* second ser. **17**, 79

Koelblod D. and van Paradijs J. (1975) *AA Suppl.* **19**, 101

Koeppen, J. (1988) *AA Suppl.* **75**, 381

Kohl J. L., Parkinson W. H. and Wothbroe G. L. (1977) *Ap. J.* **212**, L101

Kollatschny W. (1980) *AA* **86**, 308

Kondo Y., Morgan T. H. and Modisette J. L. (1977) *PASP* **89**, 675

Kraft R. P. (1960) *Stellar Atmospheres* ed. J. Greenstein (University of Chicago Press) p. 370

Krishnaswamy K. and Sneden C. (1985) *PASP* **97**, 407

Kudritzki R. P. (1987) *ESO Workshop on SN 1987a* ed. I. J. Danziger (ESO, Garching bei München) p. 39

Kurtz D. W. (1976) *Ap. J. Suppl.* **32**, 651

Kwok S., Bell M. B. and Feldman P. A. (1981) *Ap. J.* **247**, 125

Laborde G. (1961) *Ann. d'Astroph.* **24**, 89

Lambert D. L., Mallia E. A. and Warner B. (1969) *MN* **142**, 72

Lambert D. L. and Luck R. E. (1976) *Obs.* **96**, 100

Lambert D. L. and Clegg R. E. S. (1980) *MN* **191**, 367

Lambert D. L., Hinkle K. H. and Hall D. N. B. (1981) *Ap. J.* **248**, 638

Lambert D. L. (1983) *PASP* **100**, 1202

Lambert D. L. (1985) *Str. Coll.* 191

Lambert D. L., Gustafsson B., Eriksson K. and Hinkle K. H. (1986) *Ap. J. Suppl.* **62**, 373
Lambert D. L. (1987) *Ap. J. Suppl.* **65**, 255
Lambert D. L., McWilliam A. and Smith V. V. (1987) *Astroph. Space Sci.* **133**, 369
Lambert D. L. (1988) *IAU Symp.* **132**, 563
Lambert D. L. (1989) *IAU Coll.* **106**, 101
Lambert D. L., Hinkle K. H. and Smith V. V. (1990) *AJ* **99**, 1612
Lambert D. L., Heath J. E. and Edvardsson B. (1991) *MN* **253**, 610
Lambert D. L. (1991) *IAU Symp.* **145**, 299
Lamers H. J. (1972) *AA Suppl.* **7**, 113
Lamontagne R., Wesemael F. and Fontaine G. (1985) *Ap. J.* **299**, 496
Landsman W. and Simon T. (1993) *Ap. J.* **408**, 305
Lang K. R. (1992) *Astrophysical Data: Planets and Stars* (Springer)
Langer G. E., Suntzeff N. B. and Krafft R. P. (1992) *PASP* **104**, 523
Lanz T., Artru M. C., Didelon P. and Mathys G. (1993) *AA* **272**, 465
Lazaro C., Lynas-Gray A. E., Clegg R. E. S., Mountain C. M. and Zadrozny A. (1991) *MN* **249**, 62
Learner R. C. M., Davies J. and Thorne A. P. (1991) *MN* **248**, 414
Le Bertre T., Deguchi S. and Nakada Y. (1990) *AA* **235**, L5
Leckrone D. S., Johansson S. G. and Wahlgren G. N. (1991a) *NASA*, SP
Leckrone D. S., Wahlgren G. M. and Johansson S.G. (1991b) *Ap. J.* **377**, L37
Lemke M. (1989) *AA* **225**, 125
Lemke M. (1990) *AA* **240**, 331
Lemke M., Lambert D. L. and Edvardsson B. (1993) *PASP* **105**, 468
Lennon D. J., Kudritzki R. P., Becker S. T., Butler K., Eber F., Groth H. G. and Kunze D. (1991) *AA* **252**, 498
Le Squeren A. M. and Sivagnanam P. (1985) *AA* **152**, 85
Lindblad B. (1922) *Ap. J.* **55**, 85
Lindqvist M., Nyman L. A., Olofsson H. and Winnberg A. (1988) *AA* **205**, L15
Lindqvist M., Olofsson H., Winnberg A. and Nyman L. A. (1992) *AA* **263**, 183
Linsky J. L. (1980) *ARAA* **18**, 439
Linsky J. L., Bornmann P. L., Carpenter K. G., Wing R. F., Giampapa M. S., Worden S. P. and Hege E. K. (1982) *Ap. J.* **260**, 670
Little-Marenin I. R. (1986) *Ap. J.* **307**, L15
Little-Marenin I. R. (1989) *IAU Coll.* **106**, 131
Little-Marenin I. R., Benson P. J., McConahay M. M., Cadmus R. R., Stencel R. E. and Eriksson K. (1991) *AA* **249**, 465
Lloyd Evans T. (1989) *IAU Coll.* **106**, 241
Lloyd Evans T. (1991) *MN* **249**, 409
Lockwood W. (1968) *AJ* **73**, 14
Lorenz Martins S. and Codina S. J. (1992) *IAU Symp.* **150**, 403
Lucas R., Omont A., Guilloteau S. and Nguyen-Q-Rieu (1986) *AA* **154**, L12
Lucas R., Guilloteau S. and Omont A. (1988) *AA* **194**, 230
Luck R. E. (1977) *Ap. J.* **218**, 752
Luck R. E. and Bond H. E. (1982) *Ap. J.* **259**, 792
Luck R. E. and Bond H. E. (1985) *Ap. J.* **292**, 559
Luck R. E. (1991a) *Ap. J. Suppl.* **75**, 579

Luck R. E. (1991b) *IAU Symp.* **145**, 247
Luck R. E. and Bond H. E. (1991) *Ap. J. Suppl.* **77**, 515
Luck R. E. (1992) *CDS Catalogue 3139 'A Compendium of Equivalent Width Measures'*
Luck R. E. and Bond H. E. (1992) *Ap. J. Suppl.* **77**, 515
Luck R. E. and Lambert D. D. L. (1992) *Ap. J. Suppl.* **79**, 303
Luyten W. (1923) *PASP* **35**, 175

McAlliatrer H. C. (1960) *A Preliminary Photoelectric Atlas of the Solar Ultraviolet Spectrum from 1800 to 2965 Å* (University of Colorado Press)
McConnell D. J., Wing R. F. and Costa E. (1992) *AJ* **104**, 821
McGregor P. J., Persson S. E. and Cohen J. G. (1984) *Ap. J.* **286**, 609
McGregor P. J., Hyland A. R. and Hiller D. J. (1988) *Ap. J.* **324**, 1071
McWilliam A. and Lambert D. L. (1988) *MN* **230**, 573
Maeckle R., Holweger H., Griffin R. and Griffin R. (1975a) *AA* **38**, 230
Maeckle R., Griffin R., Griffin R. and Holweger H. (1975a) *AA Suppl.* **19**, 303
Magain P. (1989) *AA* **209**, 211
Malaney R. A. and Lambert D. L. (1988) *MN* **235**, 695
Malinovsky M. and Heroux L. (1973) *Ap. J.* **181**, 1009
Mantegazza L. (1991) *AA Suppl.* **88**, 255
Marlborough J. M. (1982) *IAU Symp.* **98**, 361
Mathioudakis M. and Doyle J. G. (1991) *AA* **244**, 409
Mathys G. (1990) *AA* **232**, 151
Mathys G. (1991) *AA* **245**, 467
Mathys G. and Cowley C. R. (1992) *AA* **253**, 199
Mathys G. and Lanz T. (1992) *AA* **256**, 169
Meade M.-R. and Code A. D. (1980) *Ap. J. Suppl.* **42**, 283
Meggers W. F., Corliss C. H. and Scribner B. F. (1975) *Tables of Spectral Line Intensities* (Monogr. 145, NBS, Washington D.C.)
Mendez R. H. (1991) *IAU Symp.* **145**, 375
Menten K. M., Melnick G. J. and Philip T. G. (1990) *29th Liège Coll.*, 243
Merrill P. W. and Sanford R. P. (1944) *Ap. J.* **100**, 14
Merrill P. W. (1947) *Ap. J.* **105**, 360
Merrill P. W. (1948) *Ap. J.* **107**, 317
Merrill P. W. (1950) *Ap. J.* **111**, 484
Merrill P. W. (1951a) *Ap. J.* **114**, 37
Merrill P. W. (1951b) *Ap. J.* **113**, 55
Merrill P. W. (1952) *Ap. J.* **116**, 21
Merrill P. W. and Lowen L. (1953) *PASP* **65**, 221
Merrill P. W. (1953) *Ap. J.* **118**, 453
Merrill P. W. (1955) *PASP* **67**, 199
Merrill P. W. (1956) *Lines of the Chemical Elements in Astronomical Spectra* Publ. 610 (Carnegie Institution, Washington)
Merrill P. W. (1960) in *Stellar Atmospheres* ed. J. Greenstein (Chicago University Press)
Mihalas D., Frost S. A., Lockwood G. W. (1975) *PASP* **87**, 153
Mitchell S. A. (1947) *Ap. J.* **105**, 1
Molaro P. (1987) *AA* **183**, 241

Molaro P. and Bonifacio P. (1990) *AA* **236**, L5
Moore C. E. (1945) A multiplet table of astrophysical interest *Princeton Obs. Contr.* **20**
Moore, C. E. (1950) An ultraviolet multiplet table *NBS Circ.* 488, section I
Moore C. E. (1952) An ultraviolet multiplet table *NBS Circ.* 488, section II
Moore C. E. (1962) An ultraviolet multiplet table *NBS Circ.* 488, sections III–IV
Moore Ch., Minnaert M. and Houtgast J. (1966) *The Solar Spectrum 2935 to 8770 Å.* Second revision of Rowlands preliminary table of solar spectrum wavelengths, Monogr. 61 (NBS, Washington D.C.)
Morgan W. W. (1935) *Publ. Yerkes Obs.* **7**, part III
Morgan W. W., Abt H. A. and Tapscott J. W. (1978) *Revised MK Spectral Atlas for Stars Earlier than the Sun* (Yerkes Observatory and Kitt Peak National Observatory)
Morris M., Guilloteau S., Lucas R. and Omont A. (1987) *Ap. J.* **321**, 888
Motch C., Werner K. and Pakull M. W. (1993) *AA* **268**, 561
Mould J. R. (1978) *Ap. J.* **226**, 923
Mould J. R. and McElroy D. B. (1978) *Ap. J.* **220**, 935
Mundt, R. (1984) *Ap. J.* **280**, 749
Murty P. S. (1980) *Ap. J.* **240**, 585

Naftilan S. A. (1977) *PASP* **89**, 309
Nemry F., Surdej J. and Hernaiz A. (1991) *AA* **247**, 469
Nercessian E., Guilloteau S., Omont A. and Benayoun J. J. (1989) *AA* **210**, 225
Nguyen-Q-Rieu, Graham D. and Bujarrabal V. (1984) *AA* **138**, L5
Nguyen-Q-Rieu, Epchtein N., Truong-Bach and Cohen M. (1987) *AA* **180**, 117
Nissen P. E. (1974) *AA* **36**, 57
Nordh H. L., Lindgren B. and Wing R. (1977) *AA* **56**, 1
Norris J. (1971) *Ap. J. Suppl.* **23**, 193
Nugis T. and Niedzielski A. (1990) *AA* **238**, L1
Nussbaumer H., Schmutz W., Smith L. J. and Willis A. J. (1982) *AA Suppl.* **47**, 257
Nyman L. A. (1992) *IAU Symp.* **150**, 401

O'Brien G. T. and Lambert D. L. (1986) *Ap. J. Suppl.* **62**, 899
Ohishi M., Kaifu N., Kawaguchi K., Murakami A., Saito S., Yamamoto S., Ishikawa S., Fujita Y., Shiratori Y. and Irvine W. M. (1989) *Ap. J.* **345**, L83
Ohman Y. (1936) *Stockholm Obs. Ann.* **12**, N.3
Olofsson H. (1988) *Space Sci. Rev.* **47**, 145
Olofsson H., Lindqvist M., Nyman L. A., Winnberg A. and Nguyen-Q-Rieu (1991) *AA* **245**, 611
Omont A., Lucas R., Morris M. and Guilloteau S. (1993) *AA* **267**, 490
Orlov M. Ya. and Shavrina A. V. (1990) *Astrophysics* **32**, 126

Pallavicini R., Randich S. and Giampapa M. S. (1992) *AA* **253**, 185
Parthasarathy M. (1989) *IAU Coll.* **106**, 384
Payne-Gaposchkin C. (1957) *The Galactic Novae* (North Holland)
Pedley J. B. and Marshall E. M. (1983) *J. Phys. Chem. Ref. Data* **12**, 967
Persson S. E. (1988) *PASP* **100**, 710
Pesch P. (1972) *Ap. J.* **174**, L155
Peterson D. M. and Scholz M. (1971) *Ap. J.* **163**, 51

Peterson R. C., Kurucz R.-L. and Carney B. W. (1990) *Ap. J.* **350**, 173
Petrie R. M. (1952) *Publ. Dominion Obs.* **9**, 251
Pettersen B. R. and Hawley S. L. (1989) *AA* **217**, 187
Pettersen B. R. (1989) *AA* **209**, 279
Phillips J. G. and Davis S. P. (1989) *Ap. J. Suppl.* **71**, 163
Phillips K. J. H. and Keenan F. P. (1990) *MN* **245**, 4P
Pierce K. (1968) *Ap. J. Suppl.* **17**, 1
Pilachowski C. A., Sneden C. and Hudek D. (1990) *AJ* **99**, 125
Poli A. A., Bord D. J. and Cowley R. (1987) *PASP* **99**, 623
Polidan R. S. and Peters G. J. (1976) *IAU Symp.* **70**, 59
Polosukhina N. S. and Khoklova V. L. (1978) *Pisma Astr. J.* **3**, 74
Powell A. L.-T. (1970) *Royal Obs. Bull.* **167**, 303
Praderie F., Catala C., Czarny J., Felenbok P. (1987) *IAU Symp.* **122**, 101
Prinja R. K. (1990) *MN* **246**, 392
Provost J. and van t'Veer-Menneret C. (1969) *AA* **2**, 218

Querci M. and Querci F. (1970) *AA* **9**, 1
Querci M. (1986) *The M Type Stars* ed. H. R. Johnson and F. R. Querci, NASA SP-492, p. 113
Querci M. and Querci F. (1989) *IAU Coll.* **106**, 258

Rauch T., Heber U., Hunger K., Werner K. and Neckel T. (1991) *AA* **241**, 457
Rawlings J. M. C. (1992) *IAU Symp.* **150**, 365
Reader J. and Corliss C. H. (1982) *Handbook of Chemistry and Physics* 63rd edition (CRC Press)
Rebolo R. (1991) *IAU Symp.* **145**, 85
Rebolo R., García Lopéz R., Beckman J. E., Vladilo G., Foing B. H. and Crivellari L. (1989) *AA Suppl.* **80** 135
Redfors A. (1991) *AA* **249**, 589
Redfors A. and Cowley C. R. (1993) *AA* **271**, 273
Reimers D. (1987) *IAU Symp.* **122**, 307
Reimers D., Baade R., Schroeder K. P. and Ising J. (1990) *AA* **236**, L25
Reynolds S. E., Hearnshaw J. B. and Cottrell P. L. (1988) *MN* **235**, 1423
Rich R. M. (1988) *AJ* **95**, 828
Richards W. G. and Scott P. R. (1976) *Structure and Spectra of Atoms* (Wiley)
Ridgway S. T., Carbon D. F. and Hall D. N. B. (1978) *Ap. J.* **225**, 138
Ridgway S. T. and Hall D. N. B. (1980) *IAU Symp.* **87**, 509
Ridgway S. T., Carbon D. F., Hall D. N. and Jewell J. (1984) *Ap. J. Suppl.* **54**, 177
Rinsland C. P. and Wing R. F. (1982) *Ap. J.* **262**, 201
Robinson R. D., Cram L. E. and Giampapa M. S. (1990) *Ap. J. Suppl.* **74**, 891
Roby S. W. and Lambert D. L. (1990) *Ap. J. Suppl.* **73**, 67
Rodono M. (1992) *IAU Symp.* **151**, 71
Rogerson J. B. (1989) *Ap. J. Suppl.* **71**, 1011
Rossi C., Altamore A., Baratta G. B., Friedjung M. and Viotti R. (1992) *AA* **256**, 132
Ruland, F., Griffin R., Griffin R., Biehl D. and Holweger H. (1980) *AA. Suppl.* **42**, 391
Russell S. C. and Bessell M. S. (1989) *Ap. J. Suppl.* **70**, 865
Ryabchikova T. A., Davidova E. S. and Adelman S. J. (1990) *PASP* **102**, 581

Ryan S. G. (1992) *PASP* **104**, 805
Ryan S. G., Norris J. E., Bessell M.-S. and Deliyannis C. P. (1992) *Ap. J.* **388**, 184

Sadakane K. (1976) *PAS Japan* **28**, 469
Sadakane K., Takada M. and Jugaku J. (1983) *Ap. J.* **274**, 261
Sadakane K., Jugaku J. and Takada-Hidai M. (1985) *Ap. J.* **297**, 240
Sadakane K., Jugaku J. and Takada-Hidai M. (1988) *Ap. J.* **325**, 776
Sadakane K. and Okyudo M. (1989) *PAS Japan* **41**, 1055
Sadakane K. (1991) *PASP* **103**, 355
Sadakane K. (1992) *PAS Japan* **44**, 125
Sahai R. and Howe J. (1987) *Calgary conference*, 167
Saito S., Kawaguchi K., Suzuki H., Ohishi M., Kaifu N. and Ishikawa S. (1987) *PAS Japan* **39**, 193
Samsonov G. V. (ed.) (1968) *Physicochemical Properties of the Elements* (Plenum)
Sandlin G. D., Brueckner G. E. and Tousey R. (1977) *Ap. J.* **214**, 898
Sandlin G. D., Bartoe J. D. F., Brueckner G. E., Tousey R. and van Hoosier M. E. (1986) *Ap. J. Suppl.* **61**, 801
Sanford R. F. (1929) *PASP* **41**, 271
Sanford R. F. (1940) *PASP* **52**, 203
Sanner F. (1976) *Ap. J. Suppl.* **32**, 115
Sargent W. L. and Jugaku J. (1961) *Ap. J.* **134**, 777
Sauval A. J. (1969) *Solar Physics* **10**, 319
Sauval A. J. (1972) *Contribution à l'étude des molécules dans le soleil Ph.D. thesis*, Liège
Sauval A. J., Biémont E., Grevesse N. and Zander R. (1980) *21st Liège Coll.* p.235
Sauval A. J. and Tatum J. B. (1984) *Ap. J. Suppl.* **56**, 193
Schild R.E. (1976) *IAU Symp.* **70**, 106 ed. A. Slettebak, (Reidel)
Schneeberger J., Linsky J. L. and Worden S. P. (1978) *AA* **62**, 447
Schoenberner D. and Drilling J. S. (1984) *Ap. J.* **276**, 229
Schoenberner D., Herrero A., Becker S., Eber F., Butler K., Kudritzki R. P. and Simon K. P. (1988) *AA* **197**, 209
Schild H., Boyle S. J. and Schmid H. M. (1992) *MN* **258**, 95
Schmidt-Kaler Th. (1982) *Zahlenwerte und Funktionen*, Landolt-Börnstein, Berlin, Springer, New Series, Group VI, vol. 2b, p.1
Schrey U., Drapartz S., Kaeufl H. U., Rothermel H. and Ghosh S. K. (1987) *IAU Symp.* **122**, 553
Schroeder K. P. (1985) *AA* **147**, 103
Schroeder K. P., Griffin R. E. M. and Griffin R. F. (1990) *AA* **234**, 299
Schroeder K. P. and Huensch M. (1992) *AA* **257**, 219
Schwartz P. R. and Barrett A. H. (1970) *Ap. J.* **159**, L123
Sekiguchi K. and Anderson K. S. (1987) *AJ* **94**, 129
Sellgren K. and Smith R. G. (1992) *Ap. J.* **388**, 178
Severny A. S. and Lyubimkov, L. S. (1985) *IAU Coll.* **90**, 327
Sheffer Y and Lambert D. L. (1992) *PASP* **104**, 1054
Shipman H., Liebert J. and Green R. F. (1987) *Ap. J.* **315**, 239
Simon K. P., Jonas G., Kudritzki R. P. and Rahe J. (1983) *AA* **125**, 34
Sinha K. (1991) *Proc. Austr. AS* **9**, 32
Sion E. M., Aannestad P. A. and Kenyon S. J. (1988) *Ap. J.* **330**, L55
Sion E. M., Kenyon S. J. and Aannestad P. A. (1990) *Ap. J. Suppl.* **72**, 707

Slettebak A., Collins G. W. and Truran R. (1992) *Ap. J. Suppl.* **81**, 335
Smith G. H. (1987) *PASP* **99**, 67
Smith G. H. and Dupree A. K. (1988) *AJ* **95**, 1547
Smith G. H. and Wirth G. D. (1991) *PASP* **103**, 1158
Smith L. F. and Hummer D. G. (1988) *MN* **230**, 511
Smith M. A. (1972) *Ap. J.* **175**, 765
Smith M. A. (1973) *Ap. J. Suppl.* **25**, 277
Smith M. A. (1974) *Ap. J.* **189**, 101
Smith M. A. (1981) *Ap. J.* **246**, 905
Smith V. V. and Wallerstein G. (1983) *Ap. J.* **273**, 742
Smith V. V. (1984) *AA* **132**, 326
Smith V. V. and Lambert D. L. (1985) *Ap. J.* **294**, 326
Smith V. V. and Lambert D. L. (1986) *Ap. J.* **311**, 843
Smith V. V., Lambert D. L. and McWilliam A. (1987) *Ap. J.* **320**, 862
Smith V. V. and Lambert D. L. (1988) *Ap. J.* **333**, 219
Smith V. V. and Lambert D. L. (1990a) *Ap. J. Suppl.* **72**, 387
Smith V. V. and Lambert D. L. (1990b) *Ap. J.* **361**, L69
Smits D. P. (1991) *MN* **248**, 20
Sneden C., Pilachowski C. A., Gilroy K. K. and Cowan J. J. (1988) *IAU Symp.* **132**, 501
Sneden C. (1991) *IAU Symp.* **145**, 235
Sneden C., Gratton R. G. and Crocker D. A. (1991) *AA* **246**, 354
Solf J. (1978) *AA Suppl.* **34**, 409
Sotirovski P. (1971) *AA* **14**, 319
Sotirovski P. (1972) *AA Suppl.* **6**, 85
Sowell J. R. (1989) *PASP* **101**, 101
Spinrad H. and Taylor B. J. (1969) *Ap. J.* **157**, 1279
Spinrad H. and Wing R. (1969) *ARAA* **7**, 249
Spite F. and Spite M. (1986) *AA* **163**, 140
Spite, F., Richtler T., and Spite M. (1991) *AA* **252**, 557
Spite M. and Spite F. (1990) *IAU Symp.* **148**, 372
Spite M. and Spite F. (1991) *AA* **252**, 689
Spite M. (1992) *IAU Symp.* **149**, 123
Stahl O., Mandel H., Szeifert Th., Wolf B. and Zhao F. (1991) *AA* **244**, 467
Stencel R. E., Mullan D. J., Linsky J. L., Basri G. S., Worden S. P. (1980) *Ap. J. Suppl.* **44**, 383
Stephenson C. B. (1986) *PASP* **98**, 467
Strassmeier K. G., Hall D. S., Fekel F. C. and Scheck M. (1993) *AA Suppl.* **100**, 173
Strassmeier K. G., Fekel F. C., Bopp B. W., Dempsey R. C. and Henry G. W. (1990) *Ap. J. Suppl.* **72**, 191
Striganov A. R. and Odintsova G. A. (1982) *Tables of Spectral Lines of Atoms and Ions* (Energoizdat, Moscow)
Strohbach P. (1970) *AA* **6**, 385
Struve O. and Wurm K. (1938) *Ap. J.* **88**, 84
Sun Y. L., Jaschek M., Andrillat Y. and Jashek C. (1985) *AA Suppl.* **62**, 309
Suntzeff N. B. and Smith V. V. (1991) *Ap. J.* **381**, 160
Suzuki H., Ohishi M., Kaifu N., Ishikawa S., Kasuga T., Saito S. and Kawaguchi K. (1986) *PAS Japan* **38**, 911

Swensson J., Benedict W. S., Delbouille L. and Roland G. (1970) *The Solar Spectrum from 7498 to 12 016* Mémoires de la Société Royale des Sciences de Liège, vol. hors serie 5

Swings P. and Struve O. (1941a) *Ap. J.* **93**, 455

Swings P. and Struve O. (1941b) *Ap. J.* **94**, 291

Swings P. (1952) *Mémoires Liège Inst. d'Astroph.* **8**, N.341

Swings J. P. (1973) *AA* **26**, 443

Takada-Hidai M., Sadakane K. and Jugaku J. (1986) *Ap. J.* **304**, 425

Takada-Hidai M. (1991) *IAU Symp.* **145**, 137

Takeda (1991) *PAS Japan* **43**, 823

Talavera A. (1988) *IAU Coll.* **94**, 135

Taylor B. J. (1982) *Vistas in Astronomy* **26**, 253

Taylor B. J. (1991) *Ap. J. Suppl.* **76**, 715

te Lintel Hekkert P., Caswell J. L., Habing H. J., Haynes R. F. and Norris R. P. (1991) *AA Suppl.* **90**, 327

Thackeray A. D. (1953) *MN* **113**, 211

Thaddeus P., Cummins S. F. and Linke R. A. (1984) *Ap. J.* **283**, L45

Thorburn J. A. (1992) *Ap. J.* **399**, L83

Tomkin J. and Lambert D. L. (1979) *Ap. J.* **227**, 209

Tsuji T. (1964) *Ann. Tokyo Obs.* Second ser. **9**, 1

Tsuji T. (1986a) *ARAA* **24**, 89

Tsuji T. (1986b) *AA* **156**, 8

Tsuji T. (1991) *AA* **245**, 203

Tsuji T., Iye M., Tomioka K., Okada T. and Sato H. (1991) *AA* **252**, L1

Turner B. E., Tsuji T., Bally J., Guelin M. and Cernicharo J. (1990) *Ap. J.* **365**, 569

Turner B. E. (1992) *Ap. J.* **388**, L35

Turnshek D. E., Turnshek D. A., Crane E. R. and Boeshaar P. W. (1985) *An Atlas of Digital Spectra of Cool Stars* (Western Research Co.)

Ukita N. and Morris M. (1983) *AA* **121**, 15

Underhill A. (1966) *The Early Type Stars* (Reidel)

Underhill A. B. (1973) *AA* **25**, 161

Underhill A. B. and Fahey R. P. (1973) *Ap. J. Suppl.* **25**, 463

Underhill A. B. (1977) *Ap. J.* **217**, 488

Utsumi K. (1966) *Trieste Coll.*, 49

Utsumi K. (1984) *Str. Coll.* 243

van der Hucht K. A., Stencel R. E., Haisch B. M. and Kondo Y. (1979) *AA Suppl.* **36**, 377

van Helden R. (1972) *AA Suppl.* **7**, 311

van Paradijs J. (1973) *AA Suppl.* **11**, 25

Vanture A. D., Wallerstein G. and Brown J. A. (1991) *Ap. J.* **381**, 278

van t'Veer-Menneret C., Burkhart C. and Coupry M. F. (1988) *AA* **203**, 123

van Winckel H., Mathis J. S. and Waelkens C. (1992) *Nature* **356**, 500

Varani G. F., Meikle W. P. S., Spyromilio and Allen D. A. (1990) *MN* **245**, 570

Vauclair S. (1991) *IAU Symp.* **145**, 327

Vaughan A. H. and Zirin H. (1968) *Ap. J.* **152**, 123

Venn K. A. and Lambert D. L. (1990) *Ap. J.* **363**, 234
Verbunt F., Pringle J. E., Wade R. A., Echevarria J., Jones D. H. P., Argyle R. W., Schwarzenberg-Czerny A., la Dous C. and Schoembs R. (1984) *MN* **210**, 197
Vernazza J. E., Avrett E. H. and Loeser R. (1981) *Ap. J. Suppl.* **45**, 635
Vidal-Madjar A., Fertlet R., Spite M. and Coupry M. F. (1988) *AA* **201**, 273
Viotti R. and Friedjung M. (1988) *IAU Coll.* **94**, 197
Vitrichenko E. A. and Kopylov I. M. (1962) *Isv. Crimea* **27**, 241
Vreux J. M., Andrillat Y. and Biemont E. (1990) *AA* **238**, 207

Waelkens C., van Winckel H., Bogaert E. and Trams N. R. (1991) *AA* **251**, 495
Waelkens C., van Winckel H., Trams N. R. and Waters L. B. F. M. (1992) *AA* **256**, L15
Walborn N. (1973) *Ap. J.* **180**, L35
Walborn N. (1980) *Ap. J. Suppl.* **44**, 535
Walborn N. and Panek R. J. (1984) *Ap. J.* **280**, L27
Walborn N. and Panek R. J. (1985) *Ap. J.* **291**, 806
Walborn N., Panek R. J. and Heckathorn J. N. (1985) *Future of Ultraviolet in Astronomy Based upon Six Years of IUE* (NASA/GSFC) p. 511
Walborn R. N. and Fitzpatrick E. L. (1990) *PASP* **102**, 379
Wallerstein G. and Conti P. (1964) *Ap. J.* **140**, 858
Wallerstein G. and Dominy J. F. (1988) *Ap. J.* **330**, 937
Wallerstein G. (1989) *Ap. J. Suppl.* **71**, 341
Wallerstein G. (1990) *Ap. J. Suppl.* **74**, 755
Wallerstein G., Schachter J., Garnavich P. M. and Oke J. B. (1991) *PASP* **103**, 185
Wallerstein G. (1992) *PASP* **104**, 511
Warner B. (1965) *MN* **129**, 263
Warner B. (1989) in *Classical Novae* ed. M. F. Bode and A. Evans (Wiley)
Wegner G. and Swanson S. R. (1991) *Ap. J. Suppl.* **75**, 507
Weidemann V. and Koester D. (1991) *AA* **249**, 389
Weinteb S., Barrett A. H., Meeks M.-L. and Henry J. C. (1963) *Nature* **200**, 829
Werner K., Heber U. and Hunger K. (1991) *AA* **244**, 437
Werner K. (1991) *AA* **251**, 147
Wheeler C., Sneden C. and Truran J. W. (1989) *ARAA* **27**, 279
Wiese W. L., Smith M. W. and Glennon B. M. (1966) *Atomic Transition Probabilities* (NSRDS-NBS 4)
Willems F. J. and de Jong T. (1986) *Ap. J.* **309**, L39
Willis A. J., van der Hucht K. A., Conti P. S. and Garmany D. (1986) *AA Suppl.* **63**, 417
Willson L. A. (1975) *Ap. J.* **197**, 365
Wilson R. W., Solomon P. M., Penzias A. A. and Jefferts K. B. (1971) *Ap. J.* **169**, L35
Wing R. F. (1979) *IAU Coll.* **47**, 347
Wing, R. F., Carpenter K. G., Wahlgren G. M. (1983) *Spec. Publ.* N.1 (Perkins Observatory)
Wing R. F. (1991) *Montpellier Coll.*, 275
Winnewisser G. and Walmsley C. M. (1978) *AA* **70**, L37
Woehl H. (1971) *Solar Physics* **16**, 362
Woehl H., Engvold O. and Brault J. W. (1983) *Inst. Theor. Astroph. Blindern* N.56
Wojslaw R. S. and Peery B. F. (1976) *Ap. J. Suppl.* **31**, 75

Wolf B. (1972) *AA* **20**, 275
Wolf B. (1993) *Reviews of Modern Astronomy* vol. 5, ed. G. Klare (Springer) p. 1
Wolff S. C. and Morrison N. D. (1972) *Ap. J.* **175**, 473
Wolff S. C., Heasley J. N. and Varsik J. (1985) *PASP* **97**, 707
Wright K. O. (1947) *Publ. Dominion Obs.* **8**, 1
Wright K. O. (1963) *Publ. Dominion Obs.* **12**, 218
Wu C. C., Ake T. B., Boggess A., Bohlin R. C., Imhoff C. L., Holm A. V., Levay Z. G. and Panek R. J. (1983) *NASA IUE Newsletter* 22
Wyckoff S. and Wehinger P. A. (1976) *MN* **175**, 587
Wyckoff S. and Wehinger P. A. (1977) *Ap. J.* **212**, L139
Wyckoff S. and Clegg R. E. S. (1978) *MN* **184**, 127

Yamashita Y. (1967) *Publ. Dominion Obs.* **13**, 67
Yamashita Y., Nariai K. and Norimoto Y. (1977) *An Atlas of Representative Stellar Spectra* (University of Tokyo Press)
Yerle R. (1979) *AA* **73**, 346

Zhao G. and Magain P. (1990) *AA* **238**, 242
Zhao G. and Magain P. (1991) *AA* **244**, 425
Zhou Xu (1991) *AA* **248**, 367
Zickgraf F.-J. (1988) *IAU Coll.* **94**, 125
Zirin H. (1968) *Ap. J.* **152**, L177
Zirin H. (1988) *Astrophysics of the Sun* (Cambridge University Press)
Ziurys L. M., Clemens D. P., Saykally R. J., Colvin M. and Schaefer H. F. (1984) *Ap. J.* **281**, 219
Zook A. C. (1978) *Ap. J.* **221**, L113
Zook A. C. (1985) *Ap. J.* **289**, 356
Zuckerman B., Dyck H. M. and Claussen M. J. (1986) *Ap. J.* **304**, 401

ESA Sp. **263** (1986) *New Insights in Astrophysics* (European Space Agency)
IAU Symp. **26** (1966) *Abundance Determinations in Stellar Spectra* ed. H. Habenet (Academic Press)
IAU Symp. **70** (1976) *Be and Shell Stars* ed. A. Slettebak (Reidel)
IAU Symp. **87** (1980) *Intestellar Molecules* ed. B. H. Andrews (Reidel)
IAU Symp. **98** (1982) *Be Stars* ed. M. Jaschek and H. G. Groth (Reidel)
IAU Symp. **122** (1987) *Circumstellar Matter* ed. I. Appenzeller and C. Jordan (Reidel)
IAU Symp. **132** (1988) *The Impact of Very High S/N Spectroscopy on Stellar Physics* ed. G. Cayrel de Strobel and M. Spite (Kluwer)
IAU Symp. **145** (1991) *Evolution of Stars. The Photospheric Abundance Connection* ed. G. Michaud and A. Tutukov (Kluwer)
IAU Symp. **149** (1992) *The Stellar Populations of Galaxies* ed. B. Barbuy and A. Renzini (Kluwer)
IAU Symp. **150** (1992) *Astrochemistry of Cosmic Phenomena* ed. P. D. Singh (Kluwer)
IAU Symp. **151** (1992) *Evolutionary Processes in Interacting Binary Stars* ed. Y. Kondo, R. F. Sisteró and R. S. Polidan (Kluwer)
IAU Coll. **47** (1979) *Spectral Classification of the Future* ed. M. F. McCarthy, A. G. D. Philip and G. V. Coyne (Specola Vaticana)

IAU Coll. **90** (1986) *Upper Main Sequence Stars with Anomalous Abundances* ed. C. R. Cowley, M. M. Dworetsky and C. Mégessier (Reidel)
IAU Coll. **94** (1988) *Physics of Formation of FeII lines outside LTE* ed. R. Viotti, A. Vittone and M. Friedjung (Reidel)
IAU Coll. **103** (1988) *The Symbiotic Phenomenon* ed. A. J. Mikolajewska, M. Friedjung, S. J. Kenyon and R. Viotti (Kluwer)
IAU Coll. **106** (1989) *Evolution of Peculiar Red Giant Stars* ed. H. R. Johnson (Kluwer)
Trieste Colloquium (1966) *Late Type Stars* ed. M. Hack (Osservatorio di Trieste)
Str. Coll. (1985) *Cool Stars with Excesses of Heavy Elements* ed. M. Jaschek and P. C. Keenan (Reidel)
Calgary Conference (1987) *Late Stages of Stellar Evolution* ed. S. Kwok and S. R. Pottasch (Reidel)
Third Haystack Observatory Conference on 'The Interstellar Medium' (1991) ed. A. D. Haschick and P. T. P. Ho, Astr. Soc. Pacific Conf. ser. vol. 16
23rd Liège Coll. (1981) *Upper Main Sequence Chemically Peculiar Stars* (Institut d' astophysique, Université de Liège)
29th Liège Coll. (1990) *From Ground Based to Space Borne Sub-mm Astronomy, ESA SP-314*
Montpellier Colloquium (1991) *The Infrared Spectral Region of Stars* ed. C. Jaschek and Y. Andrillat (Cambridge University Press)
NASA (1991) *The First Year of HST Observations* ed. A. L. Kinney and J. C. Blades (Space Telescope Science Institute)
Rome Colloquium (1991) *Evolutionary Phenomena in the Universe* in honour of the 80th birthday of L. Gratton ed. by P. Giannone, F. Melchiorri and F. Occhionero (Italian Physical Society)
HC (1992) *Proceedings of the 31st Herstmonceux Conference 'Elements and the Cosmos'* in honour of B. E. J. Pagel ed. M.-G. Edmunds and R. Terlevich (Cambridge University Press)
Workshop 'Elemental Abundance Analysis' (1987) ed. S. J. Adelman and T. Lanz (Institut d'astronomie de l'Université de Lausanne)

Index of elements in stars

In what follows we provide an index of the different elements studied for a given group of stars. The definitions and synonyms for the different groups are given in part three, chapter 2.

Index of molecules in stars

In what follows we provide an index of the different molecules studied for a given group of stars. The definitions and synonyms for the different groups are given in part three, chapter 2.